92nd Textile Institute World Conference (TIWC 2023)

Sustainability of the Textile & Fashion Supply Chain: Transitioning to Zero Carbon & Zero Waste

Huddersfield, United Kingdom
3 – 6 July 2023

ISBN: 978-1-7138-9696-8

Printed from e-media with permission by:

Curran Associates, Inc.
57 Morehouse Lane
Red Hook, NY 12571

Some format issues inherent in the e-media version may also appear in this print version.

Textile Institute
8th Floor, St. James' Buildings
79 Oxford Street
Manchester, M1 6FQ
United Kingdom

Phone: +44(0)161 237 1188

tiihq@textileinst.org.uk

Additional copies of this publication are available from:

Curran Associates, Inc.
57 Morehouse Lane
Red Hook, NY 12571 USA
Phone: 845-758-0400
Fax: 845-758-2633
Email: curran@proceedings.com
Web: www.proceedings.com

The Textile Institute

The 92nd Textile Institute World Conference Proceedings

3-6 July 2023
Huddersfield, UK

Sustainability of the Textile and Fashion Supply Chain - Transitioning to Zero Carbon and Zero Waste

University of HUDDERSFIELD
Inspiring global professionals

The Textile Institute

Technical Textiles Research Centre

Arts & Humanities Huddersfield

The 92nd Textile Institute World Conference Proceedings

Sustainability of the Textile and Fashion Supply Chain - Transitioning to Zero Carbon and Zero Waste

Held at The University of Huddersfield, UK
3-6 July 2023

Proceedings compiled by
Dr AJ Hebden, Dr T Smith, Dr AD Hewitt, Dr S Rana,
University of Huddersfield, Huddersfield, UK

The Textile Institute
8th Floor, St James' Buildings
79 Oxford Street
Manchester
M1 6FQ
UK

Contents

World President's Welcome

As World President of the Textile Institute I welcome you to The 92nd Textile Institute World Conference (TIWC 2023) which this year is being hosted by Conference Chairman Prof Parik Goswami CText FTI, the School of Arts and Humanities, and the Technical Textiles Research Centre (TTRC) at the University of Huddersfield, UK.

The conference theme, 'Sustainability of the Textile and Fashion Supply Chain – Transitioning to Zero Carbon and Zero Waste', is focused on identifying the current scenario as well as emerging sustainable practices and solutions throughout the whole supply chain of textile and fashion products, starting from raw materials to manufacturing processes, distribution and retailing, use by consumers and up to the end-of-life phases.

TIWC 2023 is the first international conference The Textile Institute has held since the COVID pandemic and it is a pleasure to welcome you all in person to this special event. The conference is an opportunity to network and share knowledge, discuss and make plans and to leave the conference with firm actions to transition to zero carbon and zero waste as set out by the TIWC 2023 theme.

The Textile Institute brings together designers, engineers, innovators, researchers, and practitioners from all textile and associated industries. I hope that this conference will help you to seek out new partnerships, collaborations, new friends, and work together to find solutions.

Professor Malgorzata Zimmniewska CText FTI
Textile Institute World President

Conference Chairman's Welcome

Dear Delegates,

On behalf of the organising committee, I warmly welcome you to the 92nd Textile Institute World Conference (TIWC) 2023 hosted by the University of Huddersfield.

The University of Huddersfield is dedicated to expanding the frontiers of human knowledge and aim to provide an exceptional environment for the development of researchers by encouraging multidisciplinary concentrations of research excellence such as institutes and research centres that are capable of providing first-class resources and attracting leading international collaborators. Over the last decade the University has won a number of awards and titles for its outstanding record for world-class teaching and developing students as independent learners and critical thinker. Some of the awards include the Times Higher Education University of the Year, and the first Global Teaching Excellence Award. Alongside, the University also has a growing reputation as a leader in the field of enterprise. In 2012 the University was awarded the prestigious Times Higher Education Entrepreneurial University of the Year and followed this up in 2013 and 2022 when it was awarded Queen's Awards for Enterprise: International Trade. We pride ourselves as being successful, global, and modern.

Indeed, the theme of this conference 'Sustainability of the Textile and Fashion Supply Chain – Transitioning to Zero Carbon and Zero Waste' is to address the biggest contemporary challenge that we face in today's world.

After oil and gas, the textile industry is the second most polluting industry globally and whilst initiatives such as Textiles 2030 in the UK are already causing consumers and industry to question what the future fashion-textile sector will look like as we head towards net zero, there is still a great opportunity to reimagine the wider textile landscape. Afterall, the global textile market was valued at $993.6 billion in 2021 and is expected to grow at a compound annual growth rate of 4.0% from 2022 to 2030 and it will need strong leadership and initiatives to transition towards our Net Zero goal. Contrary to the popular believe, the fashion and textile industry has a huge impact even on the UK economy. In 2021, the industry generated a £62 billion gross value-added contribution to the UK's economy, supporting 1.3 million jobs and £23 billion in total tax revenue to the UK treasury.

However, whilst the industry makes enormous contribution to the economic growth and labour market engagement, it is also costly to our climate and ecosystem. It produces 8-10% of global CO_2 emissions, consumes 79 trillion litres of water per year, lead to ~20% of industrial water pollution from textile dyeing and other chemical processing, contributes to ~35% of oceanic primary microplastic pollution and produces >92 million tonnes of waste per year. Achieving net zero by 2050 is the biggest challenge the industry face today. Every challenge brings opportunities, but it will require all stakeholders from around the world to work together.

The ethos of this conference, hence, is solution-driven involving different stakeholders from divergent fields instead of just being subject-driven. Where best to host such a conference than Huddersfield in West Yorkshire!

West Yorkshire is not only the centre of the UK textile industry, but arguably the longest established industrial cluster in the world, with the biggest concentration of textile activity in the UK. It is also a place of unparallel scenic beauty. As the popular saying, "All work and no play doesn't just make Jill and Jack dull, it kills the potential of discovery, mastery, and openness to change and flexibility and it hinders innovation and invention", so please take some time to explore Huddersfield, and its beautiful surroundings. Based at the foothills of the Pennines and River Calder in the northeastern outskirts of the town, the landscape has exceptional scenic beauty! The rivers around the town provided soft water required for the textile industry during the early Victorian era making it a prominent mill town. Castle Hill is my particular favourite spot, but there are many more to explore. Please feel free to ask me and my colleagues and you will get a lot of suggestions.

I would take this opportunity to thank my colleagues at the University of Huddersfield and from The Textile Institute for their teamwork, collaboration and tireless work for this event. I am not mentioning any names in the fear of missing someone, but you know who you are and how much I appreciate all your contribution to this event. I would also like to include and thank the UK Fashion and Textiles Association for their amazing support with the conference.

On behalf of the organising committee, we would like to thank all of the sponsoring organisations for providing their generous support. A special thanks to our media and knowledge partners.

While concluding, I would like to take the opportunity to welcome each one of you and express my gratitude to the conference keynote speakers, session keynotes, paper and poster presenters, workshop conductors and participants for their contributions who are the foundation of this conference.

I am confident that you will all find this conference interesting, and a learning platform to collaborate with like-minded colleagues. Enjoy your time at the conference and in Huddersfield!

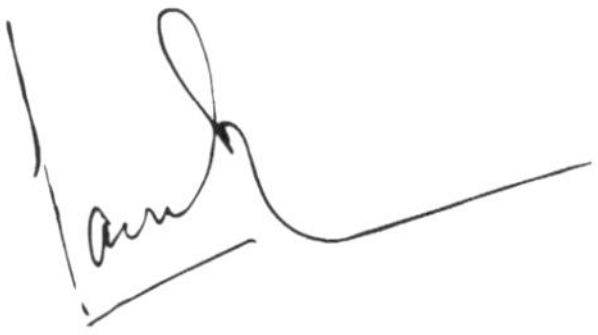

Professor Parik Goswami CText FTI
Conference Chairman

Sponsors

Thank you to all of our sponsors who made this conference possible.

Founding Sponsor

camira

Gold Sponsors

AW HAINSWORTH
QUALITY IN TEXTILES SINCE 1783

Taylor & Francis
Taylor & Francis Group

Silver Sponsor

BTMA

Bronze Sponsors

MANCHESTER 1824
The University of Manchester

natific
COLOR.SIMPLIFIED.

James Heal

General Survey

PCFC

Media Partners

ASBCI
ASSOCIATION OF SUPPLIERS TO THE BRITISH CLOTHING INDUSTRY

TIMES INTERNATIONAL
YOURS GATEWAY TO INTERNATIONAL BUSINESS
www.timesinternational.in

F2F FIBRE2FASHION

PAKISTAN textile JOURNAL
巴基斯坦纺织刊

textiles

WTiN

Committees

Conference Chair:

Prof Parik Goswami	University of Huddersfield, UK

Organising Committee:

Prof Parik Goswami	University of Huddersfield, UK
Dr Andrew Hewitt	University of Huddersfield, UK
Rachel Mountford	University of Huddersfield, UK
Kate Melaugh	University of Huddersfield, UK
Nicola Chiles	University of Huddersfield, UK
Nicola Redmore	University of Huddersfield, UK
Lorna Ingram	University of Huddersfield, UK
Dr Andrew Hebden	University of Huddersfield, UK
Dr Tim Smith	University of Huddersfield, UK
Dr Sohel Rana	University of Huddersfield, UK

TI Events Committee:

Prof Subhash Anand MBE	University of Bolton, UK
Prof Elizabeth Barnes	Manchester Metropolitan University, UK
Dr Xiaogang Chen	University of Manchester, UK
Prof Mervyn Davies	The Textile Institute, UK
Stephanie Dick	The Textile Institute, UK
Dr Kim Gandhi	The Textile Institute, UK
Robyn Ingham	The Textile Institute, UK
Helen Rowe	The Textile Institute, UK
Ian Smith	Textile Consult Ltd, UK
Prof Xungai Wang	Australian Future Fibres Research & Innovation Centre, Australia
Prof Malgorzata Zimniewska	Institute of Natural Fibres & Medicinal Plants, Poland

Programme Committee:

Prof Long Lin	University of Leeds, UK
Prof Stephen Russell	University of Leeds, UK
Prof Abhijit Mujumdar	Indian Institute of Technology, India
Amit Rawal	Indian Institute of Technology, India
Adam Mansell	UK Fashion & Textile Association, UK
Prof Prasad Potluri	University of Manchester, UK
Amanda McLaren	AW Hainsworth, UK
Alan Williams	Camira Fabrics, UK
Prof Steve Westland	University of Leeds, UK
Dr Kevin Almond	University of Leeds, UK
Prof Elizabeth Barnes	Manchester Metropolitan University, UK
Prof Tamara Galloway	University of Exeter, UK
Dr Abdel-Fattah Seyam	NC State University, USA
Dr Sanjay Gupta	World University of Design, India
Dr Rudrajeet Pal	University of Borås, Sweden

Special thanks from the Organising Committee go to Dr Charlotte Goldthorpe and Dr Jane Howroyd (for conference workshops); Dr Charlotte Goldthorpe, Dr Matthew Taylor and Katrina McLaughlin (exhibitions); Lara Booth, Kathryn Brennand, Guy McKelvey, Liz Garland and Debbie Allsop (fashion and costume shows); and all of the other incredible staff and students at the University of Huddersfield who helped to make this conference possible.

Made by Camira
in Huddersfield
Making the most environmental fabrics in the world
- from farm to fabric and everything in between.
Proud sponsors of TIWC 2023
camira
www.camirafabrics.com

Submitted Abstracts

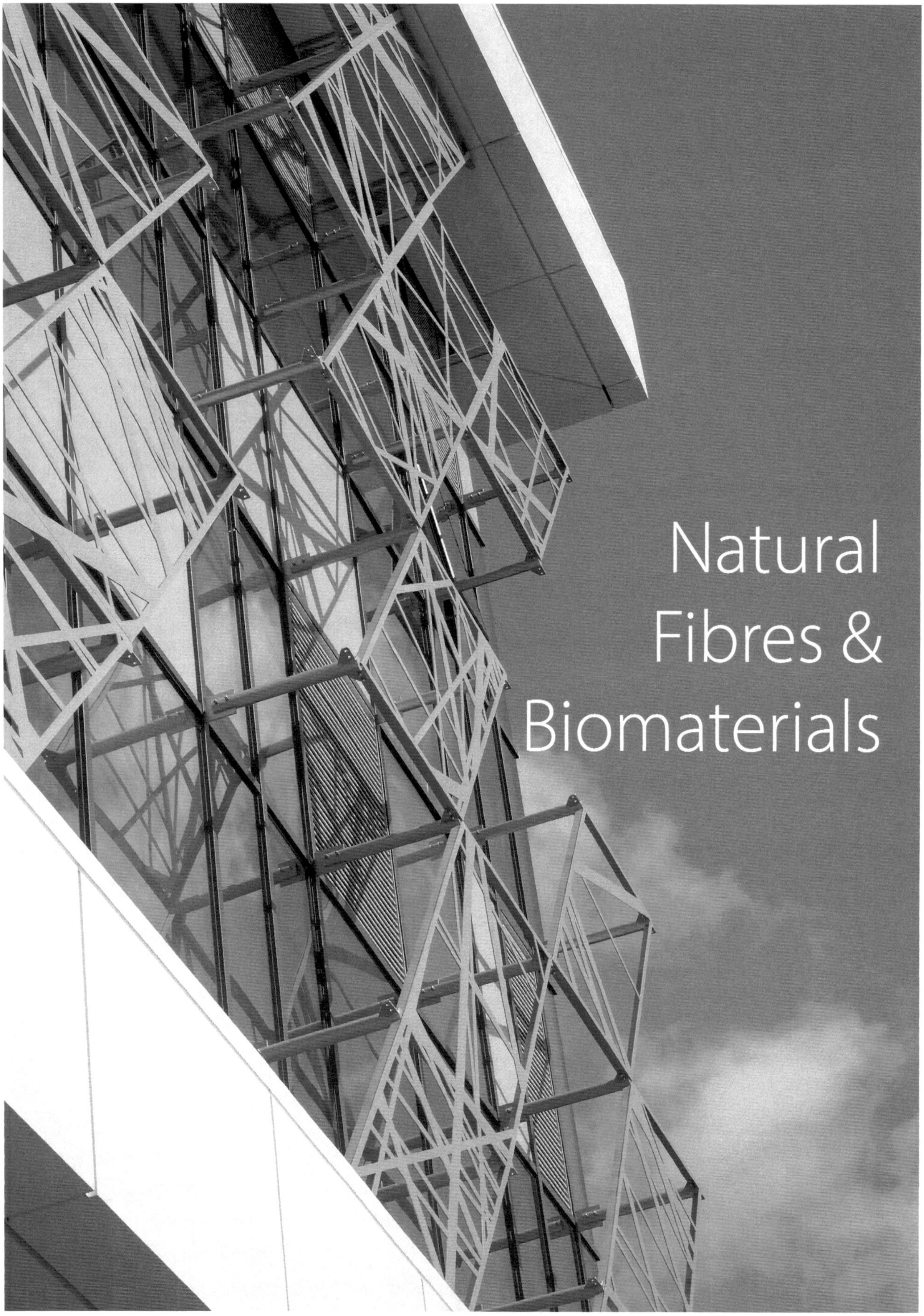

Natural Fibres & Biomaterials

ACOUSTIC AND THERMAL INSULATION MATERIALS DEVELOPED FROM WASTE BAST FIBERS AND LOCAL SHEEP WOOL

Anna Kicińska-Jakubowska[1], *Jan Broda*[2], *Małgorzata Zimniewska*[1], *Marcin Bączek*[2] *and Jerzy Mańkowski*[1]

[1]*Department of Innovative Textile Technologies, Institute of Natural Fibres and Medicinal Plants-National, Research Institute, Wojska Polskiego 71b, 60-630 Poznan, Poland*
[2]*Faculty of Materials, Civil and Environmental Engineering, University of Bielsko-Biala, Willowa 2, 43-309 Bielsko-Biala, Poland*

Abstract

This paper concerns the management of natural waste fibres. The aim of this research was the development of multifunctional acoustic and thermal insulation materials from natural protein and lignocellulosic fibre wastes, according to the rules of circular bioeconomy. For the study need, local mountain sheep wool and a mixture of bast fibre waste generated by string production were used. Insulating materials in the form of mats were produced by the needle-punching technique with different fibre contents. The basic parameters of the mats, i.e., the thickness, surface weight and air permeability were determined. The barrier properties of the mats were evaluated based on the measurement of sound absorption and noise reduction coefficients, as well as thermal resistance and thermal conductivity. It was found that mats with a higher content of the bast fibres possess a greater ability to absorb sounds, while mats with higher wool contents exhibit better thermal insulation properties. The developed mats can serve as a good alternative to commonly used acoustic and thermal insulating materials. The production of the described materials allows for more effective management of natural fibre waste generated during industrial processes and achieves the goal of "zero waste" according to the European Green Deal strategy.

Keywords

fibre waste, flax, hemp, wool, nonwoven, sound absorption, noise reduction, thermal resistance, thermal conductivity

A REVIEW OF THE UK WOOL SECTOR

Lynn Wilson[1], *Rebecca Earley*[1] *and Rosie Hornbuckle*[1]

[1]*Centre for Circular Design, University of the Arts London, 16 John Islip St, London SW1P 4JU*

Abstract

The UK wool industry is in crisis. It is reported that farmers are burning fleece and receiving as little as 50 pence, per kilo when sold at the UK market rate. Yet, whilst the regenerative and consumer health properties of wool, are widely known, fossil fuel-based fibres continue to dominate production and consumption markets. This research project, investigated the state of the UK wool sector, using a quadruple helix framework of governance, policy, industry, and society. Research methods included evidence from an extensive literature review, a wool symposium of presentations from academia and a multi-method purposeful sample of 12 qualitative industry interviews, selected using a linear supply chain framework, from sheep to product. The findings identified that, whilst wool is a devolved policy issue, there is no legislation specifically related to UK wool production and consumption exacerbated by widespread confusion about the role of industry membership organisations. Other findings identified trends in micro vertical fleece-to-product businesses and a knowledge gap across consumers, farmers, designers, and policymakers, regarding the properties and value of wool. Recommendations include the need for a systemic approach, embedded in the local context, addressing local heritage, skills, and machinery, and engaging multiple stakeholders across the value chain.

Keywords

UK wool, sustainability, regenerative fibre, wool policy, micro-vertical manufacturing, net zero, wool skills, UK wool policy

STRUCTURE-PROPERTY RELATIONSHIPS OF WOOL FROM COARSE FIBRE BREEDS OF SHEEP IN UNITED KINGDOM

Subhadeep Paul[1], Andrew Hewitt[1], Sohel Rana[1] and Parikshit Goswami[1]

[1]Technical Textiles Research Centre, Department of Fashion and Textiles, School of Arts and Humanities, University of Huddersfield, HD1 3DH, England

Abstract

Coarse wool fibres from UK sheep breeds have a very limited and low value use in terms of their application. This low price point saw countless examples of farmers burning their sheep's wool in fields as it wasn't worth the cost of transporting it for processing during the COVID-19 pandemic. This study looks to examine the structure-property relationships of these wool fibres to establish new end-use applications in technical textiles. The surface morphology and the thermodynamics and kinetics of dyeing using acid dyes were studied, which show significant differences between breeds in both physical and chemical properties, which influence sorption properties. The rate of dye diffusion to the wool fibres showed a correlation with their physical properties which tells us about the chemical reactivity of the material. These differences may be exploited for novel applications by wool industries which can have alternative and sustainable textile products used in specific sectors like automotives, households, medical and construction areas.

Keywords

Wool, dyeing, acid dye, diffusion, scales, morphology

HOW TO GUIDE DESIGNERS IN ADOPTING A MORE SUSTAINABLE, TEXTILE ORIENTED, PRODUCT DEVELOPMENT PROCESS?

Jane McCann[1] *and Wang Lu*[2]

[1]*Design and Research Consultant, Landsend, 47 Warren Road, Donaghadee, Northern Ireland, UK.*
[2]*China Women's University, 1 Yuhui E Rd, Chaoyang, China, 100101*

Abstract

Designer input has significant ecological impact from initial concept, and throughout the development process, to garment end of life. Emerging and experienced practitioners require guidance in making more responsible decisions that result in garments that are fit for purpose, from technical and aesthetic perspectives, and more sustainable in materials selection and construction methods, with transparency in supply chains applicable to both carbon footprint and ethical practice. A design-led process considers key aspects from identification of end-user requirements to the subsequent potential for; the application of recycled, regenerated and bio fibres and fabrics, garment customisation (from the capturing size and shape to cutting, embellishment and joining techniques) recently demonstrated within trade fair digital micro factories, to strategies in design for disassembly. Reference is made to academic design projects as case studies (from the China Women's University) related to performance sportswear, a sector that spans from every-day street fashion to elite sport, design for ageing, and corporate wear. In particular, the European Outdoor Industry's continuing commitment to emerging policies around eco design is recognised. An overarching aim is to promote 'less is more' in helping to highlight the need for a wider societal understanding of the 'true cost' and value of clothing.

Keywords

Design Requirements, Regenerated materials, Bio fibres, Customisation.

EVALUATION OF ANTIMICROBIAL PROPERTIES OF DENIM FABRIC FINISHED WITH PLANT-BASED NANO-EMULSIONS

Prabhuraj D Venkatraman[1], Usha Sayed[2], Swati Korgaonkar[2] and Sneha Parte[2]

[1]*Manchester Fashion Institute, Faculty of Arts and Humanities, Manchester Metropolitan University, Cavendish Street, Manchester M15 6BG, UK*
[2]*Department of Fibres and Textile Chemistry, Institute of Chemical Technology [ICT], Nathalal Parekh Marg, Matunga, Mumbai 400 019, India*

Abstract

Denim fabrics are widely used in the apparel industry as casual wear among all age groups. Recently, increased awareness among consumers toward improved lifestyles and the need for sustainable and environmentally friendly material processing led to the demand for functional finishes for denim fabrics. Hence, denim fabrics, with improved functionality and comfort, are preferred for athleisure and compete with the sportswear market. However, during an intensive activity, the human body perspires, leading to the growth of microbes developing malodor. Hence, there is a need for denim fabrics to possess antimicrobial properties suitable for athleisure wear.

This research reports the finishing of denim fabrics with herbal-based nano-emulsions derived from a range of herbs *Moringa oleifera* (drumstick tree), *Evolvulus alsinoides* (Shankhapushpi), *Milletia pinnata* (Karanja), and *Pedalium murex* Linn (Gokhru) that is widely available. Nano-emulsions were characterized for their stability, and the finished denim fabrics were evaluated for their antimicrobial properties using Gram-positive bacteria [*Staphylococcus aureus*] and Gram-negative bacteria [*Escherichia coli*]. The results indicated that the antimicrobial resistance of finished denim fabrics was >99.2% between before and after ten washes. Outcomes from this research suggest denim fabrics can be finished with durable antimicrobial properties using plant-based herbal nano-emulsions that are environmentally friendly.

Keywords

Moringa oleifera, Evolvulus alsinoides, Milletia pinnata, Pedalium murex Linn, nano-emulsion, and antimicrobial resistance.

WOOL FROM LOCAL SHEEP BREEDS - ZERO WASTE PLAN FOR LOCAL COMMUNITIES

Monika Rom[1], *Jan Broda*[1], *Katarzyna Kobiela-Mendrek*[1], *Maria Kohut*[2] *and Ingvild Svorkmo Espelien*[3]

[1]*University of Bielsko-Biala, Willowa 2, 43-309 Bielsko-Bala, Poland*
[2]*Centre of Regional Produce, Koniakow, Poland*
[3]*Selbu Spinneri, Klæbu, Norway*

Abstract

For a long time, the wool of Polish mountain sheep was wasted. This wool is double fleece and thick due to the adaptation of sheep of this breed to live in harsh conditions. Due to its properties, it has limited applicability for clothing purposes; and it is also not of interest to the existing technical textile industry in Poland. Currently, the mountain sheep population in Beskids Mountains is ~14 000, which generates ~ 56 tons of wool annually, scattered in this part of Poland. This amount is too small for the traditional industry but too much for local communities that have lost their wool processing capacity in recent decades.

Identifying good practices and organizing local sales initiatives can help turn local wool into valuable products. Our research covers the wool value chain of Polish mountain sheep, indicating the possibility of using higher-quality wool for high-value decorative products and low-quality wool for fertilizers. We pay attention to the processing process's stages, supply chain gaps, and professional competence needs. The work will present the entire zero-waste production cycle of a product made of mountain sheep wool, with an indication of the product's performance parameters.

Keywords

Wool, sustainability, value chain, zero waste strategy

Acknowledgements

The project WOOLUME is funded by Iceland, Liechtenstein and Norway through the Norway Grants in Polish-Norwegian Research. The programme is operated by the National Centre for Research and Development, contract number: NOR/POLNOR/WOOLUME/0007/2019-00

EMBEDDING LIFECYCLE ASSESSMENT INTO THE INNOVATION PROCESS FOR A NEW BIODEGRADABLE TEXTILE MADE OF PINEAPPLE LEAF FIBRE

Kate Goldsworthy[1], Laura Solomon[1], Laura Knight[1], Sanne Visser[1], Raquel Prado[2], Melissa Braithwaite[2] and Alexandra Richardson[2]

[1]*University of the Arts London, UK*
[2]*Ananas Anam Ltd, London, UK*

Abstract

This project proposed the use of pineapple leaf fibres to produce a new yarn blend as a positive climate-friendly alternative. The yarn is produced with the same agricultural waste from pineapple production that is currently used to make Piñatex, a bio-based sustainable material.

The R&D work was led by manufacturer Ananas Anam and UAL's key role was to examine the environmental impact of each proposed material and process choice at key decision hotspots in relation to the lifecycle.

The final outcome was a comprehensive study of the environmental impacts and recovery possibilities of the product with the additional mapping of existing UK industrial infrastructure for composting or recycling. These results supported the development of a new textile product which has recently been launched to the market, Piñayarn.

In this paper we will use the project as a case study to discuss:

Methods to integrate environmental assessment as part of the R&D process ensuring informed decision making throughout.

Understanding of barriers and opportunities for recovery options for the developed materials including extensive mapping of the available UK infrastructure for industrial composting.

Incorporation of data visualisation and graphic communication into the R&D process to aid interdisciplinary collaboration.

Keywords

Circular textiles, industrial composting, material R&D, process innovation, interdisciplinary collaboration

POTENTIAL OF NATURAL SUSTAINABLE BIOPOLYMERS WITH ANTIMICROBIAL PROPERTIES

Sameer Rahatekar[1], Prabhuraj D Venkatraman[2], Jonathan Butler[3], Pubudu Dias[4] and Stewart Milligan[4]

[1]School of Aerospace Transport & Manufacturing, Cranfield University, Cranfield, UK
[2]Manchester Fashion Institute, Faculty of Arts and Humanities
[3]Department of Life Sciences, Faculty of Science and Engineering, Manchester Metropolitan University (MMU), All Saints, Manchester, UK
[4]Pangaia Materials Science Limited, London

Abstract

Textiles and the fashion industry contribute toward waste generation, and according to UNCTD (UN Conference of Trade and Development), fashion is the second most polluting industry in the world. In addition, half a million tonnes of microfibre waste are dumped in the ocean annually (the majority of them is from the laundering of synthetic clothes). The industry relies on synthetic materials derived from petroleum or non-renewable resources, which has a significant environmental impact.

Biopolymers are green alternatives to petroleum-based synthetic fibres and are derived as polymers from natural resources (plant-based/animal-based/microorganisms). Biopolymers are renewable, environmentally friendly, carbon-neutral, and biodegradable. The biopolymer filaments can be blended with sustainable fibres such as Tencel (lyocell) and organic fibres for textiles and fashion.

This paper discusses various biopolymers from natural resources and how functional properties such as durability and antimicrobial properties can be enhanced. In addition, research also discusses how cellulose waste and natural biopolymers (such as seaweed) can be blended to produce a new generation of biopolymer filaments using low-impact solvents that are environmentally sustainable and suitable for various textile applications. Finally, the study highlights the benefits of such biopolymers and broader implications for the industry.

Keywords

Biopolymers, microfibre waste, cellulose, seaweed, environmental impact, and sustainable fashion.

LINEN AND HEMP FOR SUSTAINABLE TEXTILES

Malgorzata Zimniewska[1]

[1]*Institute of Natural Fibers and Medicinal Plants National Research Institute, Poznan, Poland*

Abstract

Sustainability is integral part of bio-economy development in XXI century. The principles of sustainability must be realized in human activities in terms of environment, economy and society. The significant challenge faced by textile sector is development of processes and technologies, which allow reduce water and energy consumption, effective raw materials use, implement zero waste strategy, design for recycling and from recycled materials, prevent emission of greenhouse gasses.

Among vegetable fibers, apart from cotton, linen is the most often used for textile purpose. Currently, hemp is considered to be a valuable raw materials for a sustainable economy due to its inherent pro-environmental and pro-healthy features, even the textile hemp processes are more difficult in comparison to cotton and flax due to specific fiber properties and limited access to industrial infrastructure.

The paper shows holistic approach sustainability of technological chain for both bast fibers from the plant harvesting up to final product e.g. functional hemp/linen clothing. Results of fiber and fabric properties tests, design of clothing and evaluation of the functional clothing effect on the skin of wearer are presented.

Keywords

Flax fiber, hemp fiber, fiber processing, sustainability, functional clothing

Technical Textiles & Composites

A SUSTAINABLE APPROACH: RECOVERED SILK SERICIN BIOMACROMOLECULE AND ITS REUSE AS FILMS AND NANOFIBERS FOR BIOMEDICAL APPLICATION.

Mangala Joshi[1]

[1]Department of Textile Technology Indian Institute of Technology, Delhi Hauz Khas, New Delhi 110016, India

Abstract

Sericin is a globular bio macromolecular protein which constitutes 25-30% of silk cocoon and envelops the fibroin protein which constitutes the silk fiber. In recent years silk sericin has been investigated for its unique properties in biotechnological field. These include biocompatibility, biodegradability, antibiotic and anticoagulant, antibacterial activity, antioxidant behaviour, anti-tyrosinase activity, anticarcinogenic effects and UV protective properties. Sericin promotes human skin fibroblast attachment in cell culture. It also supports cell adhesion and proliferation and plays a significant role in wound healing process. Recently several studies have been reported on sericin films, membranes, 3D scaffolds, nanoparticles, and composites, conjugated drugs in pure or blended form for cell culture, wound dressings, drug delivery applications. In this paper, our work in the area of sericin based films and nanofibrous webs having potential for biomedical application such as wound dressing etc will be presented. Nanofibres show several advantages in biomedical field especially wound dressing because of its high surface area to volume ratio, high porosity and light weight. Sericin in pure form cannot be converted to nanofibre form because of low molecular weight. Hence blending with polyurethane has been explored to exploit the good properties of both to make a good wound dressing nanofibrous membrane. In one of our studies, the feasibility of production of electrospun composite nanofibres mainly for biomedical application is reported. Biocompatible polyether-based polyurethane, natural proteinaceous polymer sericin and natural inorganic nano clay Halloysite as a drug carrier with a model drug chlorohexidine acetate were used for production of nanofibres by electrospinning technique. Sericin was extracted from Bombyx mori silk cocoons by high pressure high temperature (HT-HP) degumming. Chlorohexidine acetate (CA), an antimicrobial agent, loaded into Halloysite nanotubes at 1:1 weight ratio showed the maximum loading and was confirmed by TGA and XRD analysis. Electrospinning of polymer solution with different composition of polyurethane, sericin, CA and CA-HNTs was carried out which resulted in the formation of bead-free uniform fibres. Antimicrobial activity of nanofibrous webs was evaluated by disc diffusion method (AATCC 90) and it was found that CA and CA-HNT loaded nanofibres shows better antibacterial action against both the Gram-positive (*S. aureus*) and Gram-negative (*E. coli*) bacteria. The CA-HNT loaded nanofibres showed controlled release of CA and only 25.9 % of CA was found to be released in first 7 days. The antibacterial efficacy of the developed material and the drug release behaviour have been studied very systematically in the present composition. Hence the produced composite nanofibre can serve as an excellent material for sustained drug release for various biomedical applications.' Sericin film also has lot of potential applications but at the same time it has some drawbacks like poor tensile strength, poor thermal stability and solubility in warm water, which limits many applications of sericin especially in film form. In this present paper, a novel eco-friendly approach has been used to enhance the properties of sericin in film form through sericinmetal complex formation using alum salt. It is expected that optimum cross-linking provided by aluminium would enhance the thermal stability and would also make sericin film insoluble in warm water while retaining its other useful properties.

References

1. Mangala Joshi, M L Gulrajani and Mahadev Bar, Novel Cross-linked Sericin Film Synthesis and Characterisation, Journal of Applied Polymer Science (Impact Factor: 1.77). 02/2015; 132(5). DOI: 10.1002/app.41400

2. Tao G, Wang Y, Cai R, et al. Design and performance of sericin/poly(vinyl alcohol) hydrogel as a drug delivery carrier for potential wound dressing application. Mater Sci Eng C 2019; 101: 341–351.

3. Lamboni L, Gauthier M, Yang G, et al. Silk sericin: A versatile material for tissue engineering and drug delivery. Biotechnol Adv 2015; 33: 1855–1867.

4. Kundu SC, Dash BC, Dash R, et al. Natural protective glue protein, sericin bioengineered by silkworms: Potential for biomedical and biotechnological applications. Prog Polym Sci 2008; 33: 998–1012

COLLABORATIVE OPPORTUNITIES FOR TEXTILES IN UNLOCKING A CIRCULAR COMPOSITES INDUSTRY

Jenny Banks[1]

[1]*National Composites Centre, Bristol & Bath Science Park, Emersons Green, Bristol, UK*

Abstract

Many composite components and structures currently in use are reaching their end-of-life; but only 15% of the 110,000 tonnes of composites produced in the UK each year will be recycled or reused at end-of-life. And only 6% of this makes it into secondary applications.

Through our extensive composite recycling activities within our "Sustainable Composites" programme, we have identified that whilst fibre-reinforced composite products can be reclaimed into their constituent materials (i.e. fibre and resin), the circularity of the recovered materials is hindered by technology bottlenecks in pre- and post- reclamation processing and qualification. Some of these critical areas include: fibre sizings and re-sizing processes, automated fibre inspection/sorting, non-woven fabric production and short-fibre spinning.

Much of the foundational knowledge and technology needed to overcome these challenges already exists within the textiles sector. This presentation details the critical blockers in composite recycling and presents the research opportunities that would benefit from collaboration between the textiles and composites sectors to enable a viable circular composite supply chain.

Keywords

Sustainable composites, composites recycling, high value recyclate, eol supply chain, textile technology, collaborative opportunities, TIWC 2023

STUDY ON CERIUM OXIDE-BASED TPU NANOCOMPOSITE FILMS FOR IMPROVED WEATHER-RESISTANCE PROPERTY

Shuchita Tomar[1], *Bharti Rana*[1], *B S Butola*[1] *and Mangala Joshi*[1]

[1]*Department of Textile and Fibre Engineering, Indian Institute of Technology, New Delhi, 110016*

Abstract

Thermoplastic polyurethane (TPU) exhibits excellent flexibility, good abrasion resistance, chemical resistance, toughness, and strength. However, the weather resistance property of TPU is poor. Therefore, in order to improve the weather-resistance property, inorganic nano-fillers have been incorporated into the TPU matrix. In this work, the effect of cerium oxide (CeO_2) nanoparticles on improving the weather-resistance property of TPU film has been studied. Firstly, the CeO_2 nanoparticles were modified using 3-Aminopropyl trimethoxysilane (APTMS) to improve its dispersion in TPU matrix. Afterwards, TPU/m-CeO_2 nanocomposite films were prepared at different concentrations (3,5,7 wt %). The surface morphology, particle size, and structure of m-CeO2 nanoparticles and TPU/m-CeO_2 nanocomposite films were characterized by using transmission electron microscopy (TEM), field emission scanning electron microscopy (FESEM), and Fourier transform infrared spectroscopy (FTIR). The UVresistance property of TPU/m-CeO_2 nanocomposite films was analyzed in terms of ultra-violet protection factor (UPF) and it was found that UPF increases with an increase in CeO_2 content. The mechanical strength of nanocomposite films was also investigated. Further, neat TPU and TPU/m-CeO_2 nanocomposite films were exposed to accelerated artificial weathering for 300 h and it was observed that weather-resistance property of TPU/m-CeO_2 nanocomposite film has been improved as compared to neat TPU film.

Keywords

Thermoplastic polyurethane, cerium oxide nanoparticles, weather-resistance, nanocomposite films

UV PROTECTION AND GAS BARRIER PERFORMANCE OF ZINC STANNATE BASED TPU NANOCOMPOSITES

Bharti Rana[1], *Shuchita Tomar*[1], *S Wazed Ali*[1] *and Mangala Joshi*[1]

[1]*Department of Textile and Fibre Engineering, Indian Institute of Technology, Delhi-110016, India*

Abstract

An economically viable category of flexible elastomers is thermoplastic polyurethane (TPU). Inorganic nanomaterials incorporated TPU's diverse qualities have influenced practically every aspect of material innovation. In this study, $ZnSnO_3$ has been synthesized via hydrothermal method and incorporated in TPU matrix. Fabrication of nanocomposite films of varied thickness has been done by compression molding method. The nanomaterial has been characterized using transmission electron microscopy (TEM), field emission SEM, Fourier transform infrared spectroscopy (FTIR), X-ray diffraction (XRD), and energy dispersive X-ray (EDX). 2%-10% of as-synthesized nano $ZnSnO_3$ was added to TPU as a nanofiller. The produced nanocomposite films were evaluated for their helium gas barrier property and UV protection performance before and after exposure to UV. UV protection factor (UPF) has been improved with increased loading of ZnSnO3 and thickness of the films from about 87 for neat TPU to 285 at 10% concentration of $ZnSnO_3$. Helium gas barrier has been observed to improve by 16-18%. Moreover, tensile properties of films were also found to increase up to 6% loading of $ZnSnO_3$. Hence, $ZnSnO_3$ is proven to be act as an efficient inorganic nano-filler in the enhancement of gas barrier and UV resistance of TPU nanocomposites.

Keywords

Synthesis, Zinc stannate, TPU, Nanocomposite films, Gas barrier, UV protection

DECARBONISING BY CAPTURING EFFICIENT TEXTILE DATA ENABLING IMPROVEMENT ACTIONS IN ACHIEVING ZERO CARBON

Sajida Gordon[1]

[1]*Nottingham Trent University, 50 Shakespeare St, Nottingham NG1 4FQ*

Abstract

Textile production environmental volatility is apparent, with many activities not demonstrated openly. The importance of manufacturer environmental performances and the acquisition of this data is crucial for emissions reduction and contribution to achieving net zero. (Rasool et al, 2022). Furthermore, data associated with textiles' GHG impacts is a complex system with the possibility of capturing copious amounts at varying stages. Metrics need to be determined at first instances, this provides the raw ingredients that underpin and provide precise analytics which create data (Rennell, 2021). Data mining and optimisation become essential for, visibility and governance, however, evidently many textile producers are not participating in demonstrating visible data to assist with improvement actions. (Patel, Kathiriya 2022). This paper will aim to propose a framework for successful data capturing of textile production activities within the supply chain. Qualitative data is gathered with the use of semi-structured interviews with producers as well as overt and unstructured non-participation observation within a textile manufacturer. A detailed process is determined that identifies multiple states of carbon release and highlights lack of manufacturer obligations in advocating and participating in environmentally friendly initiatives (Kazancoglu, 2021). The framework will offer to support policy through an intergovernmental setting in capturing thorough data for a fragmented global supply network.

Keywords

Textile value-chain, textile manufacturing, carbon reduction, data acquisition.

References

KAZANCOGLU, Y., SAGNAK, M., MANGLA, S.K., SEZER, M.D., and PALA, M.O., 2021. A fuzzy based hybrid decision framework to circularity in dairy supply chains through big data solutions. Technological Forecasting & Social Change, 170, 120927.

PATEL, A.A. and KATHIRIYA, D., 2022. Cotton Crop Yield Prediction using Data Mining Technique. International Journal of Advanced Computer Science & Applications, 13 (1)

RASOOL, F., GRECO, M., and GRIMALDI, M., 2022. Digital supply chain performance metrics: a literature review. Measuring Business Excellence, 26 (1), 23-38.

Tom Rennell., 2021. What is Data vs a Metric vs a KPI vs a Report? What is Data vs a Metric vs a KPI vs a Report? (adverity.com) (Accessed 13/02/2023)

LIGNOCELLULOSIC BIOMASS AS A SOURCE OF INDUSTRIAL ABSORBENTS AND ADSORBENTS

Deepti Gupta[1]

[1]*Department of Textile and Fibre Engineering, Indian Institute of Technology Delhi, India*

Abstract

Lignocellulosic fibers are a group of natural materials that are derived from plant biomass and consist of cellulose, hemicellulose, and lignin. The use of lignocellulosic fibers as industrial absorbents and adsorbents is a growing area of research and development in the field of sustainable materials.

India has a vast variety of biomass available in the form of crop residues as well as other plant sources. A major challenge in large-scale exploitation of lignocellulosic materials indigenous to India is a lack of standardisation of methods and technology used for extracting or separating the fibres from other constituents like lignin and pectin. Traditional methods continue to be used but these do not yield standard quality of fibres. This keynote address will review the modern technologies that can be used to improve the efficiency of the fibre extraction method.

Lignocellulosic fibers have been shown to possess excellent adsorption and absorption properties due to their high surface area, porosity, and hydrophilicity. The different types of lignocellulosic fibers available and their chemical and physical properties that make them suitable for use as industrial absorbents and adsorbents. Physical and chemical methods can be used to improve the properties of LC materials. The work done by our group on extraction and use of LC fibres as potential absorbents and adsorbents will be discussed.

MESOPHASE PITCH POLYETHYLENE AND GRAPHENE BLEND FIBRES WITH IMPROVED STRAIN TO FAILURE AS POTENTIAL CARBON FIBRE PRECURSORS

Salem Aldosari[1-2], *Krzysztof Koziol*[2] *and Sameer Rahatekar*[2]

[1]*Future Mobility Institute, King Abdulaziz City for Science and Technology(kacst) Riyadh 11442, Saudi Arabia*
[2]*Enhanced Composite and Structures Centre School of Aerospace, Transport, and Manufacturing, Cranfield University, Cranfield MK43 0AL.*

Abstract

The advantage of mesophase pitch-based carbon fibres is its high modulus, but pitch-based carbon fibres and precursors are very brittle. This paper reports the development of a unique manufacturing method using a blend of pitch and Linear Low-Density Polyethylene (LLDPE)/Graphene from which it is possible to obtain precursors which are less brittle than neat pitch fibres. This study reports on the structure and properties of (LLDPE 20 wt%)/Mesophase Pitch blend precursors and graphene content ranging from 0.2wtg% to 0.5wtg% and 1wtg% fix LLDPE 20wt% and mesophase pitch 80wt%.

Fibre microstructure was determined using Scanning Electron Microscopy (SEM) which showed a two-phase region having distinct pitch fibre and LLDPE regions. Tensile testing of 20wt%LLDPE/pitch show low strain to failure (brittle), but as the percentage of graphene was increased, the strain to failure but decreased tensile strength. DSC characterisation of the melting/crystallization behaviour of LLDPE showed melting occurred around 120°C to 124°C, with crystallization between 99°C to 103°C. TGA measurements showed that for LLDPE/pitch 20wt%, 0.2wt% Graphene thermal stability was excellent to 800°C.

The Mesophase pitch and polyethylene blend showed reduced brittleness combined with excellent thermal stability, and thus is a candidate as a potential precursor for pitch-based carbon fibre manufacturing.

Keywords

Pitch, graphene, polyethylene, carbon-fibres, extrusion, blend

SCREENING FLAME-RETARDANTS FROM NATURAL SOURCES FOR BACTERIAL CELLULOSE FABRICS

Juneyoung Minn[1], *Hyunjin Kim*[1] *and Hye Rim Kim*[1]

[1]*Department of Clothing and Textiles, Sookmyung Women University, Cheongpa-ro 47-gil 100, Yongsan-gu, Seoul, South Korea*

Abstract

Textile products are used in a wide range of applications, from clothing to interiors and transportation. These ubiquitous textile products are typically flammable. Thus, textile products are often treated with flame retardants to protect from the fire accidents. However, the commonly used flame retardants have problems such as toxicity or generation of harmful chemicals, so there is an increasing demand for natural flame retardants that are friendly to the environment and human body. In this study natural materials such as turmeric, potato starch, xylitol, spinach, and banana peel were selected to impart flame retardancy to bacterial cellulose (BC) due to their flame retardancy. The natural materials were treated with ultrasonication and then dried for combustion test. The flame retardancy was evaluated based on self-extinguishment properties. In addition, the sources found to be self-extinguishing were experimented by adding the process of water bath treatment and curing process to determine the effect of treatment conditions on flame retardancy. Thus, this study has screened natural materials to enhance the flame retardancy of BC.

Keywords

Bacterial cellulose, flame-retardant, natural source, leather-like fabric

SURFACE FUNCTIONALIZATION OF TEXTILE AS A TOOL FOR FABRICATION OF SMART FABRICS

Elizabeth Amir[1]

[1]*Shenkar College, Faculty of Engineering, Department of Polymer Materials Engineering, Yeda Am 8, Ramat Gan, Israel*

Abstract

This talk will describe new strategies for chemical surface modification of textile, specifically seeking processes which are modular, inexpensive, technically simple and would not damage the fabric, while providing durable functional coatings. One of the methods is based on a two-step process that includes surface grafting of bifunctional small linker molecule, 10-undecenoyl chloride, via esterification reaction and a subsequent covalent attachment of functional materials to the double bond endgroup via UV-mediated thiol-ene click chemistry. The covalent grafting can be accomplished in a rapid manner within several minutes leading to uniform and thin functional coatings. The modified fabrics were characterized using contact angle measurements, Fourier Transform Infrared and X-ray Photoelectron Spectroscopy analysis. The developed approach allows preparation of woven, knitted and non-woven textile based on natural and synthetic polymers with controlled hydrophobicity/hydrophilicity, oil-water separation ability and pH regulation. In addition, preparation of hybrid antibacterial, antiviral and electro-conductive fabrics based on organic conjugated polymer will be presented. Polyaniline-coated fabrics were prepared via in situ polymerization of aniline monomer and the influence of the aniline concentration on the electrical resistivity of the modified fabrics was evaluated. Antibacterial properties were examined for *S. aureus* and *S. epidermidis* types of bacteria.

Keywords

Surface grafting, covalent bonding, click chemistry, functional coating

EXTRACTION OF TEXTILE GRADE BAUHINIA VAHLII FIBRE AND CHARACTERISATION OF ITS ESSENTIAL PROPERTIES

Goutam Bar[1-2] and Kavita Chaudhary[2]

[1]Department of Textile Design, National Institute of Fashion Technology, Bhubaneswar, India
[2]Department of Design, Banasthali Vidyapith, Banasthali, Rajasthan, India

Abstract

Increasing demand for bio-degradable and eco-friendly textile products to confront environmental pollution accelerates the research on bio-based fibers. Researchers are also in search of novel natural fibers for different textile products. The present paper focuses on extracting textile-grade *Bauhinia vahlii* fibre from the *Bauhinia vahlii* plant bark and analysing their essential properties. *Bauhinia vahlii* is a tropical woody climber mainly grown in the warmer region of the world. The plant bark is rich in fibre and used for making ropes and other tabletop and household products, like bags, small boxes, etc. The leaf and bark extract is also rich in natural colorant materials and has anti-microbial and antioxidant properties. So far, few attempts have been made to extract the fibre from the bark for natural fibre-reinforced composites. However, the extracted fibre strands were in a very coarser form. Textile grade *Bauhinia vahlii* fibre was extracted after several attempts by treating the *Bauhinia vahlii* plant bark with dioxane, dimethyl sulfoxide (DMSO), and varied concentrations [5%, 10%, 15%] of sodium hydroxide. Dioxane and DMSO could remove minimal lignin from the bark structure. The higher amount of lignin was dissolved, and the bark structure was softened, when treated with 10 and 15% NaOH solution at boiling for 2 hours. However, better fibre separation was achieved in the case of 15% NaOH treatment.

The extracted fibre was analysed for its physical properties, like fibre fineness, length, density, and moisture content. The extracted fibre was also analysed for its chemical composition, the effect of different solvents, crystallinity index, thermal stability, surface morphological structure, cross-sectional view, and shape factor. It was found that the extracted fibre's average fibre fineness is 15.77 microns, the moisture content is 10.5%, the density is 1.47 gm/ cm3, and the average fibre length is 20.04 mm. The extracted fibre contains 65.2% α-cellulose, 10.6% hemicellulose, 13.7% lignin, and 1.8% pectin. The fibre has a rough surface along with some fragmented impurities. A cross-sectional view of the fibre shows that fibre has an oval to round fibre cross-section with an average 0.1232 shape factor. X-ray diffraction proclaims that the extracted fibre has 56% crystallinity and a smaller crystallite size than cotton fibre. A thermogravimetric analysis confirmed that the fibre is thermally stable up to 233oC.

Keywords

Bauhinia vahlii, Crystallinity index, Chemical composition, Fibre extraction, Physical properties, Surface morphology, Thermal analysis

OPTIMUM CONDITIONS FOR THE CARBONIZATION OF KEVLAR FABRIC FOR ELECTROMAGNETIC INTERFERENCE SHIELDING

Kuldip Singh[1] *and Vijay Baheti*[1]

[1]*Department of Textile and Fibre Engineering, Indian Institute of Technology, New Delhi, 110016, India*

Abstract

This study involves the optimization of single-stage carbonization and physical activation of Kevlar fabric to form activated carbon fabric for electromagnetic interference shielding applications. The carbonization process was optimised for high carbon yield and low electrical resistivity of activated carbon fabric using Box-Behnken experimental design. For the study, three process parameters i.e., charcoal amount, nitrogen gas flow rate, and carbonization temperature were selected. The empirical equations were obtained using response surface methodology to find the relationship between the process and response parameters. An optimum amount of charcoal and nitrogen flowrate and low carbonization temperature was found suitable for higher carbon yield. Also, high charcoal amount, high carbonization temperature, and an optimum nitrogen flow rate were favourable for the low electrical conductivity of activated carbon fabric. The optimized sample for low electrical resistivity was porous and contained a higher graphitic structure. Further, this sample showed electrical resistivity of 0.0135 kΩ.cm and average electromagnetic interference shielding of 21.65 dB and 17.84 dB in the frequency range 30 MHz - 1.5 GHz and 8 GHz – 12 GHz, respectively. It was concluded that the presented carbonisation method provides a simple and easy preparation of absorption dominant activated carbon fabrics for effective electromagnetic interference shielding.

Keywords

Kevlar woven fabric, Activated carbon fabric, Carbonization, Box Behnken design, Carbon yield, Electrical resistivity, Electromagnetic interference shielding

EFFECT OF QUILTING THREAD AND PITCH ON BALLISTIC PROTECTIVE PERFORMANCE OF PARA-ARAMID FABRICS

C H Crawford[1], D J Carr[2] and C Lankester[3]

Centre for Defence Engineering, Cranfield University, Defence Academy of the United Kingdom, Shrivenham, SN6 8LA, UK.
[1]Permali Ltd., Gloucester, GL1 5TT, UK.
[2]dstl, Salisbury, SP4 0JQ, UK.
[3]Corin Group, Cirencester, GL7 1YJ, UK.

Abstract

Personal protective equipment (PPE) is designed to protect the wearer from injury and/or fatality i.e. the use of PPE increases the likelihood of survivability. Body armour, which protects the critical organs of the torso, is typically constructed of multiple layers of para-aramid or ultra high molecular weight polyethylene (UHMWPE). These layers of fabric are often quilted for ease of handling during construction of the armour and to prevent the slumping of the armour layers in the carrier. However, minimal work has been published on the effect of quilting on ballistic protection. In this work, the effects of quilting pitch (none; bias direction - 10 mm, 20 mm and 30 mm pitch), sewing thread type (cotton, polyester-cotton, para-aramid, bonded nylon) and number of layers (one, three) of a Twaron® fabric on the fragment protective performance was investigated with reference to military body armour systems. The use of polyester cotton and para-aramid sewing thread to form quilting negatively affected the fragment protective performance of the fabric used. The work reported in this paper suggested that 100% cotton or bonded nylon sewing threads should be used for quilting of body armour packs.

IN-SITU SYNTHESIS OF ZNO NPS USING NITRATE SALT

Saikat Ghosh[1], *Sohel Rana*[1] *and Parikshit Goswami*[1]

[1]*University of Huddersfield, Technical Textiles Research Centre, Department of Fashion and Textiles, Queensgate, Huddersfield HD1 3DH, UK*

Abstract

Fabrication of nanoparticles on to textile for developing a customised functional fabric through mass production is a real challenge. A probable solution of this is to develop a durable and multi-functional finishing. In this research work, nanoparticles were synthesised through in-situ pathway.

The wool fabric with twill weave, having a natural cream colour and was pre-scoured with no extra factory finishes. Analytical grade sodium carbonate and a non-ionic detergent, Ultravon JUN were purchased used for scouring of the wool fabric. The synthesis was done using metal precursor, zinc nitrate hexahydrate, with capping agent, HMTA and a coupling agent, trimethoxymethyl silane, where all the chemicals were analytical grade and used without further purification, purchased from Sigma Aldrich (UK).

Formation of nano ZnO was confirmed through SEM as well as through EDX results, which indicate the synthesis of ZnO on wool fabric. Zetaseizer also confirms the particles are synthesised in nanostructure whereas the zeta potential data indicates the aggregation tendency of the particle is less due to higher potential value.

In-situ synthesis within wool fabric, by mitigating the chemical reducing agent, helps to reduce the chemical usage in large scale. Finally, the project aims to utilise multifunctionality of nano ZnO, in automotive textile with fire retardancy, water repellency, antimicrobial properties.

OPTIMISATION OF DISENTANGLED POLYETHYLENE LAMINATE STRUCTURE FOR SOFT BODY ARMOUR APPLICATIONS

Bhupendra Singh Butola[1], *Mukesh Bajya*[1] *and Abhijit Majumdar*[1]

[1]*Indian Institute of Technology Delhi, Department of Textile & Fibre Engineering, New Delhi, India*

Abstract

Disentangled polyethylene (DPE) is a very promising material for body armour application. The effect of laminate structure, binding resin content and processing temperature on the mechanical properties and impact performance of DPE cross-ply laminates has been investigated in this research. Three structures, namely cross-ply laminates of tape (CPT), woven untwisted tape (UUT) and woven twisted tape (WTT) were prepared. Laminate made from WUT showed inferior impact energy absorption than CPT, while the WTT based laminate performed very poorly yielding the lowest energy absorption. CPT made with low resin content at higher processing temperature showed better tensile and impact resistance performances. Furthermore, to confirm the efficacy of cross-ply laminates in ballistic applications, two soft armour panels (SAPs) were prepared from WTT and CPT laminates. SAPs made from WTT got perforated by 9 mm bullet impacting at 430 m/s velocity while the SAP made from CPT laminates was able to stop the bullet and the back face deformation was 22.2 mm. Therefore, DPE tapes can be explored as a very promising material for the development of soft body armour.

Keywords

Disentangled polyethylene, Impact resistance, Laminate, Tape, Woven.

DEVELOPING SUSTAINABLE BACTERIAL CELLULOSE COMPOSITES USING MARINE POWDERS

Hyunjin Kim[1] *and Hye Rim Kim*[1]

[1]*Department of Clothing and Textiles, Sookmyung Women University, Cheongpa-ro 47-gil 100, Yongsan-gu, Seoul, South Korea*

Abstract

Bacterial cellulose (BC) is a pure cellulose material derived from bacteria. BC is known as a sustainable zero-waste fabric owing to its good moldability. Moreover, the performance of BC could be improved by incorporating additives. In this study, marine powders, namely anchovy, mussel, and shrimp powders are used to impart functionality to BC because they are rich in proteins. Marine powders could reduce waste generation because they are mostly derived from discarded marine organisms. Therefore, this study aims to develop sustainable BC composites using marine powders via physical entrapment and crosslinking methods. As a result, surface characterizations have confirmed that BC composites are physically entrapped and crosslinked with proteins from marine powders without changing the structure of BC. The durability of BC composites was analyzed in terms of flame resistance after washing, mechanical properties, softness, crease recovery, water resistance and dimensional stability. From the analyses, BC composites have sufficient durability compared to the crust cowhide leather of similar appearance and thickness with BC. This study has provided a method of producing sustainable BC composites with enhanced durability by using marine powders.

Keywords

Bacterial cellulose, marine powders, proteins, durability

A STUDY ON THE CREEP BEHAVIOUR OF OZONE TREATED JUTE FABRIC/EPOXY COMPOSITES FILLED WITH OZONIZED AND PULVERIZED SISAL FIBRES

Debarati Bhattacharyya[1] *and Vijay Baheti*[1]

[1]*Indian Institute of Technology Delhi, Research Scholar, Department of Textile and Fibre Engineering, Hauz Khas, New Delhi, 110016, India.*

Abstract

The present study deals with the ozone treatment of natural fibres mainly obtained from sisal fibres. The main advantages of performing ozone treatment of cellulosic materials are that it reduces the usage of water, energy and time, since it can physically bleach the cellulosic materials without the requirement of these precious elements of our daily lives. Also, ozonisation limits the harmful impact on our ecology, basically the chemical oxygen demand values, of the various processes. In addition, ozone treatment is also practiced in the field of natural composites, medical sectors, and in other textile processing areas. In addition, we have mentioned about an industrially scalable machine i.e. ball milling machine required to produce particles from such fibres. This machine can produce sufficient quantity of particles in a quite reasonable amount of time. Furthermore, we have also observed that by ozonising the sisal fibres, makes them more prone to breakage due to removal of lignin and other impurities along with a decrease in their surface energy, making them more brittle. Also, we have taken the SEM images of the particles and estimated the stresses required for their catastrophic fracture from Griffith theory. After preparation of the sisal particles we have made composites at different particle loadings and compared their properties with neat ozone treated jute fabric composites. Finally, a brief comparison has been made in terms of creep with composites loaded with different sisal fibre-based fillers. The results demonstrate that sisal filler loaded ozone treated jute fabric composite show better performance when compared to the other sisal fibre loaded composite.

Keywords

Cellulose; Jute fabric; Sisal fibres; Oxidation; Creep Modelling

COMMON APPAREL PERFORMANCE CHARACTERISTICS OF BACTERIAL CELLULOSE GROWN FROM KOMBUCHA

Jane Wood[1], *James Redfern*[2] *and Joanna Verran*[2]

[1]*The Department of Materials, Faculty of Science & Engineering, The University of Manchester, UK.*
[2]*The Department of Natural Sciences, Faculty of Science & Engineering, Manchester Metropolitan University, UK.*

Abstract

The textile industry is under pressure to amend many of its working practices to address their environmental impact. As various initiatives aim to implement changes in processes, there is still a need to seek alternative sources for raw materials. Textiles such as bacterial cellulose (BC), produced from microbes, are seen as potential 'environmentally friendly' alternatives.

This study aimed to find a reproducible method to manufacture BC sheets grown from a Kombucha starter culture and evaluate their performance characteristics using standard tests for tensile strength, abrasion, and moisture permeability, assessing the suitability of the material for fashion apparel applications. As BC forms as a nanofibrillar mat (similar in structure to a non-woven) and is commonly referred to as 'vegetable leather' due to its physical appearance, performance comparisons were made with animal leather, wool and cellulose non-woven textiles. The study showed that BC mats could be consistently reproduced using a black tea and glucose medium. BC was the most moisture-retentive and least breathable of all the samples tested and displayed a lower tensile strength than animal leather. However, as BC displayed superior abrasion resistance it is suggested that it could be explored as an alternative textile for use in protective apparel.

Keywords

TIWC2023, Textiles, Sustainability, Bacterial Cellulose, Kombucha, Fashion, Apparel, Performance Textiles.

Business Models & Supply Chain

'SUSTAINING FASHION' - HOW CAN DESIGN COMMUNICATION HELP? THE NEED FOR THE DESIGN IDEA TO TURN INTO THE PERFECT PRODUCT FOR BODY FIT & STYLE

Anastasia Vouyouka[1], Steven George Hayes[2] and Phoebe Apeagyei[3]

[1]*eTELESTIA CAD & Fashion Learning, www.etelestia.com*
[2]*The University of Manchester, The Department of Materials.*
[3]*Manchester Metropolitan University, Faculty of Arts & Humanities, Manchester Fashion Institute*

Abstract

A big divide exists between the aptitude for selling and real creative thinking in design and make. Retail is too powerful, exerting far too much influence on the strategic development of the fashion sector. This asymmetric power distribution makes it difficult for the technical creative dimensions to evolve, hampering small manufacturing efforts and large-scale improvements alike. Deskilling instead of upskilling results by HE attempting to serve big retailing and large student cohorts rather than designer makers and retailers being unable to provide meaningful in-house training or work-based learning.

Between consumer offerings, the current design realization process and decoupled manufacturing has resulted in a void between consumption and creation. Efforts to reconcile cooperation are failing! What the needed steps should be? How to reconcile contributing parts and cover the gap saving skills, creating them again, and save waste, the sector, and the planet? Focus group findings, and best practice examples will be explored and presented, to help add value to the new needs of the industry and education. The suggestion of a new Design Process Approach can help towards a new Business Win-Win Plan .

THE TRANSPOSING OF FASHION COMMUNICATION AND BRANDING CONCEPTS WITHIN THE METAVERSE FROM A CONSUMER PERSPECTIVE FOR STREETWEAR BRANDS.

Guy McKelvey[1]

[1]*University of Huddersfield, Queensgate, Huddersfield,*

Abstract

The aim of this study is to carry out an in depth understanding and investigation into the transposing of fashion communication, branding, and sales concepts for luxury streetwear within the metaverse analysed from the existing key target audience perspective of millennials and generation Z.

This research is focused on exploring the nascent developing and emerging concept of the metaverse in the context of fashion, which is seeing increasing consumer interest and latterly investment from fashion companies. This new and unchartered area offers opportunities and risks for companies entering it and the research focuses on looking at what is the best approach to build an effective and holistic branding strategy for fashion brands to enter the metaverse in conjunction with their current omnichannel strategy by focusing firstly; on consumer demands and appetites for fashion products in the metaverse.

Methodology

The approach of conceptualising what the metaverse is from a theoretical and commercial perspective to consumers and brands within fashion luxury streetwear through both secondary research and primary data offers a new area of exploration within fashion academia. In order to underpin this research strand and finding little current literature on the subject a small-scale exploratory study was undertaken.

Keywords

Huddersfield, TIWC2023, Future Trends, Metaverse, Fashion , Streetwear, Luxury, Retailing , Generation Z.

TRANSPARENCY IN TEXTILE SUPPLY CHAIN – WHY REAL, ACCURATE DATA IS ESSENTIAL

Frank Reidl[1]*, Andreas Roth*[1] *and Patrick Roth*[1]

[1]*Natific AG, Industriestrasse 111, 4147 Aesch, Switzerland*

Abstract

Sustainability and transparency are becoming more important, as self-regulation by the industry is being replaced by legislation from many countries. The need to provide information on the supply chain has become a legal requirement.

The supply chain for brands is diverse and global.

Consider a pair of denim jeans – cotton producer, yarn manufacturing, weaving, dyeing, finishing and garment making, all must be considered and understood. Understanding the supply chain to tier two, three and four, taking into account sewing threads, buttons, zips, transportation between sites all need to be considered and known so that the so-called FOOTPRINT and CO2 values can be calculated.

The production process has a large influence on textile sustainability and returning to the denim example, the process uses a large amount of water, chemicals, and releases a significant amount of C02 into the environment. Without a precise knowledge of these processes, conclusive ecological statements and evaluations of a product are impossible. Unfortunately, it is still common practice to just refer to available average values. Our process calculation tool provides a detailed inside into each refining step and accurate and specific values. This is essential to measure the environmental footprint of a product along the entire supply chain.

TOWARDS AN AI BASED RECOMMENDATION SYSTEM FOR SUSTAINABLE RETURN MANAGEMENT IN TEXTILE SUPPLY CHAIN

Sören Gry[1]

[1]August-Wilhelm Scheer Institute, Germany

Abstract

The steadily increasing volume of garment returns not only means an enormous increase in costs for retailers or manufacturers and an inventory risk that is difficult to calculate, but also a high environmental impact due to the garment's destruction and necessary logistics processes. Most of the existing approaches tackling those problems aim at the fundamental avoidance of returns. But many returns cannot be avoided e.g., orders for a selection of products, repairs or quality caused returns. Therefore, the aim of this research, which emerges from the project OptiRetouren, that is funded by the "Federal Ministry of Education and Research" and executed together with a German shoe manufacturer, is to appraise the potential of AI based forecasts and recommendations concerning unavoidable returns. These forecasts and recommendations are intended to enable the optimal further processing to be determined immediately and a corresponding transport or action instruction to be given. In this research the usage of an AI based recommendation system to facilitate garment second life use and to reduce product destruction is evaluated by conducting a state-of-the-art analysis followed by a conceptualization of a forecast and recommendation system.

Keywords

Sustainable return management, Second life planning, Sustainable supply chain, Artificial Intelligence, Machine Learning

REGENERATIVE AGRICULTURE APPLIED TO FIBRES, WILL THIS FLIP APPAREL'S REPUTATION FOR BEING BAD?

Charles Ross[1]

[1]RCA 216 Longwood Gate, Huddersfield, W Yorks HD3 4XF, UK

Abstract

The current drive towards better practice in the textile industry has been to reduce environmental impacts, especially post COP 26 & 27 – which have highlighted the Climate Crisis (& carbon's role in it). Meanwhile clothing has a reputation as an evil of in the world. ReGenerative Agriculture is the nirvana solution. It sequesters more carbon into the ground than the life-cycle of the product (growing, harvesting, cleaning, processing, dyeing, finishing, manufacturing, & user actions). Patagonia, The North Face, Hugo Boss, Timberland, & Eileen Fisher are embracing these fibres in their ranges for Spring Summer 2023. Even leather can become an asset of the planet. Key methodologies include using cover crops & crop rotation. Off-setting was previously used to mitigate brands' carbon footprints, this practice has now been audited to evidence better ways.

ReGenAg concentrates on Soil Health, Biodiversity (COP 15), better Irrigation, Local Trading Communities (not factory farming), & Carbon Sequestration. It works with growing textiles & producing foodstuffs for the world's population. It is easier to switch to than GOTS organic status.

This presentation will concentrate on going beyond Net Zero & reducing agricultural waste. Sustainability starts at the source; ReGenAg is ready for adopting.

Education & Policy

ADAPT OR GET LEFT: PREPARING FASHION DESIGN STUDENTS FOR THE ZERO-CARBON FUTURE

Sanjay Gupta[1]

[1]*World University of Design, Sonipat, India.*

Abstract

The fashion industry is at a crossroads. The industry has traditionally relied on creativity and artistic expression to thrive, but the rise of new technologies, shifting consumer values, and the need for sustainability is forcing the industry to adapt. The question is, will fashion designers adapt or get left behind?

The answer to this question lies in the ability of fashion design educators to prepare their students for the challenges of a zero-carbon future. The advent of technologies like VR which, integrated with other technologies, such as social media and AI, along with e-commerce is completely changing the way fashion is made, presented and retailed. These technologies are also making it possible for the fashion industry to become more sustainable, and in this way, linking innovation with sustainability.

Innovation, sustainability, and technology therefore have emerged as the three pillars of a strategy linking a zero-carbon future with the future of fashion design education. Fashion design educators will need to adapt their content and methods to ensure that their students are equipped with the necessary skills and knowledge to navigate this rapidly evolving landscape.

The urgency of this need cannot be overstated. The fashion industry must embrace innovation and sustainability if it is to survive and thrive in a zero-carbon future. The keynote address will highlight the three key disruptors in the fashion industry, and emphasize the importance of adapting to these changes. It will also demonstrate how fashion design educators can innovate and adapt to ensure that their students are not left behind in this new era of fashion design.

GROWING, HARVESTING AND PROCESSING FLAX TO INSTIGATE FIBRE, MATERIAL AWARENESS, COMMUNITY AND CHANGE

Collette Paterson[1]*, Shirley Mclauchlan*[1] *and Niki Taylor*[1]

[1]*University of Edinburgh, Edinburgh, Scotland, U.K.*

Abstract

With the climate in crisis, designers and creative thinkers have a huge responsibility and opportunity to create a positive impact on the world. In this context, the paper addresses the correlation between soil and textile, following and evaluating the growth of flax as a regenerative fibre that can seek to demonstrate the benefits of natural and local materials, drive exnovative design responses, promote considered use of materials and serve as a vehicle for community evolution and wellbeing. To that end, we explore the growing and processing of flax by hand, with the purpose of sharing the process with a broader local community to improve awareness of the connection between the textile product and its modest but rich origins. As part of this process, we engaged with designers to enable exploration through spinning and weaving with the fibre, including introducing the concept of mono-materials to develop a range of textiles qualities from flax, as well as taking a circular and zero waste approach from the outset. Preliminary results indicate positive community experience and feedback along with improved awareness toward material and colour choices. Exploratory woven samples have been created demonstrating the versatility and exciting potential of flax.

ECO-DESIGN FRAMEWORK FOR REDUCING FABRIC WASTE AND CONSUMPTION DURING FAST FASHION PRODUCTION

Abu Sadat Muhammad Sayem[1] *and Md Mazharul Islam*[2]

[1]*Manchester Fashion Institute, Manchester Metropolitan University, Manchester, M15 6BG, UK*
[2]*Department of Textile Engineering, Northern University Bangladesh, Dhaka, Bangladesh*

Abstract

Fast fashion business model is well-known for being responsible for accelerated waste generation both at post-consumption and manufacturing phases. The average marker efficiency of industrial fabric cutting process for garment manufacturing is 85%, which means average 15% of fabrics is directly scrapped during this stage. This makes all the efforts and inputs given from fibre to fabric production go fully in vain and the environmental impact and climate change contribution by this 15% of scrapped fabrics unjustifiable. Zero-Waste Pattern Cutting (ZWPC) is a recent trend in increasing fabric utilisation but is only practiced in atelier environment for making one or two garments. Our previous work demonstrated the integration of ZWPC in mass production of knitwear resulting over 95% of fabric utilisation. In this work, we demonstrated an eco-design framework for integrating ZWPC in to design and pattern cutting of woven garments in mass production scenario. As ZWPC approach demands significant changes in pattern geometry to increase fabric utilisation, it is necessary to first check through wearer trial how a garment drapes and appears on human body before finalising the design for mass production. The whole process is heuristic and based on 'error and trial' method. The physical prototyping and fit checking processes are very time and material consuming, hence not sustainable in this case. Therefore, virtual prototyping and fit analysis approach has been adopted to test the fit of eco-designed garment.

Keywords

Eco-design, fabric waste, Zero-Waste Pattern Cutting (ZWPC), marker efficiency, mass production, virtual fit analysis

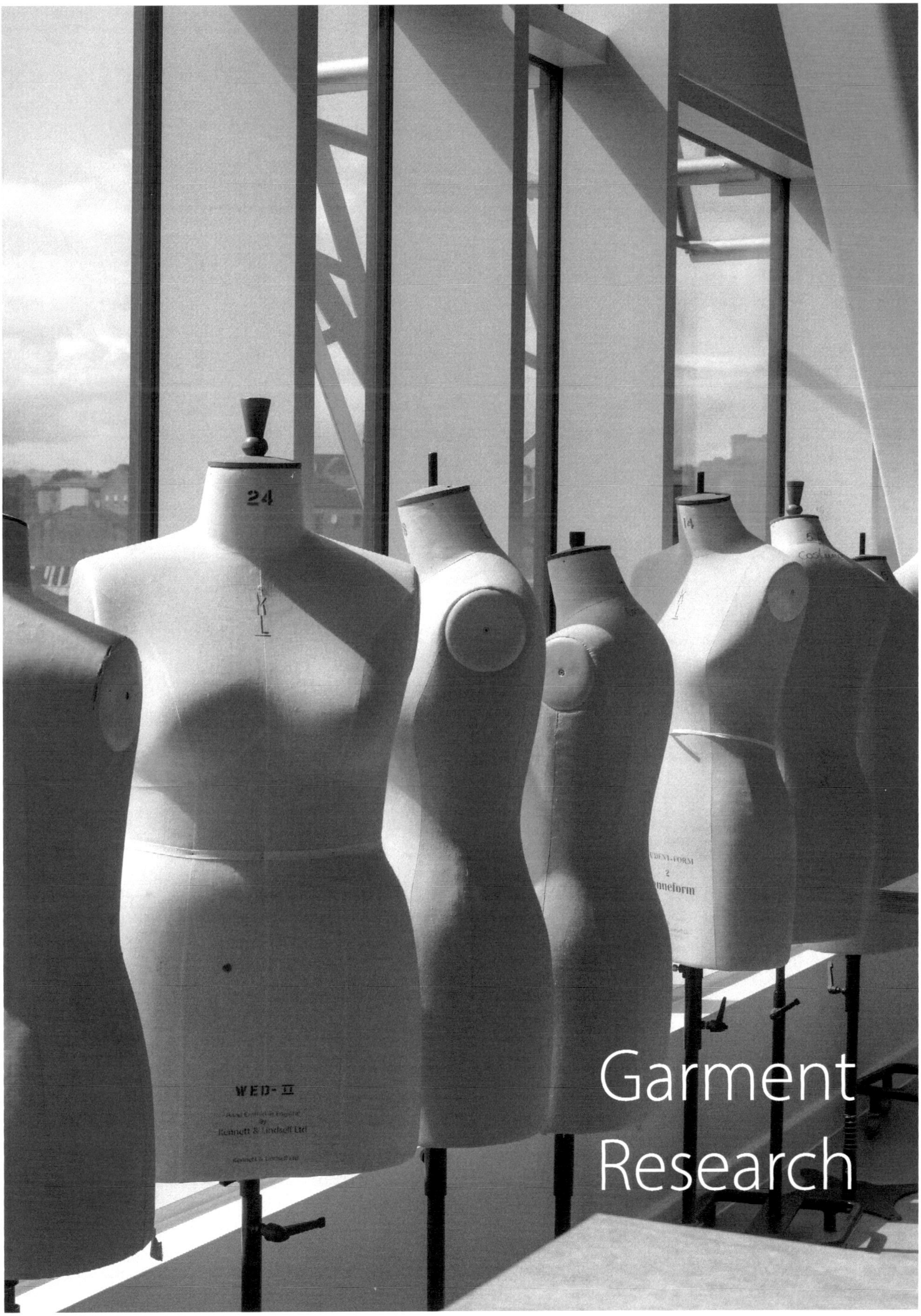

Garment Research

DEVELOPING SIZING METHODOLOGY FOR WOMEN'S SPORTS COMPRESSION GARMENTS

Kasey Hatch[1], *Kristina Brubacher*[1], *Simeon Gill*[1] *and Anura Fernando*[1]

[1]*University of Manchester, Engineering Building A, Manchester, UK*

Abstract

This research aims to improve sizing methodology for women's compression leggings, to allow the development of products that provide accurate fit and sufficient pressure delivery to the wearer. Female athletes can benefit from compression garment use, as the pressure applied to the skin and underlying tissue can improve circulation and facilitate the delivery of oxygenated blood to the active muscles during physical activity. Accurate fit is essential for these garments to function, so sizing and grading methods need to be appropriate for the target demographic. However, current sizing methods are inadequate and have no theoretical underpinning. This interdisciplinary study focuses on creating synchronicity between different research areas that are required when developing compression garments. 3d/4d body scanning, pattern drafting, mechanical testing of knitted fabrics, sizing, grading and pressure application are all critical elements within the research, and there is a drive to use technology to improve accuracy of the process. Sustainability naturally feeds into this research, with garment returns due to poor fit, causing huge environmental impact. Using technology to advance anthropometry, fabric testing and pattern creation, should enhance fit accuracy and consumer satisfaction.

Keywords

sizing, compression, garment fit, 3D scanning

ERGONOMIC DESIGN OF BADMINTON-SPECIFIC SPORTS BRAS FOR ATHLETES

Yin-Ching Keung[1], *Kit-Lun Yick*[1], *Joanne Yip*[1] *and Annie Yu*[2]

[1]*School of Fashion and Textiles, The Hong Kong Polytechnic University, Hong Kong, China*
[2]*Department of Advanced Fibro Science, Kyoto Institute of Technology, Japan*

Abstract

Today, sports bra designs for an elite level of sport with consideration of vigorous body motion, posture changes and prolonged wearing are largely neglected. Bra discomfort, displacement and fit problems negatively affect sporting performance. This study therefore aims to offer a new scientific approach that precisely identifies breast and bra displacements from movements during badminton sport.

By using high-resolution 4D scanning and motion capturing technologies, muscle movement on the back of the athlete and the associated shoulder straps displacements and dimension changes of sports bra during dynamic situations were captured and measured. Uneven changes on the skin surface of the upper body and stretch behaviour between the left and the right shoulder straps were identified.

The results suggest the need of panels separation in badminton-specific sports bra design, especially at front left strap and upper right strap at the back. The choice of materials with the desired stretch and recovery should be strategically applied. The results provide the first evidence of specific need on the separation of left and right sports bra properties for badminton-specific use.

Keywords

Ergonomic Design, Sports bra, 4D body scanning, Motion capture

FRAMEWORK FOR SAMPLE SIZE SMART GARMENT PROTOTYPING WITH SCALABILITY UNDER SUSTAINABLE PRODUCTION METHODS

A. Zidichouski[1-2]*, B. Shepherd*[2]*, J. Wood*[3] *and P. Kunovski*[1]

[1]*R&D, Kymira Ltd., Reading, UK*
[2]*Manchester Fashion Institute, Manchester Metropolitan University, Manchester, UK*
[3]*Dept. of Materials, University of Manchester, Manchester, UK*

Abstract

A two year Innovate UK Knowledge Transfer Partnership between the Manchester Fashion Institute (Manchester Metropolitan University), the University of Manchester and Kymira Ltd. was initiated to solve challenges in the prototype-to-production workflow to bring smart garment products to market. Intelligent textile developments show promise in sports science and medical diagnostic applications, though there is a gap between prototyping and mass-production. There are few manufacturers of smart garments and it is hypothesized that smart garment creation is not well-understood by mass manufacturers; conversely, prototyping methods are not in-line with standard garment making techniques. These circumstances lead to long design cycles and expensive end products for users and customers. Currently the development process is opaque, with few successful real-world examples. Standardizing aspects of wearable product development and linking researchers to mass-manufacturing resources would help achieve more reliable, cost-effective devices and reduce scaling risks. This project aims to develop design considerations, equipment, and expertise necessary to bring a new product from concept to production. Proof of concept sample garments will be prototyped, integrating better characterized electronics hardware with available technologies, whilst describing production best-practices to identify technical needs for a large-scale manufacturing partner, and considering sustainability in production processes during the design phase.

Keywords

TIWC 2023, smart garments, wearables, prototyping, manufacturing, framework, sustainability

DEVELOPMENT OF NONWOVEN GARMENT BY ADOPTING SPECIAL TECHNIQUE OF PATTERN AND 3D TECHNOLOGY

M S Cheema[1], Simeon Gill[1], Hugh Gong[1] and Tarfah Alrushaydan[1]

[1]Apparel Design and Engineering, University of Manchester, Manchester, UK

Abstract

Previous studies showed some development of nonwoven garments but could not get success in the garment industry. It was because of using inadequate traditional methods of pattern for nonwoven structure which did not improve the functionality and comfort of the nonwoven garments. This study focuses on addressing the limitations of nonwoven fabrics by developing special technique of pattern drafting based on nonwoven structural properties like shearing and flexural rigidity, tensile and surface frictional properties. Applying calculated ease at the specific parts of the body and adjust pattern as per body shape according to the physical properties of the nonwoven fabric with the help of 3D body scanning, enabled nonwoven garments to perform better in term of visual, functional, and thermal comfort in the apparel industry. Wearer trials conducted and feedback confirmed the better functionality of the nonwoven garments by adopting special technique of pattern and 3D technology and comparison with woven garments gave satisfactory results.

THE SPORTS COMPRESSION GARMENT USER NEEDS MODEL: A NEW MODEL FOR USER-CENTRED DESIGN

Kristina Brubacher[1]

[1]The University of Manchester, Oxford Rd, Manchester M13 9PL, United Kingdom

Abstract

The trend towards healthier, more active lifestyles has led to a substantial increase in sportswear consumption. The use of compression sportswear, designed to apply pressure to the underlying body, has become commonplace as athletes seek to improve their exercise performance and recovery. However, commercial sports compression garments (SCG) do not currently define the applied pressure level, likely due to large variability, meaning consumers cannot make informed purchase decisions and increasing the risk of poor functioning, unwanted garments. The aim of this research was to develop a model that can be used to identify key user needs for the design of SCGs for specific applications. The model was developed based on an extensive literature review and wear trials with commercial SCGs. Existing research widely neglects user needs and mainly focuses on pressure quantification. The developed SCG User Needs Model considers not only users' functional needs, but also expressive and aesthetic needs related to the body, garment and self-identity. The model is part of a larger SCG Design Model that considers process, user and product perspectives to improve the design of SCGs, resulting in less waste in the design phase and potentially fewer discarded garments in the use phase.

Keywords

User-centred design, sports compression garments, user needs model, waste reduction

SUSTAINABLE PROCEDURES IN COLLECTING ANTHROPOMETRIC DATA: EFFICACY IN REMOTE BODY MEASUREMENT OF SCHOOL CHILDREN FOR CLOTHING SIZING

Phoebe Apeagyei[1], *Adwoa Tweneboaah Twum*[1], *Abu Sadat Sayem*[1] *and Paula Wren*[1]

[1]*Manchester Metropolitan University, Cavendish Street, Manchester M15 6BG, UK*

Abstract

Critical aspects of sustainable development for clothing facilitate productivity, safety and wellbeing. Research hints that ill-fitting garments are wasteful, may present physiological discomfort and risks to body movement, especially for vulnerable people such the elderly and children. Commercial-sized clothing are economically manufactured using standards derived from body measurement surveys. Despite advances in body scanning technology and its automation benefits in generating anthropometric data (body measurements), the most commonly used methods in obtaining data for good clothing fit are manually applied.

Social distancing restrictions that emerged with the covid 19 pandemic have dictated real challenges in safely collecting mass anthropometric data for research and fashion practice. It is because procedures are expensive, require workstation set-ups and skilled personnel, as well as close physical proximity and body contact operations. It is crucial that obtaining anthropometric data follows an efficient and safe process. The aim and outcome of this study has hence developed and validated a novel, ethical and sustainable response to that challenge, by establishing a sustainable body measurement protocol which features a safe and remote systematic method and tools for obtaining reliable anthropometric data on children. This work has implications for academic research and professional fashion practice, especially for middle- and low-income countries.

Keywords

Sustainable body measurement, Anthropometric data, Clothing sizing, Ethics, Children

Textile Coloration & Processing

TWINNED TECHNOLOGIES FOR THE CONNECTED WORLD

Russell Thorpe[1]

[1]VeriVide Limited, Quartz Close, Warrens Business Park, Enderby, Leicester, UK

Abstract

This presentation will cover the exciting new technology being adopted by leading brands and retailers and will introduce Twin Technologies, Digital Twins and the New LED Based Viewing and Vision Based Remote Colour Assessment.

Business digitisation and transformation is key and important in both the colour development and colour approval process. Brands and retailers are actively seeking to streamline their processes into smarter, faster and more efficient ways of working. Twinned Technologies are the actual pieces of equipment and systems that streamlines the process, the New LED Lighting Cabinets and Camera Based Colour Measurement Systems which incorporates the identical quality of illumination.

Moving towards 100% LED Technology, will future proof investment and provide precise viewing and control of colour in the supply chain, this being particularly helpful in improving sustainable product quality. Digital Twins (in respect to DigiEye) are the images that represent the product, the product remains at source (manufacturing site) and colour assessment is instantly performed digitally and remotely.

Keywords

TIWC 2023, Twin Technologies, Digital Twins, LED, Vision Based, DigiEye, Sustainable

GLASS MEETS TEXTILES - THE USE OF VACUUM PLASMA COATING FOR THE FUNCTIONALISATION OF TEXTILES

Jeroen Schotsaert[1], *Jerome Jolibois*[2], *Grégory Arnoult*[1], *Hugues Wiame*[1] *and Marc Van Parys*[3]

[1]AGC Plasma Technology Solutions, a business unit of AGC Glass Europe SA, Technovation Center, Rue Louis Blériot, 12, B-6041, Gosselies, Belgium
[2]AGC Plasma Technology Solutions, a business unit of AGC Glass Europe SA, Sohnreystraße 21, 37697 Lauenförde, Germany
[3]Unitex, Hendrikslaar, 12, 9080, Lochristi, Belgium

Abstract

The use of thin film coating technologies, as magnetron sputtering and plasma enhanced chemical vapor deposition (PECVD), are well established in the glass industry since several decades for the mass production of solar control and low emissivity glasses for architectural and automotive applications. However, despite their huge potential for the functionalization of other types of surfaces and their environmental benefits, these vacuum coating technologies find currently only few applications for the industrial manufacturing of textile substrates. Major challenges are associated with the upscaling of lab scale processes for reaching high quality and uniform coatings at acceptable costs. Hence, the potential of magnetron sputtering as a powerful technology for the deposition of inorganic and metallic nano-coatings and of PECVD allowing to make hydrophobic and oleophobic coatings, remains largely unexploited in the textile industry. AGC Plasma Technology Solutions, a business unit within the multinational glass producer AGC Glass Europe SA, is bringing its experience in the field of vacuum coating technology towards the textile industry and hence supports the drive for the sustainable manufacturing of high value-added textiles. AGC is making a demonstration line available for customers to experience the technology and make prototypes for roll widths up till 1.6 meter wide. AGC's process development is currently focused on halogen-free water-repellent coatings and on metallization, while the possibilities offered by magnetron sputtering and PECVD are endless.

Keywords

Nano-coatings; thin film; magnetron sputtering; plasma enhanced chemical vapor deposition; textiles; fabrics; finishing; metallizing; coloration; water and oil repellency

SUSTAINABILITY: IMPACT ON TEXTILE LUBRICATION

Elizabeth Hornby[1] *and* *Peter Vickers*[1]

[1]*Benjn.R.Vickers & Sons LTD, 6 Clarence Road, Leeds, LS10 1ND, www.vickers-oil.com*

Abstract

A plethora of sustainability initiatives, labels and regulations is changing the global textile industry. One aspect which has been largely unexplored is the impact on textile lubrication. Lubricants which are in contact with textiles come under two broad categories: (1) processing aids for fibres and yarns; (2) textile machinery lubricants which come into contact with the material, for example needle oils for knitting machines. The presentation will discuss present trends in textile lubricant formulations, taking account of multiple, sometimes colliding factors, such as environmental aspects, technical engineering requirements, the needs of textile and finishing processes, and commercial impact (cost).

Vickers Oils, which has served the global textile industry for more than 150 years, and 30 years ago pioneered the development and adoption of "Environmentally Acceptable Lubricants" for the global shipping industry, can offer different ranges of "sustainable" products, depending on the precise drivers for each sector of the textile industry.

The Circular Economy & LCA

ENVIRONMENTAL SUSTAINABILITY OF FASHION PRODUCT MADE FROM POST-CONSUMER WASTE

Md Mazedul Islam[1] *and Bruce Montgomery*[1]

[1]*The University of Southampton, Southampton, UK*

Abstract

The fashion industry has detrimental environmental impact throughout supply chain operations that needs urgent solutions. Limited work focuses on measuring the environmental sustainability of clothing product made from post-consumer waste in the circular economy. This research aims to evaluate the environmental sustainability of fashion products made from post-consumer waste using Higg Index tool by sustainable apparel coalition. Three t-shirt products were selected that were made in a green, ordinary and subcontractor factory as a case study. Data was collected through practice based qualitative questions to manufacturing practitioners, supplemented by triangulations and scores were obtained using the Higg Index product module tool. Findings reveal significant differences in the product's environmental performances in terms of materials, manufacturing, packaging, care and end of life stages. A varying range of scores (from low to high) was identified for the three t-shirts. Product made in the green factory embracing circular economy achieved the highest score and the sub-contractor factory product obtained lowest score. Findings highlight poor and emerging sustainable practices, identify challenges, and suggest improvement in the above-mentioned categories. The research urges stakeholders to embrace sustainable practices for the fashion products to reduce environmental impact through life cycle stages and benefit the industry.

Keywords

Environmental Sustainability, Fashion Product, Post-Consumer Waste

TOWARDS A DYNAMIC DECISION SUPPORT SYSTEM FOR CONFIGURING POST-CONSUMER TEXTILE-TO-TEXTILE RECYCLING VALUE CHAIN

Rudrajeet Pal[1-2], *Erik Sandberg*[3] *and Abhijit Majumdar*[4]

[1]*Swedish School of Textiles, University of Borås, Skarabörgsvagen 3, Borås, Sweden*
[2]*Department of Industrial Engineering and Management, University of Gävle, Kungsbäcksvägen 47, Gävle, Sweden*
[3]*Department of Management and Engineering, Linköping University, Olaus Magnus väg 37, Linköping, Sweden*
[4]*Department of Textile and Fibre Engineering, Indian Institute of Technology Delhi, School of Textiles, Hauz Khas, New Delhi, India*

Abstract

Globally <13% of post-consumer textiles are being recycled, and current valorisation is hindered by challenges at product-process-supply chain levels. In future, with growing volume of low-value textile wastes, optimal configuration of textile-to-textile (T2T) recycling value chain is crucial and requires intricate decision-making. Data-driven decision- support systems (DSSs) can guide the stakeholders (sorters, recyclers, manufacturers, and retailers) on how to improve process efficacy and optimize recycling networks and align with market needs. This paper defines and develops a conceptual model underpinning a dynamic DSS for designing optimal T2T recycling value chain. Given the purpose, we explored the specific supply chain, product and process contingencies and their inter-relations via a systematic review of existing literature, 40+ interviews with stakeholders engaged with collection, sorting and recycling from Sweden, Norway, Germany, Netherlands, Switzerland, Poland and India, and multiple field visits. Initial findings suggest on designing recycling value chain and its logistics network to align and optimize demand and supply in conjunction with end market needs, improving feedstock standardization and reconfiguring the supply network for sustainability. The critical DSS factors to consider are: (i) number of supply chain tiers and members, and geographical distances in-between; (ii) inventory/stock and material flow at each tier; (iii) process yields and conversion rate based on sorting/recycling technology; (iv) demand volume and feedstock variety and variability in terms of material, colour etc, and (v) costs and prices based on market conditions. Other external factors like textile waste export-import tax, GRS standards, are also crucial.

Keywords

Post-consumer textile waste, Recycling value chain, Decision-support system, Circular Economy.

BUILDING A CIRCULAR MATERIALS REPOSITORY. CONVEYING THE COMPLEXITY OF MATERIALS FOR CIRCULARITY

Lynn Wilson[1], Martha Wood[1] and Helena McLaughlin[1]

[1]Circular Design Scotland, 19 Belford Mews, Edinburgh, EH43BT

Abstract

The concept of a circular material is an oxymoron. There is no such thing as a circular or sustainable material, only the existence of materials and circular systems. This paper presents the findings of a six-month industry research project, using qualitative industry interviews with material development companies and desk-based research, identifying new materials suitable for commercial industrial circular systems. The information is within a digital repository accessed via subscription. The findings identified an abundance of companies and projects developing new fibres, materials from bio-regenerative materials and solutions for valorising fossil fuel polymers, reducing reliance on virgin polymers. This revealed concerns about commercial viability, scalability, and overall purpose of new material development. Critically, materials designed as 'sustainable' replacements, such as Mycelium packaging, may exacerbate the climate crisis through a lack of traceability and failure to ensure full circularity. The research recommends and provides examples of new communication tools to increase investment in circular material systems.

Keywords

Circular materials, industry research, circular systems, sustainable materials, biopolymers, regenerative materials.

CLOSING THE LOOP THROUGH CIRCULAR DESIGN EDUCATION

Rachel Kelly[1], *Patsy Perry*[1] *and Sarah Moriarty*[1]

[1]*Manchester Metropolitan University, All Saints, Manchester M15 6BG, United Kingdom*

Abstract

Fashion and textiles is predominantly an extractive, linear industry which creates huge volumes of waste and we have yet to see a meaningful shift to a circular system that eliminates waste, regenerates natural systems, and keeps materials and products in use for longer. The waste hierarchy prioritises the highest value opportunities, so reducing waste at source and reuse should be implemented before recycling. Circularity requires global cooperation and an industrial-scale response supported by community and local approaches, as well as joint participation of multiple stakeholders, including higher education institutions, to encourage more sustainable design and production solutions. This study examines the transformative potential of a cross-cultural and multidisciplinary higher education initiative to encourage circular design practice in aspiring designers and engineers. A British Council-funded circular design challenge aimed at design and engineering students in China explored how circular economy principles can be a catalyst for creativity, collaboration and regenerative thinking within architecture, design and fashion practice, and the critical importance of design for physical and/or emotional durability to reduce waste. Qualitative data gathered from students and tutors informed the development of a conceptual framework for fashion and textiles which could provide directions for designing out waste rather than recycling.

Keywords

Circular design; China; design thinking; higher education

EMERGING CONCEPTS AND IMPLICATIONS OF CIRCULAR ECONOMY AND SUSTAINABILITY TERMINOLOGIES IN FASHION AND TEXTILES.

Md Mazedul Islam[1], *Bruce Montgomery*[1] *and Caroline Lo*[1]

[1]*The University of Southampton, Southampton, UK*

Abstract

While the term circular economy sustainability is gaining traction in industry and academia, there is a lack of consensus around the associated terminologies in fashion and textiles. The similarities and differences between these concepts and wider implications remain ambiguous and unclear in literature. This research aims to explore and analyze the terminologies associated with sustainability and circular economy in fashion and textiles. A systematic literature review that includes peer-reviewed journal articles and market intelligence resources published over a period from 2013 to 2023 to investigate this field. It adopts thematic and content analysis techniques. The bibliometric and contents analysis illustrates the diversity and implications of the associated concepts, terms and meanings used in the literature and how they might have evolved over time. We identify significant differences in the concepts, relationships, and implications of the terminologies across stakeholders and fashion life cycle stages. Findings argue that the emergence of such tensions triggers discussions and debates on the complex nature of the environment, society, economy, and culture, which the stakeholders must understand to facilitate a sustainable fashion industry. This review contributes to the clarification and definition of terminologies that can be utilized to understand why a uniform understanding on circular economy and sustainability is needed. It provides academics, stakeholders, and managers with a comprehensive understanding of sustainability concepts and their wider implications in fashion and textiles to stimulate further scholarly research.

Keywords

Circular Economy, Concepts, Terminologies, Fashion, Textiles, Sustainability

A CRITICAL REVIEW OF LIFE CYCLE ASSESSMENT OF FASHION PRODUCTS

Aayushi Badhwar[1], *Saniyat Islam*[1] *and Caroline Swee Lin Tan*[1]

[1]*School of Fashion and Textile, 25 Dawson Street, Brunswick, VIC 3056, Australia.*

Abstract

The extreme environmental and financial cost of extracting and using natural resources from the planet is pushing countless industries to implement circular economy action plans. Circular economy framework proposes designing out waste to lower the environmental impact. Repairing, reusing, and recycling products can reduce resource consumption while still generating economic value. Life Cycle Assessment/Analysis (LCA) is a global matrix to determine the environmental impact of a product. One of the most popular models to measure the environmental impact of a product is the HIGG MSI, which is built upon the LCA framework. HIGG index only provides a detailed account of the impact from "cradle-to-gate" which limits the model from manufacturing to retail floor. Currently, no model represents the full "cradle-to-cradle" LCA of a product that includes the product Use-Phase and End-of-Life Phase. This paper explains the significance of the inclusion of these criteria to contextualise the full impact of the garment to achieve a "cradle-to-cradle" assessment. An emphasis is placed on the HIGG MSI model to highlight these gaps in the current measurement system. The paper proposes a theoretical underpinning of how traceability platforms can assist in providing more accurate life cycle assessments.

Keywords

Circular economy, Life Cycle Assessment/Analysis, HIGG MSI, Use-Phase and End-of-Life-Phase, traceability

HARNESSING HISTORICAL TECHNOLOGIES IN THE DEVELOPMENT OF A CIRCULAR TEXTILE MODEL: THE FUTURE OF REGENERATED PROTEIN FIBRES

Hannah Auerbach George[1] *and Marie Stenton*[2]

[1]*Victoria and Albert Museum,Cromwell Road, London, SW7, United Kingdom*
[2]*London College of Fashion,20 John Prince's St, London W1G 0BJ, United Kingdom*

Abstract

The Business of Fashion Textiles and Technology (BFTT) is an AHRC-funded initiative to advance sustainability across the sector. The project, 'Rethinking Material Resources' uses a novel methodology combining expertise in archival research, textile design, and green chemistry to reimagine cross-sector industry waste as a valuable resource. The result is a transdisciplinary, systems-thinking approach which challenges the traditional design process and presents a new understanding of textile quality within a circular economy.

The historical significance of regenerated protein fibres (RPFs) as a substitute fibre during wartime is compared to the current global issue of overproduction and excess waste. Opportunities are identified for food waste to reduce the current use of petrochemical-derived fibres and textile dyes, which remain in landfill without the ability to decompose, through the matching of textile lifespan to the speed of today's fashion trends. Archival data on early RPFs was used to develop a new framework exploring the use of naturally biodegradable yet low tensile strength materials by reimagining these within a 'disposable fashion' context. Inputs vary from the elimination of hazardous chemicals used during the regeneration process, textile and garment construction techniques, and the use of sustainable coloration techniques mirroring the valorisation of waste.

Keywords

Sustainability, Circular Design, Waste, Archival Research, Speed Cycles

Defined Applications

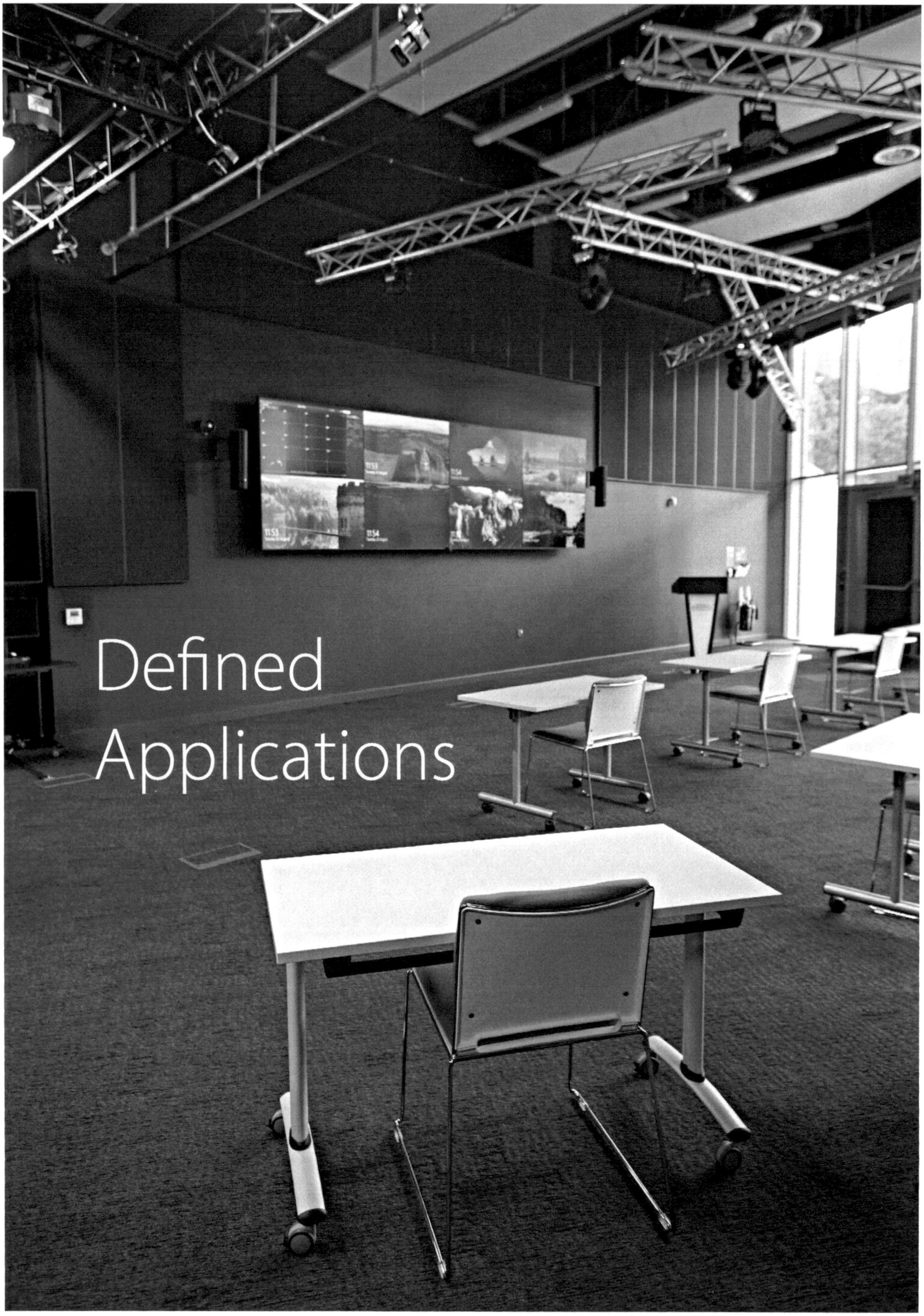

THE UNIQUENESS OF DISPOSABLE NONWOVENS FOR HYGIENE AND THE CASE FOR THEIR SEPARATE CATEGORISATION.

Adrian Wilson[1]

[1]*AWOL Media.*

Abstract

When considering the waste generated by plastics and synthetic fibres, there is perhaps now more of a case than ever before for nonwovens – or at least the disposable products based on them related to hygiene and healthcare – to be considered as completely separate to the overall textiles industry, as well as to hard plastics, paper and packaging.

The nonwovens industry has elements drawn from all of these other industries. However, the essential shortage of PPE and the panic buying of items such as diapers and wipes during the Covid-19 pandemic illustrated the essential role of many nonwoven-based disposables, This should have ensured that absorbent hygiene and medical products are kept as far away from fast fashion or non-essential disposables such as plastic bags in legislation as is possible.

At present – as has been amply illustrated in presentations from European Commission legislators and other non-industry players – all of these products are regularly being conflated.

Some points covered:

The introduction of the EU's Single-Use Plastics Directive (SUPD) in July 2021 will be seen as a watershed for the global nonwovens industry.

The 'one-size-fits-all' nature of the EU's Strategy for Sustainable and Circular Textiles aims to address the problem of fast fashion head on.

How close the system comes to transformation will depend on the level of leadership shown by key decision makers across all stakeholder groups.

Inevitably, all of the major activity by the fashion brands means there will be more competition for cellulosic fibres.

The availability of recycled PET (rPet) for the production of synthetic fibres for nonwovens is also likely to tighten considerably in the coming years.

A projected cost for getting Europe's plastics industry to net zero in greenhouse gas emissions (GHGs) by 2050 is between $5-6 trillion.

Microfibre Research

QUANTIFICATION OF MICROFIBRE RELEASE IN LAUNDERING

Alice Hazlehurst[1], *Lucy Tiffin*[1], *Mark Sumner*[1] *and Mark Taylor*[1]

[1]*University of Leeds, Leeds, UK*

Abstract

Textiles laundering is a significant source of microfibre pollution, with fibres of both natural and synthetic origin having potential for negative impacts on the environment and organisms. Quantifying microfibre release from domestic laundering is an important step in understanding the scale of the problem and to evaluate the efficacy of potential solutions. The present work explores the influence of different testing methodologies, laundering variables, and fabric variables on the quantification of microfibre release.

A review of different test methods highlighted the scale of variation in current estimates of microfibre release, and therefore, the need to utilise standardised testing methodologies for microfibre release to be reliably quantified going forwards. Such a standardised method has been used to investigate the influence of select laundering and fabric variables experimentally. Results indicated that fabric variables can have a greater influence on microfibre release than laundering variables, however a greater mass of microfibre release was also demonstrated with increasing wash liquor ratio and in-wash agitation.

Experimental data was combined with assumptions for washing load size and frequency to provide an estimate for possible annual microfibre release from untreated domestic laundering in the UK: estimated up to 87,165 tonnes of microfibre discharged in the UK annually.

Keywords

Microfibres, laundry, water pollution

Fashion & Textiles in Mourning

THE DECLINE OF FEMALE MOURNING WEAR: A CASE STUDY ANALYSIS OF FREDERIC FORSTER'S MOURNING WAREHOUSE 1849-1923 IN LEEDS, UK

Judith Simpson[1] *and Kevin Almond*[1]

[1]*School of Design, University of Leeds, Leeds, LS2 9JT.*

Abstract

The research explores the trade in UK female mourning wear by tracing the history of Frederic Forster's Mourning Warehouse, situated on Lower Briggate in Leeds 1849-1923. It also interrogates the small collection of Forster items held by Leeds Museums and Galleries (LMG). This helped to develop an understanding of why UK female mourning wear, which dominated the High Street in the late 19th century, began to disappear. The research subsequently examines what these items signify to a present-day audience and whether the need for special clothing to denote mourning has disappeared or is met by other clothing practices in a globalized world? A review of literature raises several explanations, suggesting that the cultural work done by mourning clothing is no longer required today and the symbolic efficacy of black mourning clothing disappeared as black became popular and accepted in fashionable attire. The research will test these suppositions through the object-based analysis of the items in the LMG collection and a database search of newspaper advertisements from Frederic Forster in the period 1849 – 1923. The research activities will be filmed to synthesise the insights and the film will be discussed with a selection of interview participants to explore new knowledge and understanding about the decline of the UK mourning dress trade.

Keywords

Mourning Wear, Clothing Practices, Cultural Work, Symbolic, Female, Trade

The results of this research will be published in the autumn edition of Costume: The Journal of the Costume Society Volume: 57.2.

CRAFTING SORROW: THE USE OF TEXTILES IN CONTEMPORARY ART PRACTICE TO PERFORM MOURNING

Claire Barber[1]

[1]*School of Arts and Humanities, University of Huddersfield, Huddersfield, HD1 3DH, UK*

Abstract

Someone dies, and an expectation to extend the life of clothing is turned upside down. A wardrobe of the deceased presents a conundrum for a widow when the duty of care and ongoing preservation of clothing can feel overwhelming.

This paper presents my practice-based research in which I explore and reveal experience that is imbued in the worn clothing of someone who has died, specifically through my recent work Cactus Shadow. In this work, I undertook to carefully stitch the shadow of a cactus into the deconstructed sleeve of a jacket of the deceased. It is an intervention that reflects, from a widow's perspective, on the ways the repetition of stitch is allied to the time and labour in mourning. By referring to a term used by Freud 'mourning-work' I ponder how the labour of sewing and the labour of mourning coalesce into something that is performed over time. My response emphasises the process of making work as a performative act that crafts sorrow into a material reality. This raises questions about the use of textiles as an art medium, where the histories embedded in materials and the ways things are made are fundamental to the content of the work.

Keywords

Craft, Clothing, Labour, Textile Art, Stitch, Performance, Mourning.

Submitted Papers

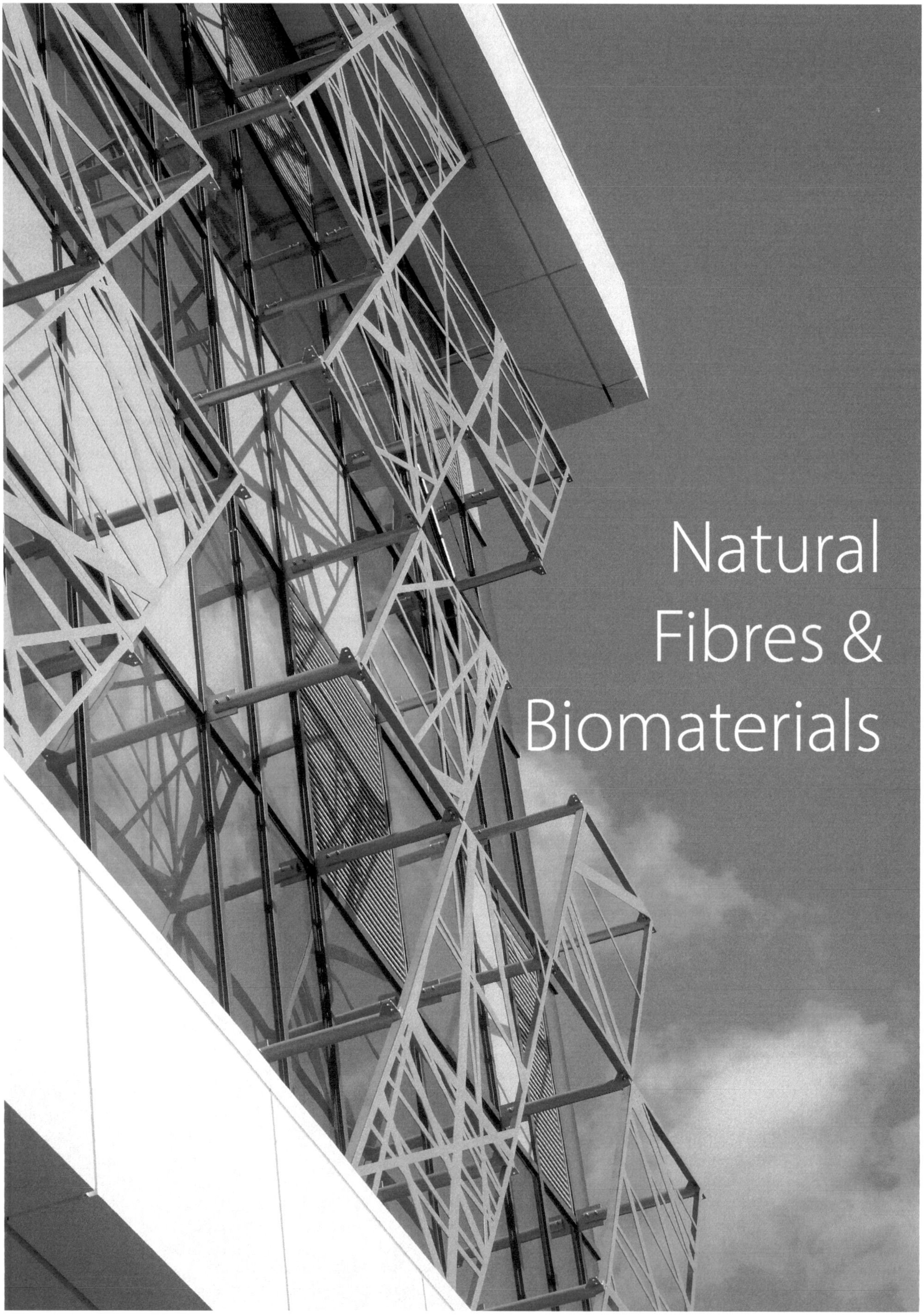

Natural Fibres & Biomaterials

FRAMEWORK FOR ENVIRONMENTALLY SUSTAINABLE FASHION AND TEXTILE PRODUCTION TO ACHIEVE UNITED NATION (UN) SUSTAINABLE DEVELOPMENT GOAL (SDG) 12

***Abu Sadat Muhammad Sayem*[1], *Julfikar Haider*[2], *Mohammad Abbas Uddin*[3], *Ayub Nabi Khan*[4], *M M Alamgir Sayeed*[5] *and Mahbub Hasan*[6]**

[1]*Manchester Fashion Institute, Manchester Metropolitan University, UK*
[2]*Department of Engineering, Manchester Metropolitan University, M1 5GD, UK*
[3]*Bangladesh University of Textiles Dhaka, Bangladesh*
[4]*BGMEA University of Fashion & Technology, Bangladesh*
[5]*Textile Physics Division, Bangladesh Jute Research Institute, Bangladesh*
[6]*Materials and Metallurgical Engineering, Bangladesh University of Engineering & Technology.*

Abstract

A framework with three key actions - identify (I), act (A), and evaluate (E) to achieve complete environmental sustainability in fashion and textile production in line with SDG 12 has been developed as a part of a Global Challenge Research Fund (GCRF) project. Called as the 'I-A-E framework', it emphasises more on zero or near-zero waste generation at source and incorporating sustainability thinking in material and process selections, rather than post-production waste management after generating huge amounts of wastes and effluents. The first step involves 'identifying' present scenarios and points of action in context of resource consumption, air pollution and greenhouse gas emissions, water pollution, and solid waste generation in product development and production. The next step is to 'act' for sustainable development, which includes incorporating (1) a zero-waste philosophy during production, (2) sustainability thinking into material selection and production processes, and (3) sustainability in resource and waste management. The third component of the framework is to 'evaluate' for celebrating and moving forward by checking the eco-indices to amend targets or set new ones. This new framework was validated through stakeholders' workshops and roundtable discussion. It is an easy-to-follow toolkit that the fashion and textile industry will be able to implement into their product development and production activities.

Keywords

Textiles, Fashion, Sustainable Development Goal, Pollution, Green House Gas, Eco-indices

Introduction

According to the United Nations Alliance for Sustainable Fashion, (UNASF) the fashion industry is responsible for about 2-8% of the global greenhouse gas (GHG) emission, which is estimated to increase by more than 60% by 2030, consuming about 215 trillion litres of water per year and causing 20% of the wastewater generation around the world, approximately 9% of annual microplastic release to the oceans and an annual material loss of US $100 billion due to underutilisation (UNASF, n.d.). The Fashion Industry Charter for Climate Action (FICCA), established under the auspices of UN Climate Change, aims to achieve net-zero emissions by 2050 and the signatories committed to reduce GHG emission by 30% by 2030 against a baseline of no earlier than 2015 to keep the increase in global warming under 1.5°C above pre-industrial levels (FICCA, n.d.).

The United Nation (UN) Sustainable Development Goal (SDG) 12 comes with 11 targets and 13 indicators (SDGS, n.d.' SDG Tracker, n.d.). Out of which, 4 targets are directly, and 2 targets are indirectly related to fashion and textile production (Table 1). A 10-Year Framework of Programmes on Sustainable Consumption and Production Patterns (10YFP) was adopted at the Rio+20 UN Conference on Sustainable Development in 2012 (UNEP, n.d.; EPA, n.d.; 10YFP Secretariat, n.d.). The 10YFP consists of six programmes: Sustainable Public Procurement, Consumer Information for SCP, Sustainable Tourism, Sustainable Lifestyles and Education, Sustainable Buildings and Construction, and Sustainable Food Systems, which provide directions for the member countries. They do not provide any direct guideline for the manufacturers. There are several frameworks to achieve sustainability and circularity in the fashion industry, which have been proposed and presented (Table 2), but they do not provide effectively appliable guidance and recommendations for the manufacturing plants, specially located in Asia. This paper presents a simple and easy to follow framework for environmentally sustainable fashion and textile production developed under the auspices of a Global Challenge Research Fund (GCRF) Project 2020/2021.

Table 1: Relationship between SDG 12 Targets and Fashion and Textile Production

		SDG 12 Targets
Fashion and Textile Production	Directly related	12.2: Sustainable management and use of natural resources 12.4: Responsible management of chemicals and waste 12.5: Substantially reduce waste generation 12.6: Encourage companies to adopt sustainable practices and sustainability reporting
	Indirectly related	12.8: Promote universal understanding of sustainable lifestyles 12.A: Support developing countries scientific and technological capacity for sustainable consumption and production

Table 2: Prevailing frameworks related to fashion industry

#	Existing Frameworks on Textile and Fashion	References
1	Framework of circular business models for fashion and textiles	Coscieme et al. (2022)
2	Apparel Industry Sustainable Strategies Framework	CSB (2022)
3	Fixing Fashion Framework	Berentson-Shaw (2015)
4	Integrated Framework for the Sustainable Performance of Fast Fashion Firms	Seo and Kim (2015)
5	Fashion Design for Sustainability: A framework for participatory practice	Williams, D. (n.d)
6	Framework for Achieving Sustainable Fashion Consumption through Collaboration	NFA (2012)
7	Design framework for enabling sustainability in the clothing sector	Santos et al. (2016)
8	Conceptual framework for sustainable fashion consumption within the circular fashion system	Vladimirova et al. (2021)
9	Common Framework for Responsible Purchasing Practices	MSI (2022)

What we know so far

Fibre production

Textile fibres have environmental footprints in many ways, especially in terms of energy demand and water depletion (Esteve-Turrillas & de la Guardia, 2017). Among the natural fibres, cotton dominates the fashion market. The major hotspots with cotton fibres are energy used in farming and usage of fertilizers or pesticides (Günther et al., 2017). Cotton cultivation requires more energy and water compared to other natural fibres such as jute and kenaf (La Rosa & Grammatikos, 2019). In case of animal fibres, wool and silk contribute to the environmental footprint. On-firm activities are accountable for two third of wool's total CO2 emission (Chen & Burns, 2006). On the other hand, for silk, cocoon processing, irrigation and drying involve environmental consequence. Synthetic fibres like nylon, polyester, acrylic, elastane, etc. also leave footprint at different processing stages (Van Der Velden et al., 2014). For regenerated cellulose fibres, energy consumption and application of hazardous substance increase environmental footprint (Rosson & Byrne, 2020).

Yarn production

Energy demand is the biggest eco-indicator for yarn manufacturing and most energy is depleted for driving machineries. Air quality (PM10 and PM2.5) is another indicator that shows impact of spinning. Air conditioning in a spinning factory can consume both energy and water (Liu, Zhu, et al., 2020). Spinning of melange yarn consumes energy as well as water, dyes and auxiliaries (Liu, Zhu, et al., 2020). Yarn count is an important factor that directly impacts environmental footprint (Van Der Velden et al., 2014).

Fabric Production

The use of non-renewable energy, generation of fibre fly, use of heavy metal containing knitting oils, generation of solid waste are the key environmental issues related to knitting, (Roos et al., 2015). It accounts for 5 MJ/kg energy consumption causing emissions such as CO_2 and particulates (Baydar et al., 2015). GHG emission from Knitting is comparatively lower than the GHG emissions from fibre production and yarn spinning processes. In one calculation it was found to be 7% of the total life cycle GHG emissions of knitted garments (BSR, 2009). Generation of cotton fibre fly in the knitting floor not only creates indoor air pollution which in turns causes acute and chronic respiratory problems for the employees but also affects the quality of the fabric due to fibre contamination (Bhowmick & Ghosh, 2008).

Weaving requires additional yarn preparation processes such as winding, warping and sizing, which increase the extent of environmental impact from weaving, when compared to knitting. Areas of concern in weaving are solid wastes from yarn, size ingredients and fabrics scraps, effluent from sizing process, particulates from sizing, warping and weaving, volatile organic compounds (VoCs) from sizing and high decibel noise from looms. According to a study done in Finland, energy consumption in weaving including singeing and sizing sub-processes is 5.4 MJ/kg of fabric (Kalliala and Talvenmaa, 2000). Higher energy requirements are reported in other studies (Muthu, 2020). From the total energy consumption in a weaving mill, 36.3% is used by the machines, 29.4% by compressors, 27.1% by air conditioners, 3.4% for illumination and 3.8% for cleaning (Koc and Cincik, 2010). Water pollution associated with sizing of warp yarns is one of the significant environmental issues related to weaving, which accounts for 30% - 70% increased chemical oxygen demand (COD) load in wastewater (Muthu, 2020). Besides sizing chemicals, lubricants and oils are used to lubricate the loom (Roos et al., 2015). Small fibrous particulates as well as VoCs are generated from the sizing process and particulates from the weaving and warping processes. Weaving machines are the main cause of noise pollution accounts for a noise level on an average of 100-120 dB (CTNP, 2019). Further sustainability issues in weaving industries involves handling of solid wastes from yarn, fabric scraps and empty beam frames, maintaining proper humidity and temperature at the weaving floor.

Colourations

The textile industry consumes a big share of global chemical stock. Hundreds of different chemicals, which fall broadly under the categories of dyes and pigment, scouring and bleaching agents, auxiliary chemicals, coatings and washing agents, are used for processing and producing textile materials. The fourth edition of the colour index published by the Society of Dyers and Colourists (SDC) lists over 34,500 dyes and pigments listed under 11,570 Colour Index Generic Names (SDC, n.d.). Uses of organic dyes during the conventional dyeing process has an adverse effect on the environment (Yu et al., 2016). Among the different groups of dyes used in textile colouration, direct and reactive synthetic dyes generally cause less environmental impact in comparison to vat or sulphur synthetic. Textile colouration is recognised as one of the largest water consuming and water polluting industry in the world (Rather et al., 2019). A large volume of effluent with deep colour, various organic and inorganic chemicals, chlorinated compounds is discharged during various textile wet processes.

Garment Production

The apparel industry is responsible for 6.7% of global carbon emissions, which is equivalent to 3.3 Giga Tonnes of CO_2 (Quantis, 2018). The growing environmental impact can be traced back to the significant rise in consumption and production of clothing items. A pair of jeans produces CO_2 equivalent of 11.5 kg of which 57% is produced during the garment production phase (Sadin et al., 2019). A women's sweatshirt has a carbon footprint of around 13.5 Kg of which 5 kg is produced during the manufacturing process (Jungmichel, 2010).

Distribution Chain

Transportation activities between manufacturing unit and retail stores are responsible for over half of the total carbon emission in fashion supply chain (Karaosman et al., 2017). Packaging and finishing items are inevitable to protect products during distribution and maintain branding. However, they in the forms of polybag, hangtags, printed levels, paper, plastic stands, boards and cartons in many cases are landfilled or incinerated after one single use.

The Framework

The framework being presented here is initially developed by the project team utilising the latest information derived through literature review and tacit knowledge and professional experience from the industry. The initial version of the framework is presented before participants (n = 30) including mid to senior level professionals from textile and apparel industry in Bangladesh, representatives from national and international non-governmental organisations (NGOs) and academics in two participatory workshops and then fine-tuned incorporating the key feedbacks from the participants. Finally, the framework is presented in a round table discussion with representative from international fashion retailers, national and international organisations to further validate it.

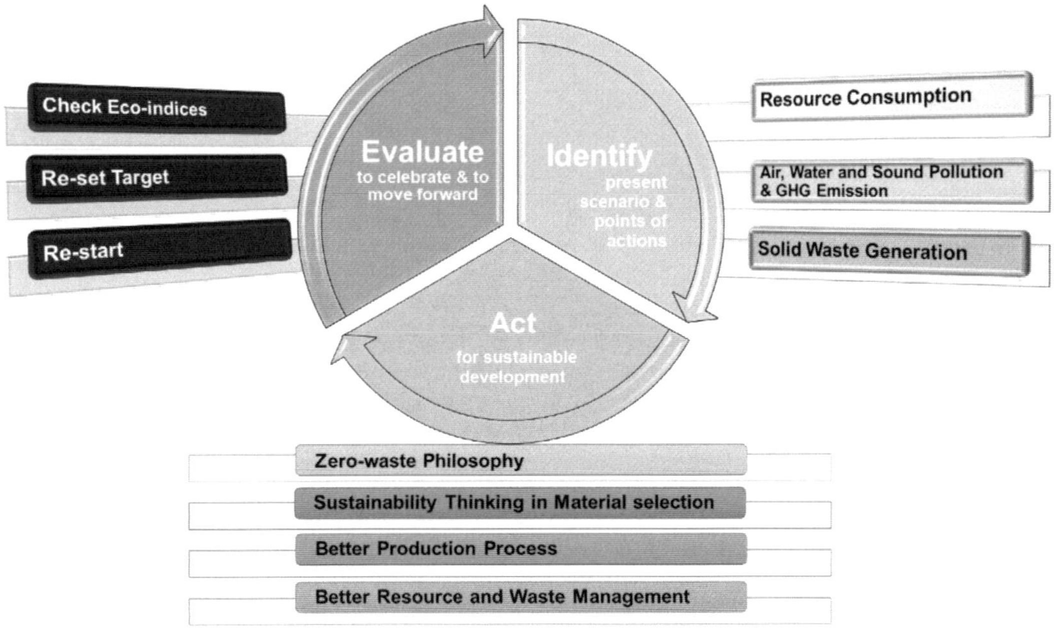

Figure 1: Proposed I-A-E Framework for textile and fashion sustainability

The proposed framework in Figure 1 recommends three key actions - identify (I), act (A), and evaluate (E) to achieve complete environmental sustainability in fashion and textile production in line with SDG 12. These key actions are elaborated as –

A. Identify of present scenario and points of actions
B. Act for sustainable development
C. Evaluate to celebrate & to move forward

Identify of present scenario and points of actions

For any producer of fibre, yarn, fabric and garment, the starting point to achieve sustainability is to first know their present status in terms of resource consumption, pollution and waste generation (Table 3). Section 2 provides a general picture of the environmental impact of fashion and textile production. The UN environment programme identified the sustainability hotspots along the textile and apparel value chain (UNEP, 2020).

Table 3: Sources of resource consumption, pollution and solid waste generation along the textile and apparel value chain

Point of Actions	Specific Areas	Production Phase
Resource Consumption	Raw material	Fibre, Yarn, Fabric, Garment
	Energy	Fibre, Yarn, Fabric, Colouration, Garment
	Water	Fibre, Yarn, Fabric, Colouration, Garment
	Chemicals	Colouration, Industrial Garment wash
Air pollution & GHG emissions	GHG	Fibre, Yarn, Fabric, Colouration,
	VoC	Fabric and Colouration
	Fly (material dust)	Yarn and Fabric
	Overall air quality (PM2.5/10)	Fibre, Yarn, Fabric, Colouration
Water pollution	Effluent	Synthetic fibre, fabric, colouration, wash
Sound pollution	Noise	Synthetic fibre, yarn, fabric, colouration, wash
Solid waste generation	Short fibre	Fibre and Yarn
	Waste yarn	Yarn and Fabric
	Waste fabric	Fabric, Colouration and Garments
	Trims & packaging	Garments and others

We recommend the use of a mapping tool to identify the environmental performance of manufacturing facility. For example, Higg Facility Environmental Module (Higg FEM) from Sustainable Apparel Coalition helps assess the Environmental Management Systems, Energy Use and Greenhouse Gas Emissions, Water Use, Wastewater, Emissions to Air, Waste Management and Chemical Management of a facility.

Act for sustainable development

Once the manufacturers identify the environmental hotspots within their production cycles, it is now time to act for sustainable development. We recommend four principles of actions. They are:

I. Adopt zero-waste philosophy
II. Incorporate sustainability thinking in material selection
III. Apply better production process
IV. Follow better resource and waste management

Adopt zero-waste philosophy

It is important to plan product development and production having a zero-waste target in mind. It is true 100% zero-waste may never be possible in textile and apparel production but having zero-waste philosophy will enable the manufacturers to gradually reduce waste and get closer to a near-zero-waste scenario. For example, recent research conducted at Manchester Metropolitan University, UK demonstrated that the application of Zero-Waste Pattern Cutting (ZWPC) approach into mass production of selected apparels could achieve over 98% of fabric utilisation, while in traditional industrial practice over 85% is considered impossible (Ramkalaon and Sayem, 2020). Previously, the practice of ZWPC was thought to be applicable only in atelier environment for designing and making one or two garments at a time. Implementation of this approach in the industry needs a new mindset and a new approach in apparel design and pattern cutting, and the use of digital pattern cutting and marker making techniques including 2D/3D tools.

Similar to zero-waste, zero-discharge approaches need to be identified for fibre, yarn and fabric productions. Here, the "Road to Zero" initiative from ZDHC is worth mentioning (ZDHC, n.d.). However, this initiative is still too far away to reach the small and medium scale manufacturers in least developed and developing countries.

Incorporate sustainability thinking in material selection

The use of sustainable and renewable materials in fibre production and other textile applications such as dyeing can play a major role in promoting sustainability in textile and fashion industry (Sayem and Haider, 2019). The use of 100% recyclable and biodegradable materials must be ensured not only in fibre making but also in other areas such as colouration, accessories, packaging etc. Product Life Cycle design and life cycle analysis are important at the early stage of product development can help in minimising the environmental impact and ensure material circularity. We recommend the use of any reliable lifecycle analysis (LCA) tool like Higg Index - Material Sustainability Index (HIMSI) from SAC, Eco Index tool by Outdoor Industry Association, OpenLCA, CCaLC, Granta EduPack software etc. by the product developers, designers and purchase team to make sure more and more sustainable materials can be fed into the production process.

Apply better production process

Here we present the better alternatives that can be employed in the production of fibre to garments.

a. Fibre and Yarn

- Organic agricultural products have been reported to substantially minimise the growth stage impacts on the environment (Baydar et al., 2015).
- Naturally coloured cotton has better pest and disease resistance than the conventional cotton (Chen & Burns, 2006). Furthermore, the use of coloured wool from sheep can help minimise the necessity of wool colouration (Chen & Burns, 2006)
- Crop rotation technique can minimise the damage caused by harmful living organisms and substance, thus can reduce environmental footprints via countering the necessity of pesticides (Kazan et al., 2020).
- Despite having some environmental impacts, recycling of fibres from thrown away wastages can spare the elevated impacts associated with the production of virgin fibres (Esteve-Turrillas & de la Guardia, 2017).
- To bring energy efficiency, apart from process optimisation, replacement of current machineries with energy efficient ones can be carried out (Hasan et al., 2019; Malinauskaite et al., 2019; Sa et al., 2018).

b. Fabric

In case of knitting

- Adoption of advanced technologies such as seamless knitting and 3D fabric production, would certainly reduce the wastage of energy resources and carbon footprint by removing post knitting operations, reducing material waste, shortening downtime as well as lowering transport emissions (Chakraborty, 2016; Power, 2012).
- Energy efficient developments in needle technology for high speed circular knitting in order to save energy and resources (Power, 2012).
- Use of semi-combed yarn, as combed yarn consumes more energy than the carded yarn of same count both in knitting and weaving (Koç & Kaplan, 2007).
- Use of eco-friendly lubricants as machine oil reduce the pollution load of wastewater stream (Power, 2012).
- Separating the knitting machine by polythene wrap would be an effective measure to prevent fibre contamination from cotton fly in the knitting floor. Improvement in the atmospheric cleaning process inside the knitting room can bring substantial change but involves with initial cost as well as running cost (Bhowmick & Ghosh, 2008).

In case of weaving

- In sizing, chitosan would be an excellent, economic as well as ecologically compatible alternative to synthetic sizing agents for cellulosic yarns, which is biodegradable and can also be recycled from the wastewater (Stegmaier et al., 2008).
- Keratin from chicken feathers, an accessible and environment friendly sizing agent, can be used to size the polyester or polyester/cotton blended warp yarns and effectively contributes to negligible amount of total ammonia and nitrogen in wastewater but easily biodegradable in the activated sludge (Reddy et al., 2014). Starch based warp yarn sizing extracted from potato would be an economic as well as eco-friendly sizing agent in replacing synthetic sizing agents.

- Over the last 20 years, textile machinery manufacturers especially the weaving machine manufacturers have made considerable progress in lowering down the noise level by adopting mechatronics, PLC operating systems, active noise control mechanism, blocking and absorbing airborne sound as well as vibration damping.
- Air suction system, blowing system and proper ventilation can maintain proper humidity, temperature and remove air contaminants from weaving section (CTNP, 2019).
- Rate of yarn breakage connects with machine speed and yarn count. Therefore, speed of machine needs to be under control to minimize yarn breakage rate and reduce solid waste (Power, 2012).
- Empty cone and empty beam frames should be reused in the weaving process (Alkaya & Demirer, 2014)

c. Colouration

- Uses of energy efficient machineries and process optimisation can reduce environmental pollution during fabric coloration
- Water-free dyeing technique such a supercritical fluid dyeing should be made available by making the technology patent universal (Sayem and Rossbach 2003).

d. Garments

- Apply ZWPC technique in mass production of garment to maximise fabric utilisation (Ramkalaon & Sayem, 2020).
- Apply tools for sustainable fashion product design & development to assist designers (Kozlowski et al., 2019).
- Build sustainable fashion supply chain to embed sustainability through every stage of the fashion and textile production (Shen, 2014).

e. Distribution Chain

- Using biodegradable and less impactful materials for packaging and finishing item.
- Redesigning packaging to ensure recycling could be an approach to avoid landfilling of packaging after single use.

Follow better resource and waste management

- Wastewater treatment and recycling treated water in production system and in domestic use like for use in toilet flush.
- Rainwater harvesting for industrial use.
- Use of renewable energy can reduce the energy consumption in the textile industry
- Installation of wind-powered turbo-ventilators on production plant roofs
- Use of direct solar energy for fibre drying and water heating in the textile industry.
- Small fabric cut offs can be broken down into individual fibres and reprocessed.

This has to be kept in mind that the list provided above is a non-comprehensive list and it is expected that novel alternative approaches, techniques and processes will appear time to time on the market. The implementing party (the producer and manufacturer) will need to keep an open eye always in search of better alternatives for them.

Evaluate to celebrate and to move forward

At retailers and manufacturers level, the effluent characterising parameters, such as chemical oxygen demand (COD), dissolved oxygen (DO), biochemical oxygen demand (BOD), total dissolved solids (TDS), total suspended solids (TSS), the potential of hydrogen (pH) and colour index are commonly tested and the acceptable limits of these parameters are included in several national and international guidelines (Shamsuzzaman et al, 2021). We recommend the evaluation of following product eco-indices following LCA methodology as per BS EN ISO 14040: 2006 and PAS 2050:2011 (Ccalc, 2013) and include them in guidelines of local authorities' control parameters:

- carbon footprint (or Global warming potential);
- water footprint;
- acidification potential;
- eutrophication potential;
- ozone layer depletion potential;
- photochemical smog; and
- human toxicity potential

In addition, we also recommend the mapping of environmental performance of manufacturing facility as mentioned in the Section 3.1. Once measured, textiles and apparel manufacturers will be able to judge the environmental performances of their facilities and products. This will provide them a benchmark to set target for next year and re-start with I-A-E for next year.

Conclusion

The framework was developed primarily targeting the textile and apparel industry in Bangladesh, which is the partner country of our GCRF project. However, this is applicable to all textile and apparel manufacturers located at any part of the world. Effective implementation is the key to the success of any framework. It has been revealed by the workshop participants that a combined effort from all stakeholders will need to come together to implement the proposed framework. The textile and apparel manufacturers will not be able to implement this alone unless different parties like fashion retailers, local industry bodies like Bangladesh Garments Manufacturers and Exporters Associations (BGMEA) and the relevant government and non-government organisations get involved. We recommend setting up working groups involving all stakeholder similar to the FICCA working groups for implementation of the proposed framework.

Acknowledgement

We thankfully acknowledge "Research England" for the GCRF funding and the Research and Knowledge Exchange (RKE) Directorate of Manchester Metropolitan University (MMU) for facilitating the project. The authors are thankful to all research assistants (RAs) namely Anubhav Malik and Anna Romachney from MMU; Humayra Akhter Himu, Tanvir Mahady Dip, Bristi Sarker and Nusrat Jahan from Bangladesh University of textiles (BUTEX); and Rayed Barkat and Mohammad Iftekhar Rahman from BUFT for their contribution to the project activities.

References

10YFP Secretariat (n.d.), the 10 year framework of programmes on Sustainable consumption and production, https://sustainabledevelopment.un.org/content/documents/944brochure10yfp.pdf (Accessed on 22 May 2023)

Alkaya, E., & Demirer, G. N. (2014). Sustainable textile production: A case study from a woven fabric manufacturing mill in Turkey. Journal of Cleaner Production, 65, 595–603. https://doi.org/10.1016/j.jclepro.2013.07.008

Baydar, G., Ciliz, N., & Mammadov, A. (2015). Life cycle assessment of cotton textile products in Turkey. Resources, Conservation and Recycling, 104, 213–223. https://doi.org/10.1016/j.resconrec.2015.08.007

Bhowmick, N., & Ghosh, S. (2008). Role of yarn hairiness in knitting process and its impact on knitting room's environment. WSEAS Transactions on Environment and Development, 4(4), 361–372.

Berentson-Shaw, J. (2015), Fixing Fashion: A Framework to Make Sustainable Fashion the Industry Standard, Huffpost, https://www.huffingtonpost.co.uk/jess-berentsonshaw/sustainable-fashion_b_8393590.html (Accessed on 22 May 2023)

BSR. (2009). Apparel Industry Life Cycle Carbon Mapping. Business for Social Responsibility, June 2009, available online at https://www.bsr.org/reports/BSR_Apparel_Supply_Chain_Carbon_Report.pdf (access date 22 May 2023)

Ccalc (2013), Carbon calculation, http://www.ccalc.org.uk/about.php (access date 22 May 2023)

Chakraborty, S. (2016). A Detailed Study on Environmental Sustainability in Knit Composite Industries of Bangladesh. American Journal of Environmental Protection, 5(5), 121. https://doi.org/10.11648/j.ajep.20160505.13

Chen, H. L., & Burns, L. D. (2006). Environmental analysis of textile products. Clothing and Textiles Research Journal, 24(3), 248–261. https://doi.org/10.1177/0887302X06293065

Coscieme, L.; Saskia Manshoven, Jeroen Gillabel, Francesca Grossi and Lars F. Mortensen (2022), A framework of circular business models for fashion and textiles: the roleof business-model, technical, and social innovation, SUSTAINABILITY: SCIENCE, PRACTICE AND POLICY, VOL. 18, NO. 1, 451–462, https://doi.org/10.1080/15487733.2022.2083792

CSB (2022), Apparel Industry Sustainable Strategies Framework, Center for Sustainable Business, NYU Stern, https://www.stern.nyu.edu/experience-stern/about/departments-centers-initiatives/centers-of-research/center-sustainable-business/research/return-sustainability-investment-rosi/apparel-industry-sustainable-strategies-framework (accessed on 22 May 2023)

CTNP (2019), TEXTILE VALUE CHAIN. Control Techniques for Noise Pollution in Textile Industry. (2019). Available online at https://textilevaluechain.in/in-depth-analysis/articles/textile-articles/control-techniques-for-noise-pollution-in-textile-industry/ (access date 22 May 2023)

EPA (n.d.), 10-Year Framework of Programmes on Sustainable Consumption and Production, https://www.epa.gov/international-cooperation/10-year-framework-programmes-sustainable-consumption-and-production

Esteve-Turrillas, F. A., & de la Guardia, M. (2017). Environmental impact of Recover cotton in textile industry. Resources, Conservation and Recycling, 116, 107–115. https://doi.org/10.1016/j.resconrec.2016.09.034

FICCA (n.d.), The Fashion Industry Charter for Climate Action, UNFCCC secretariat (UN Climate Change), available online at https://unfccc.int/sites/default/files/resource/Industry%20Charter%20%20Fashion%20and%20Climate%20Action%20-%2022102018.pdf (last accessed 04 July 2021).

Günther, J., Thevs, N., Gusovius, H. J., Sigmund, I., Brückner, T., Beckmann, V., & Abdusalik, N. (2017). Carbon and phosphorus footprint of the cotton production in Xinjiang, China, in comparison to an alternative fibre (Apocynum) from Central Asia. Journal of Cleaner Production, 148, 490–497. https://doi.org/10.1016/j.jclepro.2017.01.153

Hasan, A. S. M. M., Rokonuzzaman, M., & Tuhin, R. A. (2019). energies Drivers and Barriers to Industrial Energy E ffi ciency in Textile Industries of Bangladesh. Energies, 12, 1775. https://doi.org/10.3390/en12091775

Jungmichel, N. (2010). The carbon footprint of textiles. Systain Consulting, Berlin, Germany.

Kalliala, E. and Talvenmaa, P. (2000), Environmental profile of textile wet processing in Finland, Journal of Cleaner Production, Volume 8, Issue 2, pp. 143-154, https://doi.org/10.1016/S0959-6526(99)00313-3.

Kazan, H., Akgul, D., & Kerc, A. (2020). Life cycle assessment of cotton woven shirts and alternative manufacturing techniques. Clean Technologies and Environmental Policy, 22(4), 849–864. https://doi.org/10.1007/s10098-020-01826-x

Karaosman, H., Morales-Alonso, G. and Brun, A. (2017) "From a systematic literature review to a classification framework: Sustainability integration in fashion operations," Sustainability (Switzerland), 9(1). doi: 10.3390/su9010030.

Koç, E., & Çinçik, E. (2010). Analysis of energy consumption in woven fabric production. Fibres and Textiles in Eastern Europe, 79(2), 14–20.

La Rosa, A. D., & Grammatikos, S. A. (2019). Comparative life cycle assessment of cotton and other natural fibers for textile applications. Fibers, 7(12). https://doi.org/10.3390/FIB7120101

Liu, Y., Huang, H., Zhu, L., Zhang, C., Ren, F., & Liu, Z. (2020). Could the recycled yarns substitute for the virgin cotton yarns: a comparative LCA. International Journal of Life Cycle Assessment, 25(10), 2050–2062. https://doi.org/10.1007/s11367-020-01815-8

Malinauskaite, J., Jouhara, H., Ahmad, L., Milani, M., Montorsi, L., & Venturelli, M. (2019). Energy efficiency in industry: EU and national policies in Italy and the UK. Energy, 172, 255–269. https://doi.org/10.1016/j.energy.2019.01.130

MSI (2022), Common Framework for Responsible Purchasing Practices, The MSI Working Group on Purchasing Practices. https://www.fairwear.org/stories/the-common-framework-for-responsible-purchasing-practices-out-now/

Muthu, S. S. (2020). Assessing the Environmental Impact of Textiles and the Clothing Supply Chain. In The Textile Institute Book Series (2nd ed.). WOODHEAD PUBLISHING.

NFA (2012), Framework for Achieving Sustainable Fashion Consumption through Collaboration, Nordic Initiative, Clean and Ethical, Nordic fashion Association, https://www.bsr.org/reports/nice-consumer-framework.pdf

Power, E. J. (2012). Sustainable developments in knitting. International Journal of Business and Globalisation, 9(1), 1–11. https://doi.org/10.1504/IJBG.2012.047519

Quantis (2018), Measuring Fashion Report, available online at https://quantis.com/wp-content/uploads/2018/03/measuringfashion_globalimpactstudy_full-report_quantis_cwf_2018a.pdf (accessed on 22 May 20203).

Ramkalaon, S., & Sayem, A. S. M. (2020). Zero-Waste Pattern Cutting (ZWPC) to tackle over sixty billion square metres of fabric wastage during mass production of apparel. The Journal of The Textile Institute, 1–11.

Rather, L. J., Jameel, S., Dar, O. A., Ganie, S. A., Bhat, K. A., & Mohammad, F. (2019). Advances in the sustainable technologies for water conservation in textile industries. Consumption, Footprint, and Life Cycle Assessment, pp. 175-194. https://doi.org/10.1016/b978-0-08-102633-5.00010-5

Reddy, N., Chen, L., Zhang, Y., & Yang, Y. (2014). Reducing environmental pollution of the textile industry using keratin as alternative sizing agent to poly(vinyl alcohol). Journal of Cleaner Production, 65, 561–567. https://doi.org/10.1016/j.jclepro.2013.09.046

Roos, S., Sandin, G., Zamani, B., & Peters, G. (2015). Environmental assessment of Swedish fashion consumption.

Rosson, L., & Byrne, N. (2020). Comparative gate-to-gate life cycle assessment for the alkali and acid pre-treatment step in the chemical recycling of waste cotton. Sustainability (Switzerland), 12(20), 1–15. https://doi.org/10.3390/su12208613

Sa, A., Thollander, P., & Rafiee, M. (2018). Industrial energy management systems and energy-related decision-making. Energies, 11(10). https://doi.org/10.3390/en11102784

Sandin, G., Roos, S., Spak, B., Zamani, B. & Peters, G. Environmental assessment of Swedish clothing consumption — six garments, Sustainable Futures. Mistra Future Fashion, G.Sandin-Environmental-assessment-of-Swedish-clothing-consumption.MistraFutureFashionReport-2019.05.pdf

Santos, A. D.; Fabrizio Ceschin, Suzana Barreto Martins and Carlo Vezzoli (2016), A design framework for enabling sustainability in the clothing sector, Latin American Journal of Management for Sustainable Development, Vol. 3, No. 1, pp 47-65, https://doi.org/10.1504/LAJMSD.2016.078615

Sayem, ASM., Haider, J., 2019. An Overview on the Development of Natural Renewable Materials for Textile Applications'. In Hashmi, S. (eds.) Reference Module in Materials Science and Materials Engineering, Elsevier.

SDC (n.d) Colour Index: Fourth Edition Online, Society of Dyers and Colourists (SDC), UK, https://colour-index.com/assets/files/upl/CI_leaflet.pdf

SDGS (n.d.) Goal 12, Ensure sustainable consumption and production patterns, https://sdgs.un.org/goals/goal12

SDG Tracker (n.d), SUSTAINABLE DEVELOPMENT GOAL 12, Ensure sustainable consumption and production patterns, https://sdg-tracker.org/sustainable-consumption-production

Seo, Min Jeong and Kim, Minjeong (2015), "A New Integrated Framework for the Sustainable Performance of Fast Fashion Firms", International Textile and Apparel Association (ITAA) Annual Conference Proceedings. 132. Accessed from https://core.ac.uk/download/pdf/128974465.pdf

Shamsuzzaman, M., Md. Abul Kashem, Abu Sadat Muhammad Sayem, Adnan Maroof Khan, Sayed Md. Shamsuddin, Md Mazedul Islam (2021), Quantifying environmental sustainability of denim garments washing factories through effluent analysis: A case study in Bangladesh, Journal of Cleaner Production, 290, 125740, ISSN 0959-6526, https://doi.org/10.1016/j.jclepro.2020.125740.

Shen, B. (2014). Sustainable fashion supply chain: Lessons from H&M. Sustainability (Switzerland), 6(9), 6236–6249. https://doi.org/10.3390/su6096236

Stegmaier, T., Wunderlich, W., Hager, T., Siddique, A. B., Sarsour, J., & Planck, H. (2008). Chitosan - A sizing agent in fabric production - Development and ecological evaluation. Clean - Soil, Air, Water, 36(3), 279–286. https://doi.org/10.1002/clen.200700013

UNASF (n.d.), Homepage available at https://unfashionalliance.org/ (last accessed 04 July 2021).

UNEP (n.d), 10YFP - 10 Year Framework of Programmes on Sustainable Consumption and Production Patterns, https://www.unep.org/explore-topics/resource-efficiency/what-we-do/one-planet-network/10yfp-10-year-framework-programmes

UPEN (2020), Sustainability and Circularity in the textile value chain, Global Stocktaking, UNEP, available online at https://www.oneplanetnetwork.org/sites/default/files/unep_sustainability_and_circularity_textile_value_chain_1.pdf (accessed on 22 May 2023)

Van Der Velden, N. M., Patel, M. K., & Vogtländer, J. G. (2014). LCA benchmarking study on textiles made of cotton, polyester, nylon, acryl, or elastane. International Journal of Life Cycle Assessment, 19(2), 331–356. https://doi.org/10.1007/s11367-013-0626-9

Vladimirova, K., Iran, S., Barber, J., Blazquez, M., Burcikova, M., Henninger, C. E., Johnson, E., Joyner Martinez, C., Laitala, K., Maldini, I., McNeil, L., Niinimaki, K.,

Onthank, K., Plonka, M., Sauerwein, M., Wallaschkowski, S. (2021). Conceptual framework

for Sustainable Fashion Consumption within the Circular Fashion System. International Research Network on International Fashion Consumption, https://sustainablefashionconsumption.org/projects/sfc-framework/

Williams, D. (n.d), Fashion Design for Sustainability: A framework for participatory practice, Centre for Sustainable Fashion, University of the Arts London, https://ualresearchonline.arts.ac.uk/id/eprint/14350/1/LeNSin_full_paper._dilys%20williams.pdf

Yu, M., Li, W., Wang, Z., Zhang, B., Ma, H., Li, L., & Li, J. (2016). Covalent immobilization of metal-organic frameworks onto the surface of nylon - A new approach to the functionalization and coloration of textiles. Scientific Reports, 6(February), 1–9. https://doi.org/10.1038/srep22796

ZDHC (n.d.). Roadman to Zero, https://www.roadmaptozero.com/

THE LIVING TEXTILES ECOSYSTEM: A LOCAL TEXTILE BIOFABRICATION STRATEGY.

Jane Scott[1], Ben Bridgens[1], Dilan Ozkan[1] and Romy Kaiser[1]

[1]*Hub for Biotechnology in the Built Environment, School of Architecture Planning and Landscape, Newcastle University, NE1 7RU*

Abstract

Historically UK textile ecosystems have celebrated their regional identity, evolving around the specific knowledge and skills required by particular materials such as wool in Yorkshire and cotton in Lancashire. The Living Textile Ecosystem utilises the biotech sector emerging in the North East of England, positioning textiles as a pivotal bio-fabrication strategy for the development of new materials and construction methods to support the transition of the building sector to net zero. Textiles provide unique attributes including high biocompatibility, the ability to support scale-up as composites and scaffolds, and the potential to provide intelligent material systems which adapt to changing environmental conditions.

This paper provides an initial scope for the Living Textiles Ecosystem, using the example of biohybrid mycelium-knit composites, mapping the interrelationships between materials and technologies within this new interdisciplinary field. The focus is the BioKnit prototype and The Living Room, both of which are composed of 3D knitted fabric, fungal mycelium and cellulose based bulk substrate materials. The materials and technologies used to create these living textile architectures are deconstructed, with consideration of both local material sources and manufacturing capabilities.

Two approaches to enable the commercial development of textile biohybrids are proposed: the adaptation of current industrial processes for alternative material inputs, and the development of new large-scale knit processes adapted to emerging biofabrication technologies.

Keywords

Knitting, Mycelium, Ecosystem design, repurposing waste, adapting industrial process, biofabrication.

Introduction

Could textile materials and technologies offer alternative production techniques for the construction industry? Transformative approaches are certainly required as the construction industry looks towards ambitious goals for net zero. Currently building materials are responsible for 11% of global carbon emissions, and an additional 28% of global carbon emissions are generated during building operation (United Nations, 2020). Future building materials are therefore required that not only dramatically reduce embodied carbon but also provide enhanced functionality that can address the problem of operational energy use. For example, materials that can improve the way that a building performs to reduce the need to operate energy intensive heating, cooling and ventilation systems. Finally, it is essential that materials are developed within circular systems, so that materials are reused or recycled at end of life.

Biotechnologies are beginning to offer innovate alternatives to challenge traditional building materials, suggesting that growth could replace construction in future buildings (Dade-Robertson, 2021). Mycelium-based composites, composed of mycelium (the root network of fungus) and bio-based substrates such as sawdust and food waste, have been developed for commercial application in interiors and architectural prototypes (Jones et al 2020). Whilst most composites are formed using rigid moulds to produce bricks and blocks, researchers are developing alternative production systems such as 3D printing (Elsacker et al, 2022), soft moulds (Scott et al, 2021, Yogiman et al, 2020) and robotic manufacture (Elsacker et al, 2021). The Living Textiles Research Group based in the Hub for Biotechnology in the Built Environment at Newcastle University is focused on biohybrid strategies which bring together mycelium with textiles, specifically the application of knitted fabrics as a scaffold for growth and a mechanism to enable the scale-up of biotechnologies for the built environment (Fig 1).

Figure 1: Detail illustrating the integration of mycelium and substrate with permanent knitted formwork (visible as navy threads) (Image credit, Dilan Ozkan).

This paper describes two approaches developed by the Living Textiles Research Group to enable scale-up of mycelium composites to architectural scale textile biohybrids. The first approach utilises industrial knitting technology to manufacture BioKnit, composed of complex 3D modules which are assembed into a preform suitable for use as a permanent knitted formwork. The second approach applied hand knit methodologies in the development of a freehand knitted canopy, produced off machine, integrating shape and form into one continuous textile to act as formwork for a second largescale prototype. Each textile requires different material inputs, different knitting and control systems and alternative mechanisms to grow the mycelium composite. This paper analyses how the two approaches can be positioned to develop a Living Textiles Ecosystem, reflecting on the potential to adapt current industrial knitting processes for alternative material inputs, and the development of new processes tailored to emerging biofabrication technologies.

Biofabrication of Mycelium Composites

To grow a rigid composite material, mycelium spawn (a mixture of growing mycelium and a nutrition source such as millet or rye) is mixed with a bulk substrate material. Substrates include cellulose rich materials such as sawdust or wheat straw that provide nutrition for mycelium as well as aggregate in the composite mix. The mixture is left to grow in a controlled environment for 7-10 days. During this time the mycelium colonises the substrate, binding all the materials together. After the growth period the mycelium is rendered inactive through heat treatment (Elsacker et al, 2021) or by air drying (Scott et al, 2021). Mycelium composites take the shape of the mould after growing.

BioKnit: Industrial Knitting for Mycelium Biohybrid Textile Composites

BioKnit was designed by our group in 2022 and demonstrated how knitted fabric could be used as a scaffold for mycelium growth in the biofabrication of a 1.8m high freestanding structure (Scott et al, 2022, Agraviador et al 2022). BioKnit was composed of seven 3D modules knitted on a 12gg Shima Seiki SSR industrial knitting machine. The modules we assembled by hand into a single preform, creating permanent textile formwork for a mycelial architecture (Fig 2). In the development of the prototype a series of biofabrication protocols were established for textile biohybrids. This included the specific mixture required for use with knitted textiles (Kaiser et al, 2023), the mycelium inoculation process created specifically for the creation of robust yet slender tubular arches (Scott et al, 2022) and the external support system using a custom-made environmental chamber (Agraviador et al, 2022).

Figure 2: BioKnit self-supporting structure on exhibition in The OME, Newcastle University (Image credit: HBBE).

Adaption of Current Industrial Knitting Processes for Biotechnology

The design of BioKnit exploited the production capabilities of the Shima Seiki System, using multiple materials to make shaped preforms that combined panels of mesh, interconnecting tubular arches and multiple branching forms to create the overall geometries (Fig 3). To achieve the scale required for BioKnit, knitting was undertaken at the full needle-bed width. The complexity afforded by machine technology necessitated the use of yarns with superior knitting performance for fabric production. Through initial testing, wool was identified as providing excellent biocompatibility for mycelium biofabrication, therefore 2/30nm merino wool (with superior mechanical properties for complex knitting on 12gg machines) was selected for production of the knitted modules. Linen was used for the intarsia mesh panels, however linen was not used to host mycelium growth. Waste was minimised in the production process of the knitted modules because each panel was knitted to shape, with no postproduction required. Minimal seaming points were created so that each module was joined at only four narrow edges, reducing the impact of seams in the structure (Scott et al, 2021) and the time required for seaming.

Figure 3: The knit design process; from Shima Seiki programme (left) individual module (centre) and assembled preform (right). (Image Credit, HBBE)

One major challenge working with knitted fabric for mycelium biofabrication is the natural extensibility of the material which allows the soft knit/mycelium structure to deform during filling and growth. This is a problem both for design and fabrication, reducing accuracy in simulations and challenging the structural performance of the resulting material. For BioKnit, extensibility was reduced by knitting at maximum tightness using 2 ends of yarn at a reduced stitch setting. The limited the extensibility, in combination with the catenary arch design solution (Agraviador 2022), transformed this manufacturing challenge into the opportunity to generate complex form.

Machine Specifications

Industrial Knitting machines have a maximum knitting width of between 25cm and 200cm, this is determined by the needle bed width. These machines are designed for efficiency in knitwear production, rather than production of panels at an architectural scale. The Shima Seiki SSR112 machine at Newcastle University has a maximum knitting width of 114cm limiting the production width of individual modules. Here again a creative design approach incorporating multiple intersecting modules, and internal and external shaping during knitting enabled the team to adapt the conventional technology for a new application. However, in working towards building scale components, machine technology needs to adapt for significantly larger fabric widths to minimise postproduction processing.

The Living Room: Hand-Knit methods for Biohybrid Textile Composites

In 2023 a second mycelium biohybrid knitted prototype, *The Living Room,* was grown (Scott and Bridgens, 2023). The Living Room is part of More with Less: Reimagining Architecture for a Changing World, the inaugural exhibition at the Farrell Centre, Newcastle-upon-Tyne, which offers new visions for architecture in the face of the climate emergency (Fig 4). The Living Room consists of an exposed knitted formwork on the interior and a smooth mycelium plaster on the exterior, to create a freestanding architecture reminiscent of a thin-shell vaulted structure. The ambition for *The Living Room* was to focus on local materials as resource for the structure, thinking about ways to design sustainable architecture that is fundamentally different from conventional construction methods. As such The Living Room is a bio-fabricated architecture grown using by-products from regional industry and locally available waste materials. It is composed of Herdwick wool, a mix including sawdust and waste from the paper-making industry, both sourced from the North of England.

Figure 4: The Living Room, on Exhibition at More With Less, the Farrell Centre, Newcastle. (Image Credit, Ben Bridgens).

Adaptation of Freehand knit Processes for Biotechnology

This prototype prioritised reducing the environmental impact of the material input. The decision to use local waste and by-products from other industries required a move away from conventional knitting yarns, and therefore conventional industrial knitting technologies. Instead, the knitted textile component was itself scaled up by creating 5cm long knitted stitches using a freehand knitting process. The 'yarn' used for freehand knitting was composed of Herdwick wool tops. Herdwick sheep are a traditional breed that live in the hills of the Lake District and produce a coarse wool that is less suitable for fine yarns, and unsuitable for manufacture using the 12gg Shima Seiki industrial knitting machine. Wool tops are a bulk of fibres 3cm in diameter produced in long lengths that have been scoured and combed prior to spinning. Wool tops were sourced from Europa Wools in Yorkshire. Initial lab-based experimentation determined that the coarse nature of the wool, and the untwisted bundles of fibres provides an excellent interface for mycelium growth (Fig 5). The knitted canopy was knitted by hand, in one piece, using a circular knitting technique.

Figure 5: Details illustrating integration of mycelium growth (white) through free-hand knitted wool fabric. Early experimental sample (left) and final structure (right). (Image credit, Ben Bridgens)

Research was undertaken to identify bulk cellulose waste for the substrate mix required for mycelium growth. Hardwood sawdust from a local sawmill and by-products from papermaking were identified as suitable for the requirements of the composite. These materials are unusual in that they have no commercial value and do not feed into other industrial processes. However, further work is required to assess the quantities of these materials that are available in line with the bulk material requirements for the construction industry.

The Living Room was produced by tensioning the knitted canopy inside a 4m x 4m x 2m scaffolding frame. The form was knitted into the structure during the free-knitting process using a pattern derived from a computation model of the form. As the edges were tensioned into place the three-dimensional form emerged within the scaffolding frame. The mycelium was pre-grown in lab conditions prior to inoculation of the substrate mix. The application process was also undertaken by hand to facilitate the application of the correct thickness of paste. The thicknesses had been determined by structural modelling of the proposed form. For this prototype the environmental chamber was constructed using the scaffolding frame. Heat, light, and humidity were controlled and monitored over a two-week growing period. After growing, the biohybrid structure was dried for 10 days and then all the external support system was removed.

The Living Room was constructed by hand, using craft techniques and the tacit knowledge of materials. This expertise was particularly critical at four key stages; knitting the canopy, tensioning the form, mixing the mycelium substrate, and applying the mycelium plaster. For example, control of stitch length within the knitting process was determined by hand. It was essential to knit the canopy with a high stitch density and to reduce extensibility as much as possible so that it didn't deform under the mycelium composite load. Control of stitch length was monitored and maintained manually as the loops were formed. Without significant experience of working with knitting and mycelium, the scale of this build would have been extremely challenging. It is significant that in moving towards a more sophisticated integration of knitted fabrics for mycelium biohybrids, The Living Room required the knowledge of materials and processes that underpin textile thinking and craft methodologies.

Discussion

The two structures, BioKnit and The Living Room are examples of textile biohybrids that utilise the design and manufacturing opportunities afforded by knitted fabric to grow transformative architectural forms from mycelium. Two approaches to enable the commercial development of textile biohybrids are proposed through this research. The first approach looks to current manufacturing technologies and proposes techniques to adapt current industrial knitting for alternative material inputs. In this approach the constituent fibres require good biocompatibility, and the yarns must retain the strength and flexibility required when knitted at low stitch lengths and high stitch densities.

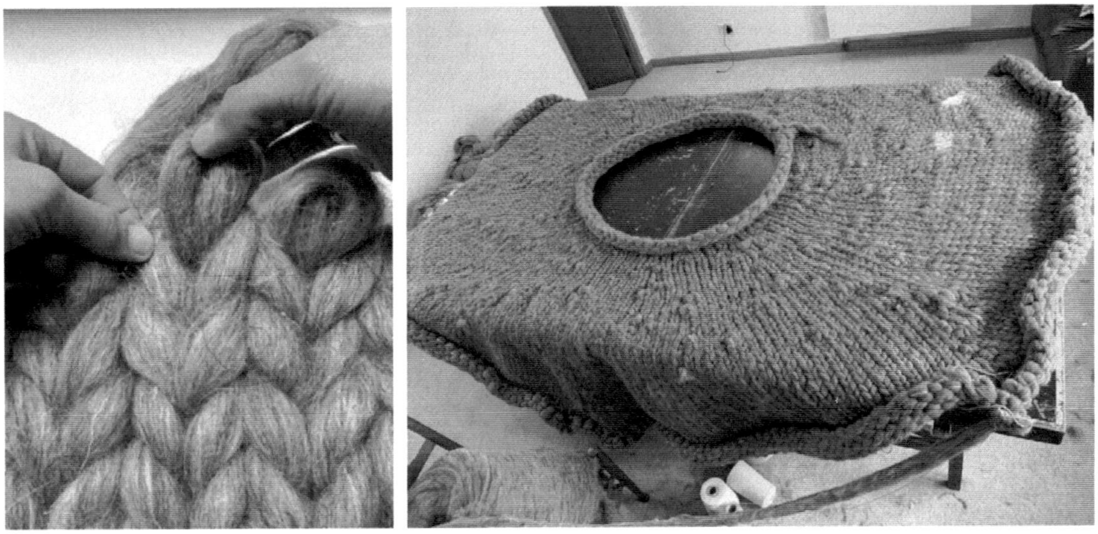

Figure 6: Freehand knitting process; stitch length is controlled manually (left). Canopy is produced as a large continuous textile using circular knitting process (right). (Image credits, Jane Scott).

Whilst complexity of the knitting structure and 3D form can be programmed into the fabric at the design stage, machine controls, including takedown and stitch length benefit from careful adaptation during sampling to make sure that faults do not occur while knitting. This first method promotes a modular system which can be scaled through repetition of modules and a secondary assembly process joining multiple modules into a cohesive preform.

The second approach looks beyond current industrial fabrication technologies to new scales of textile making designed for biotechnology at the scale of the built environment. This approach integrates the unique form-finding and form-making capabilities of knitting to propose the design and fabrication of bespoke canopies that act as permanent formwork for mycelium biohybrid composites. The advantage of this approach is in the ability to integrate structure and three-dimensional form within a continuous material with no disruption to material behaviour and performance due to seams and assembly. In addition, the use of loosely organised fibres within wool tops enables excellent mycelium growth through the textile leading to enhanced integration of the biohybrid composite. By taking this technique off machine the freedom to work to any size specification can be considered, and new ways to automate this approach are currently under review within The Living Textiles Research Group.

Both approaches enable change in multiple interconnected areas of the textile and materials processing industries in the North of England. Underpinning this work is an ambition to develop high value products for architecture and construction from waste resources readily available in the North of England. Through large-scale prototyping and the development of bespoke structures for public exhibition such as BioKnit and The Living Room, materials and processes can be tested at scale and critiqued in a public forum.

Conclusions

BioKnit and The Living Room provide two examples of how textiles and biotechnology can come together to produce alternative materials and fabrication processes with potential for application in the built environment. BioKnit uses industrial knitting technology to produce modules and preforms that can be tailored to bespoke projects and applications. It demonstrates the flexibility of industrial knitting as a production method for large scale bespoke textiles. In this example the 3D knit capabilities of industrial knitting machines are scaled up to generate preforms suitable for architectural applications. The environmental benefits of this approach include the ability to repurpose current manufacturing capabilities for new applications and the ability to produce preforms locally using UK knitting capacity.

An alternative approach is presented in The Living Room. This research project demonstrates how industrial by-products and waste from one industry can be appropriated for use in another sector. The wood and paper waste act as a food source for mycelium and as it grows the materials are bound together by the mycelium. The textile acts as the permanent formwork to generate the form of the architecture. In this example textiles and biotechnology begin to enable circular processes where waste from one industry can provide resource for another. This can benefit regional industries and demonstrate a means to dramatically reduce the environmental impact of construction. Direct benefits of this production process utilised for The Living Room include a reduction in transportation costs as all materials are sourced and grown locally, as well as a reduction in industrial waste, both of which have a positive economic as well as environmental impact for local industries.

The underlying conclusions from this research are that in the development of mycelium textile composites for architecture, changes to material inputs will necessitate changes to the biofabrication processes and resulting biohybrid textiles. However, the development of two large-scale prototypes demonstrate that knitting processes and technologies are highly adaptable for new biofabrication techniques. The success of both methods in the design and biofabrication of self-supporting structures at an architecture scale demonstrate how textile technologies can be adapted for different material inputs and alternative applications. The work undertaken within this investigation focuses on mycelium-knit composites, however this work provides the foundations for the development of a broader living textiles ecosystem. An extensive programme of research is now required to assimilate the emerging ecosystem into a low impact industry for sustainable construction.

Acknowledgements

This research was developed in the Hub for Biotechnology in the Built Environment (HBBE) and for the More With Less exhibition at the Farrell Centre, Newcastle-upon-Tyne,UK. HBBE is funded by Research England and is a joint initiative between Newcastle University and Northumbria University

References

Agraviador, A. Scott, J. Elsacker, E. Kaiser, R. Hoenerloh, A. Topcu, A. Bridgens, B. 2022. BioKnit: the coordination of computation with material investigation in the design of biohybrid textiles towards architectural integration. Proceedings, ACADIA 2022.

Dade Robertson, M. 2021. Living Construction. Routledge: Oxen

Elsacker, E. Peeters, E. De Laet, L. 2022. Large-scale robotic extrusion-based additive manufacturing with living mycelium materials. Sustainable Futures, Vol 4, https://doi.org/10.1016/j.sftr.2022.100085

Elsacker, W. Sondergaard, A. Van Wylick, A. Peeters, E. De Laet, L. 2021. Growing Living and Multifunctonal Mycelium Composites for Large Scale Formwork Applications using Abrasive Wire-cutting. Construction and Building Materials. 328.122732. https://doi.org/10/gh56vx.

Jones, M. Mautner, A. Luenco, S. Bismarck, A. Sabu, J. 2020. Engineered mycelium composite construction materials from fungal biorefineries: A critical review, Materials & Design, Vol:187, https://doi.org/10.1016/j.matdes.2019.108397.

Kaiser, R. Scott, J. Bridgens, B, Elsacker, E. 2023. BioKnit: development of mycelium paste for use with permanent textile formwork, Frontiers in Bioengineering and Biotechnology [in press].

Scott J, Gaston E & Agraviador A (2021), Configured Knitting: Grafting as an assembly process for knitted architecture. in V Stojaković & B Tepavčević (eds), Proceedings The 39th Conference on Education and Research in Computer Aided Architectural Design in Europe: eCAADe 2021 Towards a New, Configurable Architecture. Education and Research in Computer Aided Architectural Design in Europe (eCAADe), pp. 473-482, Novi Sad, eCAADe 2021

Scott, J et al. 2022 Knitted Cultivation, Textiling a Multi-Kingdom BioArchitecture in ed Hyvsjel and Cruz, Structures and Architectures, CRC Press (2-10)

Scott J, and Bridgens B, Ozkan D, Kaiser R, Agraviador A (2023), The Living Room, More With Less Exhibition, The Farrell Centre, Newcastle upon Tyne.

United Nations Environment Programme: Global Alliance for Buildings and Construction. 2020. GLOBAL STATUS REPORT FOR BUILDINGS AND CONSTRUCTION Towards a zero-emissions, efficient and resilient buildings and construction sector

Yogiaman, C., Pambudi, C.P., Jayahankar, D.K., Tracy, K.J. 2020. Knitted Bio-Material Assembly, Proceedings of ACADIA 2020: Distributed Proximities: 58-66.

ASSESSING THE ENVIRONMENTAL IMPACT OF RECYCLED WOOL TEXTILES

Antonia Vyrkou[1], Owais Siddique[1], Paraskevi Fotoglou[2], Ian Burn[2] and Athanasios Angelis- Dimakis[1]

[1]School of Applied Sciences, University of Huddersfield, Huddersfield HD1 3DH

[2]Camira, The Watermill, Wheatley Park, Mirfield WF14 8HE

Abstract

Following the prevailing paradigm of industrial circularity, and inspired by the Textiles 2030 Roadmap, the stakeholders involved in the UK textile value chains (i.e. farmers, manufacturers, and retailers) need to transform the methods they are using to supply resources and produce, use, and dispose of clothing and textiles. More specifically, the manufacturing businesses need to be able to tangibly quantify the material impacts of the products they produce, taking measures to alter their production processes and raw material choices to reduce impacts, following the circular design standards and promoting reprocessed pre-consumer waste and recycled fibres. On this basis, Camira Fabrics plans to launch a new fabric made of recycled wool. The objective of this paper is to quantify the environmental impact of this new product and to assess its performance in comparison with existing textiles, made entirely out of virgin wool. The results have shown that the environmental impact of the new fabric decreases with an increased share of recycled wool. All three main indicators (carbon footprint, water footprint and fossil fuel depletion) will be reduced from 29% to 87% depending on the share of the recycled wool (from 25% to 100%), compared to a similar product made of virgin wool.

Keywords

Recycled Wool, Textiles, Life Cycle Assessment

Introduction

According to the Global Textile Market Size & Share Report, 2022-2030, in 2021 the global textile industry was valued at USD 993.6 billion with a prediction of an annual growth of about 4% in the years 2022 to 2030 (Grand View Research, 2022). Approximately 13.3% of the total income of the industry came from wool-based products that cover a wide range of applications (clothing, upholstery, carpeting etc) and another 39% was the cotton textile revenue. This growth, however, came with a huge price in terms of water consumption and environmental impact. It is estimated that the production of a pair of jeans consumes 3781 litres of water and emits 33.4 kg of carbon equivalent (The World Bank, 2019). In general, the textile and fashion industries are responsible for 10% of the annual global carbon emissions and 20% of the wastewater production (European Parliament, 2022).

In recent years, following the worldwide effort of industries and consumers to become more environmentally friendly and socially conscious, the textile industry had been motivated to develop better production processes focusing on innovative technologies, recycling and reuse of products, by-products, and resources. A helpful tool in the effort to identify and assess the processes with the greatest environmental impact is life cycle assessment (LCA), giving valuable information about the impact of the different stages of production (sheep rearing, wool spinning, fibre dyeing etc). While performing LCA for wool products and textiles is becoming a common practice, there are a few particularities and obstacles that need to be considered compared to LCAs for other products, as detailed in the recently published "Wool innovation action plan to support a circular and sustainable future" (KTN, 2023).

Wool is one of the products that come from sheep farming (e.g. milk, meat etc) following the same processes and using the same resources. The overall environmental performance can be affected by fluctuating values of those products and a critical assumption that needs to be made is the allocation of the farming impacts to the different products/by-products. Different methods of impact allocation can produce different result, especially since farming is usually one of the major contributors to the overall environmental impact. It has been estimated that approximately 45% of the total greenhouse gas emissions of the wool textile industry in the UK were contributed by sheep farms (for greasy wool production) due to livestock maintenance (Thomas et al., 2012).

Another big obstacle for LCA practitioners is the lack of data or inaccurate data in a regional level as the differences between sheep breeds, farming practices, climate and landscape can affect the environmental impact. This can lead to assessments with a high level of uncertainty, hence the collection of representative and accurate data for different regions, production and processes and consumer practices is as a major challenge as is crucial for the reliability and accuracy of the life cycle assessments.

Camira is recognised as a leading textile company, known for environmental innovation and product stewardship. The company innovated a new category of natural fibre fabrics, made from wool blended with harvested nettles, hemp and flax, as a result of its work on a UK government funded project in 2005 – 2008 called STING (Sustainable Technology in Nettle Growing). Camira has also designed and manufactured recycled polyester fabrics for over 25 years using both pre-consumer and post-consumer PET waste. The drive to develop products using recycled wool is a response to a need to find circular solutions for the unavoidable wool waste, which is generated as part of its own manufacturing, especially from yarn cone remnants from manufacturing transport fabrics and waste selvedges. Camira's furniture manufacturer customers are also showing an interest in fabric take-back schemes to recycle their upholstery waste or chair covers into new fabric.

LCA and Environmental Product Declaration (EPD) are becoming increasingly demanded in the commercial interiors sector by architects and designers who are working with end-user clients or for public tenders. They not only highlight the tangible environmental impacts, but also allow for a level of comparative evalution, although there are many variables which can make this difficult.

The objective of this paper is to assess the environmental performance of the iinouiio production line, a novel product introduced by Camira based on recycled, and to compare its impacts with a similar product, which is manufactured with virgin wool. It also examines the implications of increasing the amount of recycled content and the consequent reductions in environmental impacts.

Methodology

The methodological approach used to carry out the research follows the LCA framework, as detailed in the ISO14040/44:2006, and includes the following four steps:

- Goal and Scope Definition - The definition of the study objective and the selected system boundaries
- Life Cycle Inventory (LCI) – Formulation of a list of all inlet and outlet flows that the system exchanges with the environment.
- Life Cycle Impact Assessment (LCIA) - Selection of impact categories, classification, and impact calculation.
- Interpretation of results – Identification of significant issues and contributors and formulation of conclusions based on the goal and scope of the study.

The software used to assess the environmental impact of the processes was SimaPro 9.2 Academic License and the most recent version of the ecoinvent Database (3.7.2). Although the comparison mainly focuses on carbon footprint, water footprint, and fossil fuel depletion, twelve (12) midpoint impact indicators have been calculated (

Table 1).

For most of the indicators, the CML-IA method is used, a LCA methodology developed by the Centre of Environmental Science (CML) of Leiden University in The Netherlands. This method is an update of the CML 2 baseline 2000 and corresponds to the files published by CML in August 2016 (version 4.7). The CML-IA (baseline) method elaborates the problem-oriented (midpoint) approach, and it is one of the most common methods used in a midpoint impact Life Cycle Assessment.

The only exception is the water depletion indicator, for which the AWARE (Available WAter REmaining) method is used, which is the recommended method from WULCA (a Life Cycle Initiative group project on the assessment of use and depletion of water resources within life cycle assessment) to assess water consumption impact assessment in LCA. In May 2016, the method was also endorsed by the EU Joint Research Centre. The implementation in SimaPro corresponds to AWARE version 1.2c. AWARE is to be used as a water use midpoint indicator representing the relative available water remaining per area in a watershed after the demand of humans and aquatic ecosystems has been met. This method assesses the potential of water deprivation, to either humans or ecosystems, building on the assumption that the less water remaining available per area, the more likely another user will be deprived.

Table 1. Environmental Impact Indicators

Impact Category	Impact Acronym	Unit
Abiotic depletion	ADP	kg Sb_{eq}
Abiotic depletion (fossil fuels)	ADP (FF)	MJ
Global warming	GWP	kg $CO_{2\,eq}$
Ozone layer depletion	ODP	kg CFC-11_{eq}
Human toxicity	HTP	kg 1,4-DB_{eq}
Fresh water aquatic ecotox.	FAETP	kg 1,4-DB_{eq}
Marine aquatic ecotoxicity	MAETP	kg 1,4-DB_{eq}
Terrestrial ecotoxicity	TETP	kg 1,4-DB_{eq}
Photochemical oxidation	PCOP	kg $C_2H_{4\,eq}$
Acidification	AP	kg $SO_{2\,eq}$
Eutrophication	EP	kg $PO_4^{-3}{}_{eq}$
Water Depletion	WDP	m^3

Goal and scope definition

The goal of the analysis is to assess the environmental impact of the iinouiio production line, using as a functional unit 1 m^2 of the final product. The analysis has followed a cradle-to-gate approach, from sourcing the raw materials until the point that the product leaves the industry gates for distribution.

The final product is manufactured by combining waste wool, which is sourced from a different Camira production line (Vigor), and virgin wool from different suppliers across the globe. The system boundaries include a series of processes for each one of the two resources. The waste wool is collected at the point of production, transported back to the UK, where it is shredded, blended with New Zealand virgin wool, and spun into a new yarn (which will be referred to hereafter as the recycled wool yarn). The virgin wool, on the other hand, is sourced from several global wool producing regions, scoured, dyed and spun, with 15% of nylon, into a yarn (which will be referred to hereafter as the virgin wool yarn) in different countries in Europe, and transported back to the UK. The two yarns are weaved together at a Camira plant in Huddersfield and the woven product is then finished via a series of processes (dry cleaning, scouring, tentering, cropping, pressing, and rolling).

LCI assumptions

All the values for the Life Cycle inventory have been provided by the corresponding industrial facilities. The weight of the final product is 564 gsm. A detailed breakdown of the numerical values is not provided in the manuscript, since some of these values are sensitive private information, but can be provided at request to any interested individual. When heat is required, it has been assumed that a typical natural gas boiler is used, with 80% efficiency, if no further information has been provided. Regarding transportation, average distances between countries have been used for the estimate of the corresponding impacts. Finally, the virgin wool used is sourced from sheep raised for meat production. Shearing is undertaken for animal welfare.

Table 2 presents the background systems used for the analysis and the corresponding entries of the ecoinvent database, for retrieving the characterization factors. The ecoinvent database does not include region specific factors for wool production, As stated earlier, the Wool Innovation Action Plan identified as a key issue for the assessment of woollen products the lack of reliable regional data on farming systems. To overcome this problem, different characterisation factors were used only for the global warming potential and the fossil fuels depletion based on the findings of Wiedemann et al. (2015). Moreover, as expected. there are no values in the database for the characterisation factors related to the production line of the recycled wool. For that purpose, the environmental impact is estimated using the indicators related to the Vigor production line, as estimated by a Camira internal project.

Table 2. References for the background systems

Resource	Emission factor reference
Wool	Sheep fleece in the grease {RoW}| sheep production, for meat | APOS, U Sheep fleece in the grease {RoW}| sheep production, for wool | APOS, U
Nylon	Nylon 6-6 {RER}| market for nylon 6-6 | APOS, U
Electricity	Electricity, low voltage {BG}| market for | APOS, U Electricity, low voltage {GB}| market for | APOS, U Electricity, low voltage {IT}| market for | APOS, U
Heat	Heat, district or industrial, natural gas {RER}| market group for | APOS, U
Compressed air	Compressed air, 600 kPa gauge {RER}| market for compressed air, 600 kPa gauge | APOS, U
Steam	Steam, in chemical industry {RER}| market for steam, in chemical industry | APOS, U
Transportation	Transport, freight, lorry, unspecified {RER}| market for transport, freight, lorry, unspecified | APOS, U Transport, freight, sea, container ship {GLO}| market for transport, freight, sea, container ship | APOS, U
Sodium bicarbonate	Sodium bicarbonate {GLO}| market for sodium bicarbonate | APOS, U
Soap	Soap {GLO}| market for | APOS, U
Water	Water, decarbonised {GB}| market for water, decarbonised | APOS, U

In terms of using recycled wool, there are at least 12 different options available to choose from with regards to modelling recycling in LCA and the consequent boundary setting (Ekvall et al., 2020). The two that were chosen to be implemented in this report are the "simple cut off" method and the "cut off with economic allocation" method.

The cut off method is perhaps the easiest and simplest one. It allows boundaries to be chosen before or after recycling and each product accounts for its associated processes. It is commonly referred to as the 100/0 method, as it gives 100% of the impact of the virgin materials to the first product and 0% to the processes/products where recycled materials have been used. This method is recommended for the application to EPDs. The cut-off method also removes any risk from the environmental impact assessment, as the impacts are only linked to the production line that causes them, irrespective of any potential future use (Frischknecht, 2010).

On the contrary, the economic allocation method essentially considers the waste stream to be recycled as a by-product of the original production line. This means that an estimate of the waste stream's economic value is required, which will be compared with the price of the initial product. The environmental burdens of the initial cradle-to-gate production line are then allocated according to the ratio between the initial product price and the estimated value of the waste stream. This method can be applied in cases where the two streams (initial product, waste) are different since it is not affected by changes in the material properties. However, it can be sensitive to changes in the economic values of scrap costs. In our case, though, where the waste provider is the same company as the waste receiver, it is safe to allocate a percentage of the value of the initial product, without a lot of uncertainty.

Results

Five different scenarios have been assessed, by changing: (a) the ratio between recycled wool yarn and virgin wool yarn in the final product; and (b) the ratio between recycled wool and virgin wool in the recycled wool yarn. The five scenarios are summarized in Table 3. Scenario 0 (S0) describes a product manufactured with 100% virgin wool. It has to be noted that this is an imaginary product, i.e. a product that is not currently manufactured by Camira, and has only been introduced for comparative purposes, as the baseline. The remaining four scenarios (S1–S4) describe alternative configurations with gradually increasing percentage of recycled wool in the final product. Scenarios 1 and 2 are the ones that describe better the current state-of-the-art of the company and are the most probable to be implemented. It is also assumed that the weight of the product does not change significantly by varying the percentage of the recycled wool.

Table 3. Assumptions for scenario formulation

	S0	S1	S2	S3	S4
Percentage of recycled wool in the recycled wool yarn	0%	50%	100%	100%	100%
Percentage of recycled wool yarn in the final product	0%	50%	50%	75%	100%
Percentage of recycled wool in final product	0%	25%	50%	75%	100%

Table 4. Environmental Impact Indicators

Indicator	Scenario 0	Scenario 1	Scenario 2
ADP	6.6E-05	4.8E-05	2.9E-05
ADP (FF)	52.05	36.88	21.70
GWP	8.56	6.07	3.58
ODP	4.1E-07	3.1E-07	2.2E-07
HTP	2.56	1.80	1.05
FAETP	2.98	2.03	1.07
MAETP	4752.54	3044.86	1337.17
TETP	0.12	0.08	0.03
POCP	2.0E-03	1.4E-03	8.6E-04
AP	8.9E-02	6.6E-02	4.2E-02
EP	4.0E-02	2.8E-02	1.6E-02
WDP	2.80	1.71	0.62

Table 4 presents the environmental impact indicators for the first three scenarios, using the cut-off method to incorporate the impact of the recycled wool, whereas Figure 1 illustrates the comparison between all five scenarios for the three main indicators (global warming potential, water depletion potential and fossil fuels depletion). All absolute values are expressed in the units of the indicator (as detailed in Table 1) per m^2 of the final product.

Clearly, and as expected, the environmental impact of the new fabric decreases with an increased share of recycled wool. From the three main indicators assessed, GWP and WDP continue to decrease until the share between recycled and virgin wool yarn is 100:0 (S5), where ADP (FF) remains relatively constant for percentages of recycled wool yan greater than 50% (S3). More specifically, GWP will be reduced between 29% and 81%, WDP will be reduced between 39% and 87%, ADP (FF) will be reduced between 29% and 59%, for S1 and S5, respectively.

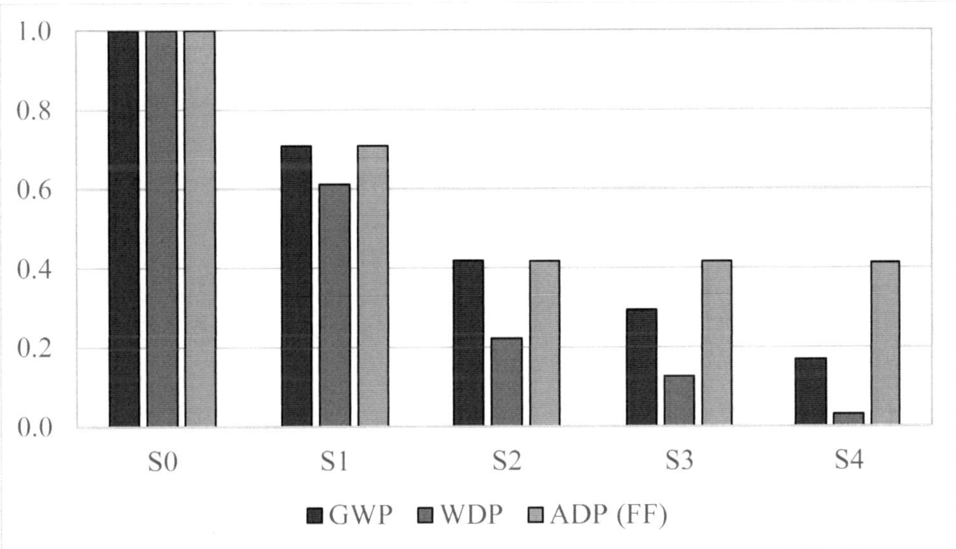

Figure 1. Environmental impact comparison for scenarios S0-S4 (relative values with S0 indicators equal to one)

Figure 2. Environmental impact breakdown per stage for S1

Currently, the most realistic option for the initial launch of the new product is Scenario 1, which incorporates 25% of recycled in the final product. For that reason, it was decided to perform an environmental impact breakdown per stage, in order to identify the environmental hotspots, i.e. the stages with the highest contribution to the environmental impact. Overall, across all indicators the main contributors are the virgin wool supply, as expected, and the spinning process, which is mostly due to the electricity consumption of this stage.

To assess the significance of the recycling allocation method, two alternative scenarios have been formulated, using the economic allocation method. Scenario 5 (S5) assumes that the economic value of the waste stream equals to 10% of the initial product stream, which is a realistic estimate, whereas Scenario 6 (S6) assumes that the economic allocation share reaches 25%. It can be seen from Figure 3, that even in the case where the waste wool is assumed to carry forward a share of the environmental impact of the initial product, the environmental performance of the new fabric remains better that the 100% virgin wool fabric in all three main indicators for S5 (with a reduction of 12%, 22% and 25% for WDP, ADP (FF) and GWP respectively), whereas for S6 only WDP is increased by 28%.

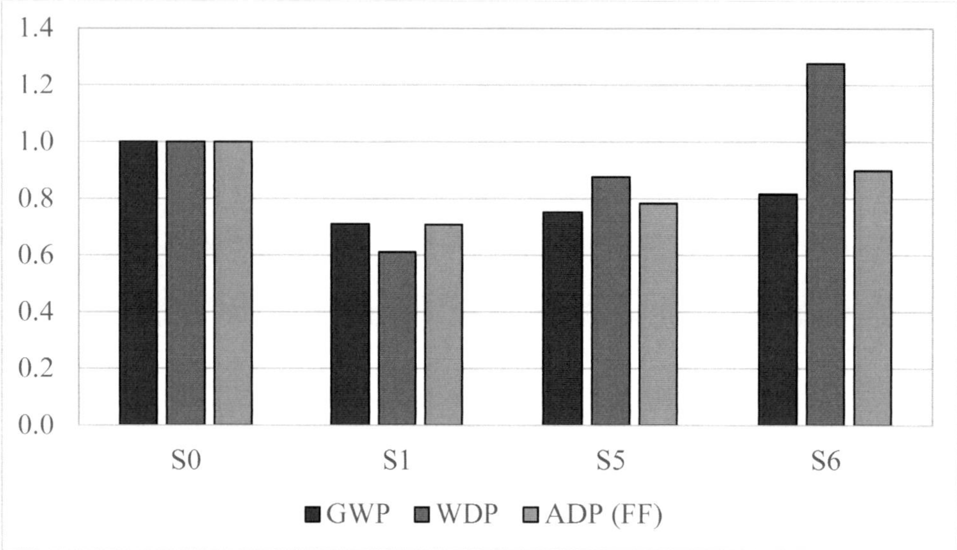

Figure 3. Environmental impact comparison for S0-S1 vs. S5-S6 (relative values with S0 indicators equal to one)

Discussion

The analysis has revealed the new product made of a combination of recycled and virgin wool has better environmental performance compared to a 100% virgin wool garment, when ignoring the environmental impact of the textile to be recycled. Since sheep farming is the major contributor to the environmental impact of a woollen garment, a reduction of the virgin wool leads to a reduction in all major indicators. That reduction can be either continuous, with an increasing share of recycled wool (e.g. for carbon and water footprint) or can reach a plateau, after incorporating 50% of recycled to the product (e.g. fossil fuel depletion).

The results are slightly worse when a different allocation method is chosen. Even in this case, though, a 10% impact allocation of the original product to the recycled wool does not alter the overall, with the environmental performance of the recycled being better compared to the 100% virgin wool garment for all indicators.

References

Ekvall T., Anna Björklund, A., Sandin, G., Jelse, K., Lagergren, J., Rydberg M. (2020) *Modelling recycling in life cycle assessment*. Report No C551, IVL Swedish Environmental Research Institute, ISBN 978-91-7883-226-2

European Parliament (2022) *The impact of textile production and waste on the environment.* Available from: https://www.europarl.europa.eu/news/en/headlines/society/20201208STO93327/the-impact-of-textile-production-and-waste-on-the-environment-infographic (Accessed: 01 March 2023)

Thomas B., Fishwick M., Joyce J., van Santen A. (2012) *A Carbon Footprint for UK Clothing and Opportunities for Savings*. Available from: https://docplayer.net/27216227-Final-report-a-carbon-footprint-for-uk-clothing-and-opportunities-for-savings.html (Accessed: 01 March 2023)

Frischknecht, R. (2010). "LCI modelling approaches applied on recycling of materials in view of environmental sustainability, risk perception and eco-efficiency" International Journal of Life Cycle Assessment 15, 666–671 https://doi.org/10.1007/s11367-010-0201-6

Grand View Research (2022) *Global Textile Market Size & Share Report, 2022-2030.* Available online at: https://www.grandviewresearch.com/industry-analysis/textile-market (Accessed: 01 March 2023)

KTN (2023) *Wool innovation action plan to support a circular and sustainable future.* Available online at: https://iuk.ktn-uk.org/programme/circular-economy-innovation-network/https://www.grandviewresearch.com/industry-analysis/textile-market (Accessed: 10 May 2023)

The World Bank (2019) *How Much Do Our Wardrobes Cost to the Environment?* Available online at: https://www.worldbank.org/en/news/feature/2019/09/23/costo-moda-medio-ambiente (Accessed: 01 March 2023)

Wiedemann, S.G., Ledgard, S.F., Henry, B.K., Yan M.J., Mao N., Russell S.J. (2015). *"Application of life cycle assessment to sheep production systems"*. International Journal of Life Cycle Assessment 20, 463–476 https://doi.org/10.1007/s11367-015-0849-z

ARE TEXTILE GRADUATES ADEQUATELY PREPARED FOR TODAY'S RAPIDLY CHANGING WORKPLACE?

Janet Wilson[1]

[1]*Textile Trainer and Consultant, Hampshire, UK*

Abstract

Against the background of a rapidly changing economic and commercial environment worldwide, and with a general shift towards more sustainable ways of living, future textile graduates are likely to need an increasingly wide range of skills throughout their professional careers.

There are current skills gaps in nearly every area of the textile industry both in the UK, and other countries too. Graduates also often lack the entrepreneurial skills required for working in SME's or for starting their own businesses.

This paper examines the key skills required in the workplace (manufacturing, retailing, SME'S, and start-ups) and reviews the subjects currently taught by university textile courses in the UK. Findings identify the key relationship between textile design and textile technology and include observations made in over 35 years of teaching in higher education. Distinction is made between Education and Training.

Some recommendations for the education and training of future generations in the textile sector are given.

Keywords

Education, Training, Textile Design, Textile Technology.

Introduction

Currently in 2023, rapid changes are occurring in many economic and commercial sectors throughout the world as markets seek to recover and reposition themselves following the covid pandemic of 2020-2022. In the UK there are the added pressures of a general cost-of-living crisis and new challenges to trade following Brexit. As the general public has less disposable income, the retail environment has been severely impacted. In 2021, PwC (PricewaterhouseCoopers 2021) reported that 17 500 chain stores closed in 2020 alone. Environmental concerns are also driving changes within society as a whole with a general shift towards more sustainable ways of living. As a result, shifts and changes are occurring in all sectors of society including the jobs market.

Now would seem a timely moment to review the subjects currently taught by university textile courses in the UK to see whether textile graduates are adequately prepared for today's rapidly changing workplace.

It is likely that future textile graduates will need an increasingly wide range of skills throughout their professional careers including a range of entrepreneurial skills that will help them to navigate the changing commercial environment. Graduates need far broader skills than ever before and university degree level courses are the most obvious and best place to deliver these skills.

Background

Since the advent of what has become known as fast fashion in the 1990's, the majority of clothing and textile goods consumed by Western nations in particular have been mass-produced in vast quantities in far away places, mainly China, and shipped around the world. This is the direct cause of several of the environmental problems that the world is facing today. Apart from the direct impact on the world's natural resources (water, land and air), it causes massive over-consumption and produces an enormous amount of waste that accumulates so quickly that we are unable to dispose of it all safely. In addition, with rising costs generally, it is increasingly likely that this will soon become uneconomic. This manufacturing model is unsustainable, harmful to the environment and it is no longer viable.

It would be more sustainable to manufacture products locally and in smaller quantities, thus providing new employment opportunities. This would incur less waste and be far kinder to populations and the environment. Individual countries could then concentrate their overseas trade on those unique items, which they alone can produce according to the nature of their climate and the talents of their population.

The key skills required by textile graduates

The transition to more sustainable manufacturing models is heavily dependent on the knowledge and skills of the new graduates entering the industry. Currently in 2023, there are skills gaps in nearly every area of the textile and fashion industries, both in the UK and many other countries too and as a result poor, ill-informed or unconsidered choices for clothing and furnishing products are often made. There is also much misinformation in the industry generally.

During my time of teaching short courses at the London College of Fashion, (from 2006-2020), and at the London Institute and the London College of Distributive Trades prior to this, I trained people working in the textile and fashion industries in product knowledge and participants came from all over the world. The majority of participants worked in fashion retail including buyers and merchandisers from all the major UK store groups and independents, and also people working in various positions in SME's. In addition there were many entrepreneurs who were setting up businesses and also participants from overseas working in many different sectors of the fashion and textile industry. Several times there were delegates from countries such as India, Bangladesh or Portugal whose father either had a textile or clothing factory in those countries and who were studying business degrees at UK universities. They came to gain an overview of textile processes. Very occasionally there were students studying on current fashion and textile courses.

In over 20 years of teaching textile product knowledge I observed that there are skills gaps amongst many textile and fashion graduates working in the industry and that they attended short courses to fill the gaps in their knowledge. Although some came to expand their knowledge in specialist areas such as quality and testing for example, most participants lacked basic product knowledge generally, for example some did not understand the difference between a fibre and a fabric and thought that cotton was a fabric. In addition it was apparent that many graduates lacked the entrepreneurial skills required for working in SME's or for starting their own businesses.

Skills gaps have been acknowledged by other sectors of the industry too. Recently the Chair of the London & Southern Section of the Society of Dyers and Colourists contacted me to say that their members had been discussing this issue and to see if we could do anything about it.

In order to see where improvements could be made, an overview of some of the skills required in the workplace was made. This included assessing the requirements of retail companies such as John Lewis and M&S and manufacturing companies such as Heathcoat Fabrics (for fabrics) and Fashion Enter (for clothing). In addition the course content in a range of different university courses in the UK was examined.

Industry

The company requirements tended to be very general, for example the key competencies that John Lewis Partnership are looking for in each employee are: 'delivering results, leading and developing, passionate retailer, team player and vision and creativity'. Other competencies cited elsewhere included:

- Specialist Skills eg. commercial, creative and digital.
- Innovative or resourceful.
- Personable and punctual.
- Great listening skills.
- Take responsibility for success in their field.
- Demonstrate problem solving.
- Demonstrate leadership skills.
- Decisiomaking ability.
- Solve complex challenges
- Convert opportunities into value for each individual customer.
- Motivated.
- Team Player.

In 2018, the keynote guest speaker of the London & SE England Section of the Textile Institute was Jenny Holloway of 'Fashion Enter', a clothing factory and sampling studio founded in North London by Jenny in 2006. Jenny had to set up her own training academy, as she was unable to find people with the right skills. This is now an award-winning social enterprise, which has trained over 650 people in one year, with technical qualifications such as stitching and industrial patterns. Fashion Enter manufactures 10,000 garments a week employing over 100 staff with a yearly turnover of £3m. They manufacture to the highest standards for clients such as M&S and ASOS. Fashion Enter is a centre of excellence for training based on the government apprenticeship scheme. Today Fashion Enter is the largest training provider for level (3-5) technical skills working with a total of 33 retailers including M&S, New Look, Jigsaw and many SME organisations too. It is also the preferred training provider for Savile Row Bespoke (Moulds, 2017).

University courses

To give an overview of the subjects taught on university textile courses in the 2020's, the course content of a range of different university courses in the UK was examined. In view of the large number of textile and fashion courses available, research was concentrated on BA Textile Design courses in particular. Online prospectuses were examined and course leaders were contacted.

As expected, various aspects of textiles are taught in the different institutions and the course content varies greatly. In general, some courses are more technically orientated and some more design orientated. However it was not easy to find information to assess the extent and depth to which the various subjects are covered. As universities are self governing and have the power to award their own degrees and construct their own courses, there is enormous variation in the competencies of graduates.

The study of textiles is at the interface of Technology and Design. An understanding of both the technical and aesthetic aspects of the subject are required. From personal experience there are gaps in graduate knowledge in both of these areas: the technical knowledge of materials and more widely an understanding and appreciation of the aesthetic aspects of textiles and clothing. Some of the more technical-orientated courses would benefit by knowing more about design and some of the more design-orientated courses would benefit by having more technical information. Both areas are discussed below.

Technical materials knowledge

If the textile and fashion industries are to transition to more sustainable manufacturing models, it is vital that graduates have a good technical understanding of textile materials. Such knowledge will help to prevent markets being flooded with significant amounts of poor quality textiles and clothing products that have a short life span and cannot be disposed of appropriately at the end of their life.

The future success of such a transition is largely dependent on people working in the textile and clothing sectors having a good technical understanding of how to produce good quality textile materials and the characteristics that make them last longer. Importantly they need to be able to communicate this message to consumers. Consumers must be educated to appreciate clothing that lasts longer and that this is something that they should aspire to.

A basic materials training should be included in all degree-level textile courses and although studied to a lesser extent in design-based courses it should nevertheless be a notable component. It is recommended that all textile courses include the following basic materials training in their courses:

- An understanding of the raw materials used to make clothing and furnishings, including:
  - The properties of fibres and their characteristics.
  - Whether goods made from them can be recycled,
  - How they biodegrade and their impact on the environment.
  - Different yarns and their characteristics.
- Fabric knowledge including:
  - Different types of fabric, their characteristics and end uses.
  - The different fabric construction methods.

- Colouration methods and their environmental impact.
- The features that bring 'quality' to a fabric and make it long lasting.

Aesthetics

In addition to having a good knowledge of textile materials, it is also important to the future success of more sustainable manufacturing models that people working in the textile and clothing sectors have a good appreciation of the aesthetic features of textiles and clothing products.

Some practical, hands-on training in the visual arts should be included in all degree-level textile courses, and although studied to a lesser extent in technology-based courses, it should nevertheless be a notable component.

Practice of the visual arts gives rise to a natural appreciation for the aesthetic appeal of different products and the skill involved in their production. It is important to understand why some products have an innate appeal over others and what makes them desirable.

With the wide availability of fast-fashion goods dominating the markets for the last 30 years or so, consumer-purchasing decisions have been dominated by price. To a large extent consumers have lost their appreciation of design and craftsmanship and their recognition of what constitutes quality. They have lost their sense of value. This issue has been further exacerbated by the decline in teaching arts and technology subjects in schools. A recent report found that the number of younger people taking up Design and Technology qualifications had declined drastically in England, with 21.8% in 2020 entering a Design and Technology GCSE, compared with 44.2% in 2009 (Tuckett 2017).

On the other hand, some companies are beginning to promote the idea that well-designed products are more sustainable than others and designers who are aesthetically accomplished are better able to educate the consumer to appreciate the benefits and added value that these products bring.

An appreciation of the aesthetics of textile products is best achieved by practical participation in arts-based activities. The value of this area of study is neither widely understood nor appreciated. Design is a visual language that can only be explored in depth by practical means and like any other language it may be taught and learnt through the hands-on repetition and practice of working with various materials. Drawing and painting underpins the learning process and visual expertise gained here is transferred into the practical construction and decoration of textiles. The practical making of fabrics and clothing cultivates an appreciation for these items and a sense of their value.

The theoretical study of most subjects predominant in our education system today, engage neural pathways in the left hemisphere of the brain. The practice of any design discipline however engages different neural pathways in the right hemisphere of the brain. This is a different learning experience that is concerned with the actual doing of something rather than theory or ideas. The learning experience is assimilated so that it becomes an integral part of our being. With practice, one gradually develops a sophistication of design and imagery that is evident when compared with similar material produced by an untrained person. One only attains mastery after many years of practice and one is able to execute works of art and textiles to a high degree of excellence combining both artistic and technical knowledge. The result can only be experienced and it is a feast for the eyes much as music is a feast for the ears.

The activities of any of the creative arts such as drawing, painting, music and any of the performing arts for example, are all right-brained activities. To do any creative task, one needs to be in a creative state of mind and it takes time for the mind to readjust from its habitual left-brained state to this different state of consciousness.

When I first started teaching textile design on the BSc (Hons.) Textile Design course at Huddersfield University (1986-1996), then Huddersfield Polytechnic, one of my first tasks was to request the rearrangement the course timetable. The technical subjects were randomly placed in amongst spaces allocated to design, and usually only one hour or so long. The students barely had time to get out their paint brushes, let alone transition into the different mindspace required to create visual designs. The mind needs to be still and clear for this process to take place.

Also in 1986 at Huddersfield Polytechnic, I was asked to teach 'Colour and Design' to a small group of apprentices on day release from local textile companies studying for a BTEC HNC qualification. They came to college one day a week to study a range of different subjects of which this was one. As a designer, rather than doing this theoretically, I chose to set practical colour exercises asking them to explore different aspects of colour theory by painting different colour sequences in small squares. As the apprentices embraced this new way of working, the intense silence of their concentration was palpable and with practice they produced some inspiring artworks. Importantly they drew great satisfaction and confidence from their achievements.

Examples of these exercises include: contrast of hue; dark/light contrast; cold/warm contrast; complementary contrast; simultaneous contrast, successive contrast and contrast of extension.

As with music and sound, different colour and design combinations evoke feelings and moods and so it is important that the colours of the textiles that we surround ourselves with, both clothing and home furnishings, resonate with us favourably. For example it is possible to increase ones subjective feeling of warmth in a room by (2ºC-5ºC) if it is decorated in warm colours rather than cool colours.

In my group of apprentices, there was one lad in the class who was unable to see the after-image of *successive contrast* and got very worried. His tendency to left-brained thinking blocked him from doing so.

In another example related to music, I have a good American friend who is a good pianist and musician and she and her husband like going to concerts. Her husband is a mechanical engineer who works in the car industry. After a recent concert he told her that he thought that the musicians had made mistakes because they hadn't followed a logical sequence of notes, or the ones he was expecting. His left-brained way of thinking had blocked him from experiencing and appreciating the uniqueness of the musical piece.

Colour is just a small part of the visual language. There is also shape and form, texture and surface and a variety of different materials. All these aspects are normally taught during a one-year, pre-degree foundation course, or in school, where students are taught the a-b-c of the visual language in preparation for their degree studies. However students who do not attend a foundation course may miss out on this significant experience.

Entrepreneurs

There is a saying that is sometimes attributed to Einstein and sometimes to others, that states that 'no problem can be solved from the same level of consciousness that created it'.

Apart from fostering an aesthetic appreciation of textiles, practice of the visual arts also helps to produce an entrepreneurial way of thinking and the textile and fashion industries badly need more entrepreneurs to pioneer more sustainable manufacturing models and tackle the environmental problems that the world is facing,

It is sometimes thought that it is solely natural talent that creates an entrepreneur and that such skills cannot be taught. Certainly some people are more gifted in this way than others, however the practice of the visual language is all about exploring the relationships between things. It is a wide view in which the eye is trained to see and experience the connections between things. This requires intense concentration and gradually leads one into a more meditative state of mind. It is literally a quantum leap. It is in this state of mind that insights occur and new innovations and discoveries are sometimes found.

Using the legendary story of Newton's apple falling as an example. Had Newton been thinking about a mathematical problem, or what he was going to have for supper that night, it is highly unlikely that he would have formulated his theory of gravity. Only when one's mind is clear or free from thoughts can one enter a different mode of consciousness and it is here that insights can occur. It is likely that solutions to some of the problems that the world is currently facing may be found in this state of mind.

(*Entrepreneur: Someone who starts their own business, especially when this involves seeing a new opportunity. Cambridge Online dictionary).

(*Quantum Leap: A great improvement or important development in something. Cambridge Online dictionary).

The difference between education and training

At an event organised by the TI London & SE England section in 2019 in which a fellow committee member interviewed two former LCF students who had set themselves up as women tailors in Savile Row, a current LCF student asked the question, would it be better for a fashion student to go straight into an apprenticeship rather than do a college degree? The interviewer gave the answer that a university degree is the most essential part of ones broader education and that this can be followed by the specialist training that one receives during an apprenticeship, an industrial placement or some other form of work-related experience (TI event 2017).

Training can and should be a continuing activity that takes place throughout ones working career. It usually involves adding to one's factual knowledge or receiving updates on new developments. Occasionally it might be to learn a new skill.

A university education should by definition be a broad-based experience, designed to broaden the mind and outlook of the student and introduce them to new subjects, concepts and experiences of which they have been previously unaware.

On the first day of teaching Colour & Design to my group of apprentices at Huddersfield Polytechnic in 1986, when I gave out the paint brushes, the face of one lad became contorted in an expression of panic. That is my definition of 'education'. It opens one up to new realms and experiences that are sometimes outside of one's comfort zone and that can sometimes be uncomfortable.

Since the passing of the Further and Higher Education Act of 1992, all the former polytechnics were awarded university status and became independent, self-governing and self-funding institutions (Lewis 1991). This gave them the power to award their own degrees and construct their own courses, thus joining pre-1992 universities in a unified system (White Paper 1991).

Although greater participation and the diversity of courses available is to be celebrated, it is not easy to find information to assess the extent and depth to which the various subjects are covered. Evidence of the teaching of some of the basic core subjects for both textile and fashion degrees is not always apparent and it is suspected that in some places it is overlooked. Potential new students cannot be expected to know what subjects are essential to their studies when they apply to university courses in this field.

In addition, university courses are facing increasing competition from a wide range of online courses that can be taught from anywhere in the world. It is likely that they will need to reassess the value and advantages of what they offer over online training courses and communicate this message to potential new students.

It is recommended that an independent, impartial organisation is established in order to scrutinise the quality of university textile and fashion courses. Perhaps this could be a collaborative initiative of the Textile Institute and the Society of Dyers and Colourists, maybe involving other organisations like the UK Fashion & Textile Association for example. The Textile Institute already ratifies some university textile courses.

Since the former polytechnics became self-governing universities in the 1990's, their courses have been subject to less outside scrutiny than previously, when their courses used to be inspected by the Council for National Academic Awards (CNAA). Abolished in 1993, the CNAA was an independent organisation that oversaw the quality of degrees awarded in the then non-traditional institutions (polytechnics and art colleges), and maintained standards in line with traditional universities. The CNAA inspected and conducted regular appraisals every 3 or 4 years of all courses nationwide. The inspections were conducted by teams of experts in relevant disciplines, usually from other similar institutions.

Recommendations

In summary, the following recommendations are made.

(1) It is recommended that the following basic materials training is included in all degree-level textile courses and although studied to a lesser extent in design-based courses it should nevertheless be a notable component.

- An understanding of the raw materials used to make clothing and furnishings, including:
  - The properties of fibres and their characteristics.
  - Whether goods made from them can be recycled,
  - How they biodegrade and their impact on the environment.
  - Different yarns and their characteristics.
- Fabric knowledge including:
  - Different types of fabric, their characteristics and end uses.
  - The different fabric construction methods.
- Colouration methods and their environmental impact.
- The features that bring 'quality' to a fabric and make it long lasting.

(2) In order to foster an appreciation of the aesthetic benefits of textile products, it is recommended that all degree-level textile courses include some practical, hands-on training in the visual arts within their courses and although studied to a lesser extent in technology-based courses, it should nevertheless be a notable component.

(3) It is recommended that an independent, impartial organisation is established in order to scrutinise the quality of university textile and fashion courses. Perhaps this could be a collaborative initiative of the Textile Institute and the Society of Dyers and Colourists, maybe involving other organisations like the UK Fashion & Textile Association for example. The Textile Institute already ratifies some university textile courses.

References

Lewis, Michael S. 1 January 1991. The Polytechnics: A Peculiarly British Phenomenon. Oxford University.

Moulds, Josephine. Traditional Industry Cutting its Cloth for the Modern Era. The Times Business Pages. 17 July 2017.

PricewaterhouseCoopers (PwC) 14 March 2021. Over 17 500 chain stores closed in 2020 with the impact of the pandemic yet to be felt. www.pwc.co.uk/press-room/press-releases/over-17-500-chain-stores-closed-in-2020-with-the-impact-of-the-pandemic-yet-to-be-felt.html

Textile Institute, London & SE England Event. 20 November 2019. In Conversation: Alan Cannon Jones with Antonia Ede and Deema Abi-Chanin.

Tuckett, Sam. 17 July 2017. A Spotlight on Design and Technology Study in England: Trends in Subject take up and the teacher workforce.

White Paper: Higher Education: A New Framework (1991) London: Her Majesty's Stationery Office May 1991.

Technical Textiles & Composites

EXPLORING THE POTENTIAL OF TEXTILE SIZING FOR ELECTROSPUN PCL/PLA NANOFIBERS CORE-SHEATH YARN FOR TISSUE ENGINEERING SCAFFOLDS

Anna Doersam[1,2], *Olga Tsigkou*[1,2] *and Celina Jones*[3]

[1]*The Henry Royce Institute, Royce Hub Building, The University of Manchester, Oxford Road, Manchester M13 9PL*

[2]*Department of Materials*[2], *School of Natural Sciences, The University of Manchester, Oxford Road, Manchester M13 9 PL*

[3]*Manchester Fashion Institute, Manchester Metropolitan University, Cavendish St, Manchester M15 6BL*

Abstract

Tissue engineering aims to replace, retain or restore diseased or damaged tissues by developing scaffolds that provide a temporary support structure for cell adhesion. Textile technologies have shown unique advantages in mimicking the fibrous extracellular matrix (ECM) and biomechanical properties of human tissues. Nanostructured textile scaffolds hold great promise due to their high surface area, controllable porosity, and favourable cellular response. Electrospinning is commonly used to produce fibers in nanoscale from natural/synthetic polymers, which can closely mimic the fibrous part of the ECM in human tissue.

The nanofiber sheath of core-sheath yarns can be damaged by mechanical stresses during textile production, such as weaving. Sizing, a preparatory process that applies a thin protective coating of adhesive polymeric material, is commonly used to improve yarn strength, abrasion resistance, and reduce fiber friction. This study aims to explore the potential of textile sizing on biopolymeric nanofibrous PCL-PLA core-sheath yarn used to fabricate textile scaffolds for tissue engineering applications due to the limited abrasion resistance of the nanofiber sheath. A variety application techniques and sizing formulas were investigated to produce a homogenous protective layer.

The nanostructure of untreated yarn shows relatively uniform and bead-free nanofibers. A homogenous polymeric layer was obtained by using a modified size box to apply the size on the yarns surface as observed by SEM. Further research, especially on the development of suitable sizing recipes for PCL nanofibers, and a suitable sizing process is suggested. The results show the potential of textile sizing on nanofiber core-sheath yarn to improve the abrasion resistance of the nanofibrous sheath for fabricating textile scaffolds.

Keywords

Textile scaffolds, Tissue Engineering, Nanofibrous Yarn, Textile Sizing

Introduction

Tissue engineering (TE) aims to repair, restore, or replace diseased or damaged tissues by developing scaffolds that provide a temporary support structure for cell adhesion and growth. A wide variety of approaches using textile techniques for TE have been established such as electrospun non-wovens, woven, knitted or braided structures, as illustrated in **Figure 1** (Kun et al., 2019). Textile technologies have shown unique advantages in mimicking the fibrous extracellular matrix (ECM) and biomechanical properties of human tissues (Jiang et al., 2021). In addition, textile-based scaffolds are customisable and can be designed with a range of porosities, mechanical properties, complex shapes, and structures and can be produced by a variety of biocompatible polymers (Anand, 2010, Gorgieva et al., 2018, Kun et al., 2019).

Electrospun nanofibers have structural similarities to native ECM fibrils and adjustable properties to regulate cellular activities (Wu et al., 2022). Current manufactured electrospun nanofiber mats are of great interest but there is little emphasis on the development of scaled, 3D structures, which are more clinically relevant (Doersam et al., 2022, Khorshidi et al., 2016, Wang et al., 2013). Key challenges of electrospun nonwovens include inadequate mechanical strength, low porosity that impedes cell infiltration, high inconsistencies in scaffold properties and performance and limited scalability (Khorshidi et al., 2016). (Li et al., 2018) suggested the manufacture of nanofiber-covered multi-strand filaments, which not only have the biological properties of nanofibers but also have the mechanical properties of the core yarn. Nanostructured textiles, meaning textile structures made from nanofibrous yarn, can create 3D structures with desirable characteristics for repair and regeneration in various tissues (Wu et al., 2022). Various electrospinning modifications have been explored over the last two decades to fabricate nanofiber yarn. Apart from the great challenges in producing pure nanofiber yarns, their physical properties such as low tensile strength and high elasticity, as well as their limitation in length, challenge their suitability for textile manufacturing (Yang et al., 2018, Zhou et al., 2010).

Nanofiber core-sheath yarn structures consisting of a multifilament or microfiber core and a nanofibrous sheath, combine mechanical strength while enhancing microstructural properties for cell attachment and growth (Zhou et al., 2010, Fakhrali et al., 2014, Cai et al., 2020). The resulting nanotextiles exhibit superior properties due to their surface structure and biological properties (Joseph et al., 2018). However, the mechanical stresses and strains applied on the yarn through textile production processes, such as weaving, could damage the nanofiber sheath of the yarn. To further impart strength, improve the yarn abrasion resistance, and reduce fiber friction, sizing of yarn is commonly used as an essential preparatory weaving process. A thin protective layer of adhesive polymeric coating, such as PVA and carboxymethyl cellulose, is applied. The protective layer helps to withstand the mechanical stresses and strains applied on the yarn during weaving.

In this study, a nanofibrous core-sheath yarn with polylactic acid (PLA) multifilament and polycaprolactone (PCL) electrospun nanofiber coating has been fabricated. Due to the limited abrasion resistance of the yarn during scaffold fabrication, which resulted in uncoated areas, textile sizing as a technique to improve the abrasion resistance of a nanofibrous-covered core yarn for textile processing, from a textile engineering perspective is a promising approach. Due to the different yarn properties, the size formulation, and the sizing conditions, it is often challenging and almost impossible to determine the correct formulation for each yarn type and structure. In fact, sizing of a PCL-PLA nanofibrous core-sheath yarn, or pure nanofibrous yarn has not been reported in the literature before. The primary objective of this study is to improve the abrasion resistance of the nanofibers sheath of the PCL-PLA hybrid yarn for greater processability.

The effectiveness of sizing is determined by a variety of parameters, including process parameters, yarn properties, size formulation, and de-sizing efficiency (Ibrahim et al., 1997, Gandhi, 2020). The most used process in the textile industry is to wind single-end warp yarns onto a creel and pass them through a heated-size bath, which is made up of sizing agents, water, and additives. The sized yarns are guided through a series of rollers to remove excess sizing material and to ensure an even coating, then dried before being wound on a warp beam.

Commonly used sizing agents are water-soluble natural starch and polyvinyl alcohol (PVA) (Limeneh and Tesema, 2019). Ideally, the right amount of sizing material will form a thin protective polymer coating around the yarn, without decreasing the yarn's elasticity or resulting in a high stiffness (Kovačević et al., 2020). Gelatinisation and foam forming of the size bath will result in uneven size pick-up and form rigid films on the yarn (Kovačević et al., 2020). In addition, the applied size should be easily and completely removable from the sized fabrics. Firstly, to recover the nanostructured surface of the yarn and secondly to limit cytotoxicity of the sizing agents. Desizing commonly uses enzymes to break down starch-based sizing agents. It is worth mentioning that desizing with boiling water to remove PVA damages nanostructured surface (Fakhrali et al., 2014).

In general, due to the different yarn properties, the size formulation, and the sizing conditions, it is often challenging and almost impossible to determine the correct formulation for each yarn type. For the purpose of this study, three different sizing methods were designed to explore even coating techniques. The following 3 techniques were designed for this purpose: Airbrush spray coating, a set-up in an environmentally controlled oven with a drying area, and lastly a sizing box set-up with an adjustable gap to remove excess sizing material, an integrated temperature-controlled plate and a controlled yarn take-off (**Figure 2**).

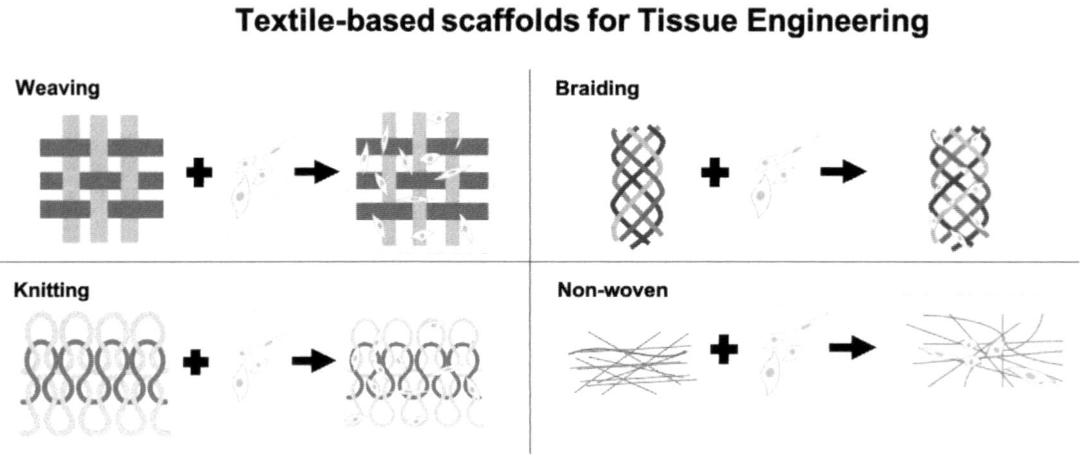

Figure 1 Schematic of tissue engineering scaffolds adopting textile structures for cell attachment and growth

Materials

Fabrication of nanofiber core-sheath yarn

A conjugated electrospinning setup was used to develop a PCL-PLA core-sheath nanofiber-structured yarn. The PCL pellets (Mn = 80000) was purchased from Sigma Aldrich, UK. Two oppositely placed needles (22GA 0.5", 0.41mm ID, 0.71mm OD) were applied with positive and negative voltages to generate the nanofibers which were then wrapped onto the PCL microfiber to produce the core–sheath yarn.

Sizing agents

A number of commercially available sizing agents were used for the preliminary testing of their suitability for non-commercial biopolymer yarn. Sizing agents used were Elvanol™ T (Kuraray Japan); Solamyl 9560 and Amitrolit 8169 (Agrana Staerke, Germany) and Emsize® E18 and Emsize® CMS60 (Emsland-Stärke GmbH, Germany). Elvanol™ T is a copolymer of polyvinyl alcohol and is biodegradable. It's a low add-on sizing agent with high film strength, abrasion resistance, and excellent adhesion, used alone or in combination with starch and additives. Elvanol™ T is readily dissolvable in hot water, without the need for costly enzymes and easily removed.

Solamyl 9560 and Amitrolit 8169 are biologically degradable hot and cold soluble modified potato starches. Emsize® E18 is a high amylose modified pea starch developed for synthetic yarns or blended yarns of cotton and synthetics. Emsize® CMS60 is a carboxymethyl potato starch which is soluble in cold water. All sizing agents do not require enzymatic degradation for removal.

Table 2 Sizing agents & recipes for sizing PCL/PLA core-sheath yarn

Type	Sizing recipe	Quantities
Type 1	Emsize® CMS60	1,7g
	Emsize® E18	0,8g
	Water	50ml
Type 2	Elvanol™	2,7g
	Solamyl 9560	1,3g
	Water	50ml
Type 3	Elvanol™	2,7g
	Amitrolit 8169	1,3g
	Water	50ml
Type 4	Elvanol™	5,5g
	Water	50ml

Method

Size bath preparation and sizing techniques for coating electrospun PCL-PLA core-sheath nanofiber yarn

The size bath was prepared by adding specific amounts (5-11wt%) of solid granules of one or combined sizing agents to an agitated deionise water (diH_2O) bath (~50ml), stirring the solution for 30 min to wet the granules, and then gradually heating to 90°C for 90 min followed by a cooking time of 30 min at 90°C. **Table 2** includes the prepared sizing recipes.

Three different techniques have been developed to apply the sizing evenly to the yarn in a process that is as controlled as possible, which are shown in **Figure 2.** The airbrush spray coating setup, as shown in Figure 2(A), included a commercial airbrush (Harder & Steenbeck Ultra Airbrush) with a 0.4mm spray nozzle and 5 ml substrate cup. The airbrush was set at a distance of 15cm from the yarn in a fume hood with the air pressure set at 3.5 bar.

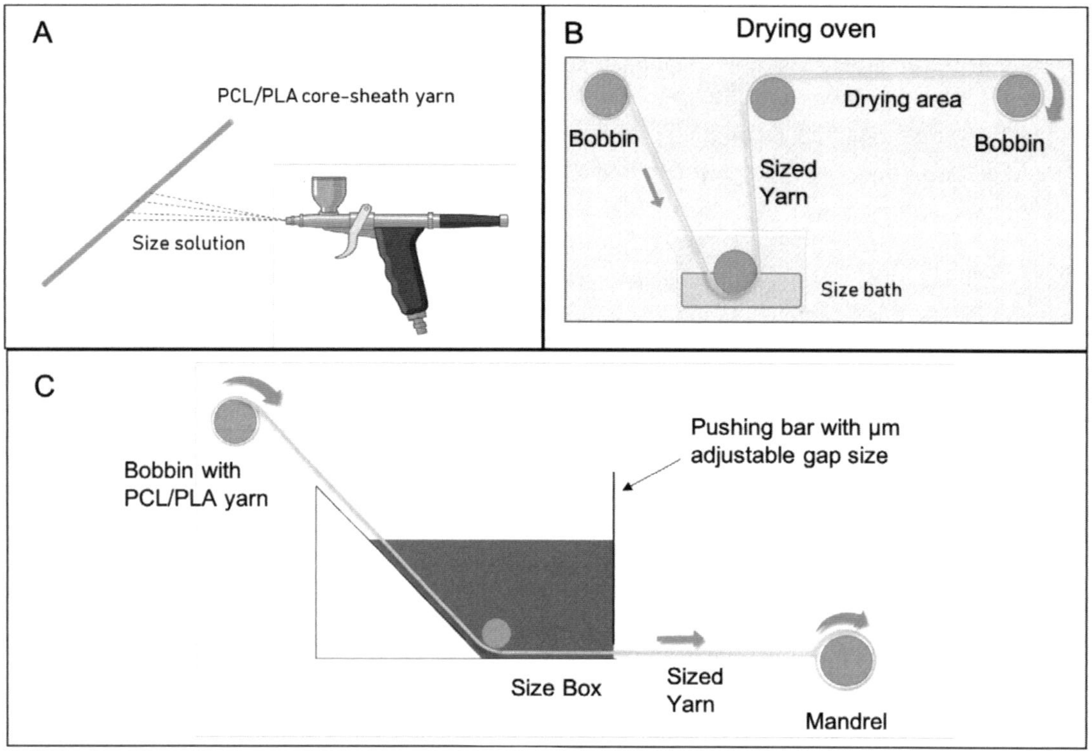

Figure 2 Schematics of different sizing techniques for sizing PCL/PLA core-sheath yarn (A) Airbrush spray coating, (B) environmentally controlled setup in an oven with a sizing bath and integrated drying area, (C) modified size box set-up with controllable take-up reel speed and adjustable pushing bar

Figure 2 (B) shows the environmentally controlled size bath set-up in a drying oven, which is set at 55°C. The yarn is supplied from a bobbin and guided by rollers through a sizing bath before passing onto a drying area. When dry the sized yarn is then wound on a small bobbin. Figure 2 (C) shows the design of a sizing process using a size box with an adjustable gap size of a pushing bar and an integrated temperature-controlled plate, the nanofibrous yarn is drawn and guided at a constant speed through a size bath solution and wound on a mandrel which was set at 1rpm. The yarn is coated with the size when pulled through the bath. The sized PCL-PLA core-sheath yarn is kept in the environment until dry.

Characterisation

The morphological structures of the untreated PCL/PLA nanofiber core-sheath yarn as well as sized yarns were observed by a scanning electron microscope (SEM, Tescan Vega 3) with an accelerating voltage of 10 kV after gold-palladium spouting. The chemical analysis of sized and untreated core-sheath yarns was performed using Fourier transforms infrared spectroscopy (FTIR) over the range 4000 – 400 cm^{-1} to demonstrate the presence of the polymeric sizing film on the PCL-PLA yarn surface.

The equipment yarn-on-yarn abrasion testing equipment used in this study complies with ASTM D6611-16 – Standard Test Method for Wet and Dry Yarn-on-Yarn Abrasion Resistance. A length of yarn is twisted in contact with itself between three rollers. A weight is attached to one end of the yarn to apply a prescribed tension. The other end is drawn back and forth through a stroke at a defined speed until the yarn fails due to abrasion upon itself within the twisted region. In deviation from the standard, the test was not performed until yarn breakage, rather the morphology of the different yarn treatments after 25 and 50 cycles on untreated yarn and 25 cycles on sized yarn was examined and twisted two times.

Results and discussion

Nanofiber Yarn morphology

The SEM images of hybrid nanofiber yarns are demonstrated Figure **3**. Figure 3 (a) shows a clear core-sheath structure and smooth bead-free nanofiber surface. Also, it can be observed that the yarn has good uniformity in diameters.

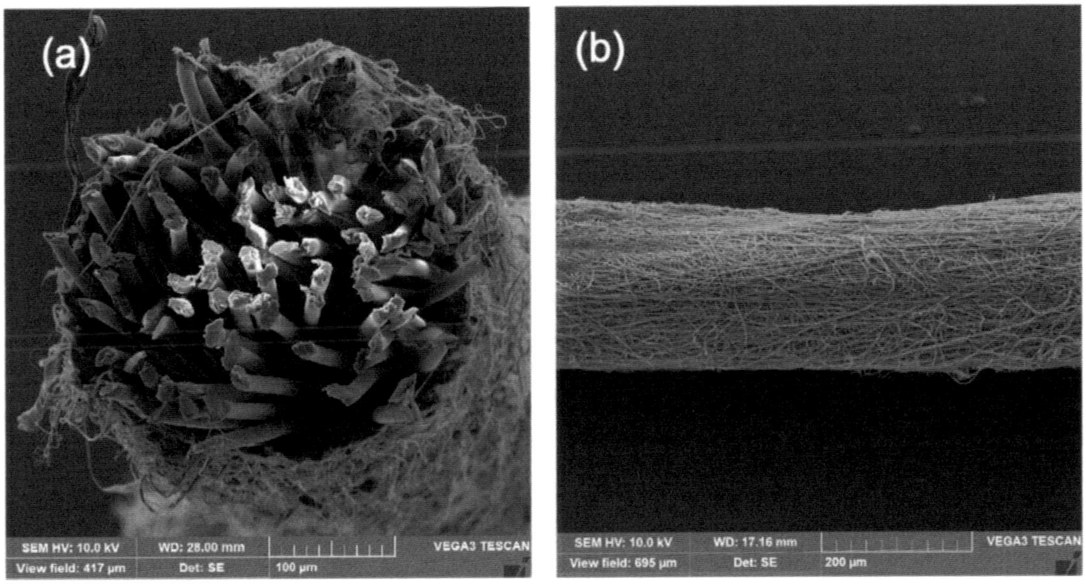

Figure 3 SEM Images of a PCL/PLA core-sheath yarn (a) cross-section with a clear multifilament PLA and a PCL sheath (b) top view of the nanostructured yarn surface

Sizing process and quality assessments of the sized yarns

Three different techniques were developed to apply the sizing evenly to the yarn in a process that is as controllable as possible. Airbrush spray coating is an inexpensive method and has been utilized in nanotechnologies to evenly coat surfaces (Koo et al., 2016). Adopting this process resulted in an uneven polymeric film with beads and droplets on the yarn. Moreover, due to the viscosity of the size solutions, not all recipe types were suitable. In addition, this process would need further optimisation to allow a continuous coating process.

The sizing bath setup in an environmentally controlled oven resulted in a discontinuous process due to the short distance of the drying area compared to the required drying time of the yarn before it could be wound onto the bobbin. The sizing technique of a modified sizing box with an adjustable gap and mandrel speed to wound the yarn gave the most controllable process. The gap size allowed for excess size to be removed similar to squeezing rollers used in industrial setups without having to risk damage by compressing the nanofiber structure of the yarn. This process would also allow for a temperature-controlled size bath by the integrated plate.

The aforementioned size box technique was used to size the yarn with prepared size recipes type 1-4. **Figure 4** shows the morphology of the polymer film on the nanofiber surface of PCL-PLA core-sheath yarn. For all recipes, it is visible that the nanostructured yarn surface is coated with a polymeric film. However, there are clear differences in the uniformity and thickness of the coating.

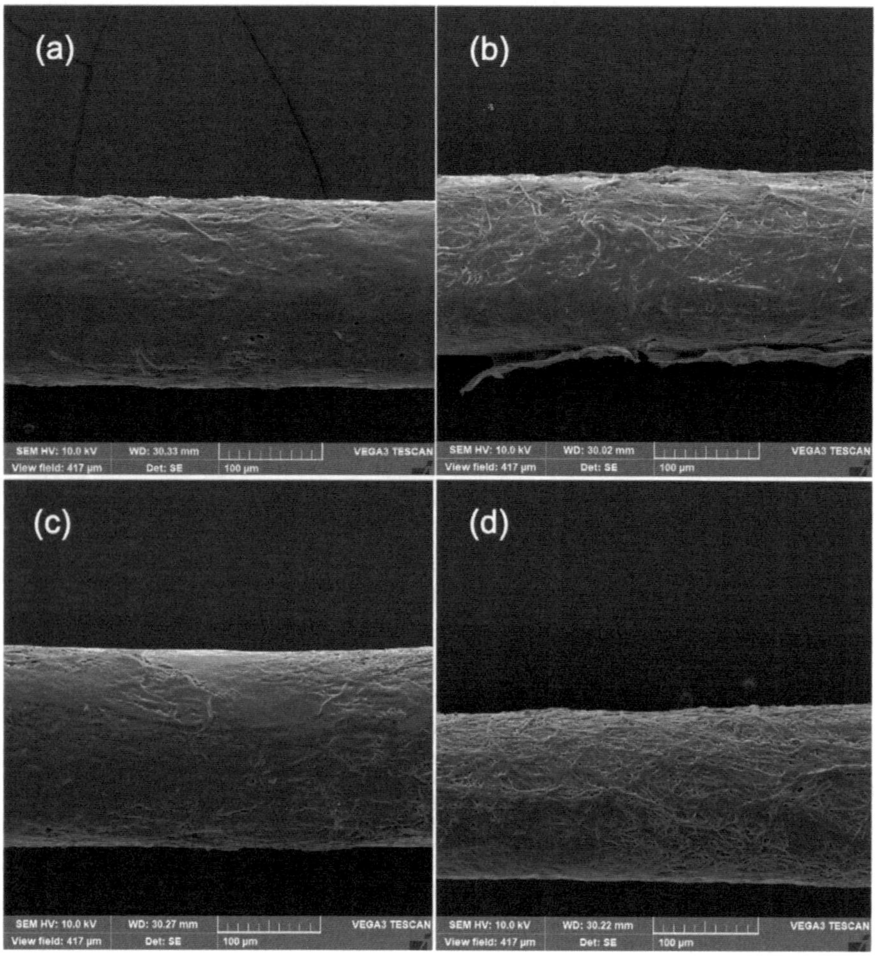

Figure 4 SEM Images of sized yarn with size type recipes 1-4 using a designed size box technique (a) CMS50/E18, (b) ElvanolTM, (c) ElvanolTM / Amitrolit 8169 and (d) ElvanolTM / Solamyl 9560 all after 25 cycles.

Analysis of FTIR spectra

The FTIR spectra of the sized yarn with type 1-4 size formulas on PCL/PLA nanofiber core-sheath yarn is shown in **Figure 5.** The FTIR spectra of the untreated yarn show a peak at 1727 cm^{-1}, attributed to the stretching mode of the C=O, and peaks at 2867 cm^{-1} and 2946 cm^{-1} that are ascribed to the stretching of the C–H bond which is characteristic of PCL. The peaks at 1268 and 868 cm^{-1} appear in the spectrum of PLA (Dadras Chomachayi et al., 2020).

The spectra of these sized samples exhibited several stretches and shifts in peaks associated with the starches. The intensity of the absorption peak in the range of 3500-3300 cm^{-1} slightly increased with for type 1,2 and 4 of the size blends which corresponds to the O-H stretching vibration of the starch, and the peak width indicated the hydrogen bonding of starch. The lower band intensity at 1750 cm^{-1} was attributed to the stretching vibration of C=0. The ratio of the band intensity at 1047cm^{-1} and 1022cm^{-1} represented the ratio of ordered crystalline regions to amorphous regions of starch, which increased with the ratio of pea starch (E18) to potato starch, including CMS60, Amitrolit, Solamyl (Ji et al., 2018). The Elvanol^{TM}, co-polymer PVA, coating, significantly reduced the peak 1241 of PCL.

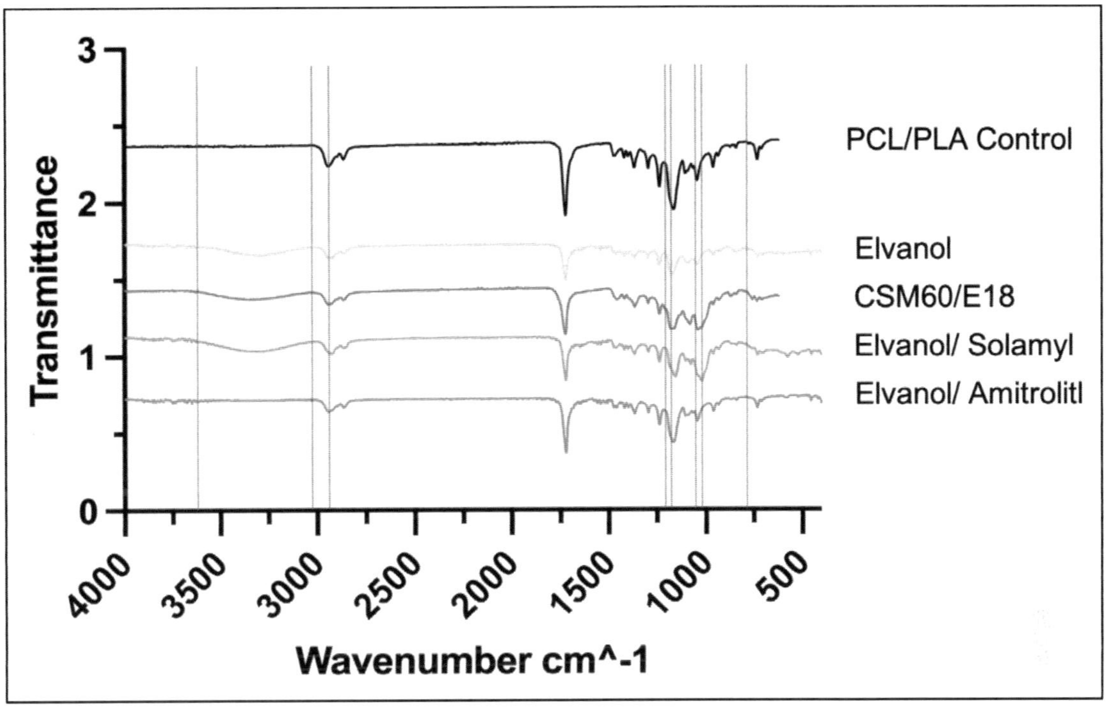

Figure 5 FTIR spectra for wavelengths from 4000 to 400 cm^{-1} of sized and untreated PCL/PLA nanofiber core-sheath yarn. The peak shift at wavelength of 1600 cm^{-1} corresponds to the presence of potato starch. The reduced peak intensity at 1727cm^{-1} and 2946cm^{-1} confirms a higher presence of polymeric film on the PCL/PLA yarn.

Mechanical abrasion resistance of sized yarn

Figure 6 shows the wear and damage during yarn-on-yarn abrasion of the nanofibers sheath. Untreated PCL-PLA yarn shows some wear after 25 cycles in the abrasion area and after 50 cycles the nanofibrous sheath is fully destroyed showing the inner core. The sizing materials add polymeric film on the fiber surface which reduces hairiness and friction to protect the nanofiber sheath structure. The abrasions resistance can therefore be a measurable characteristic of the effect of sizing to improve the mechanical properties of the yarn for subsequent manufacturing processes. Except for type 3 sizing formula, all other sized yarns show visible abrasion and damage to the nanofiber sheath. The type 3 sizing formula shows no visible damage on the surface compared to the other sizing formulas. If comparing Figure 6 (b) and (f), the morphology of the yarn is less hairy and smoother of the sized compared to the untreated yarn. Overall, the abrasion testing demonstrates that sizing with a suitable size formula and technique has the potential to improve the abrasion of nanofiber core-sheath yarn.

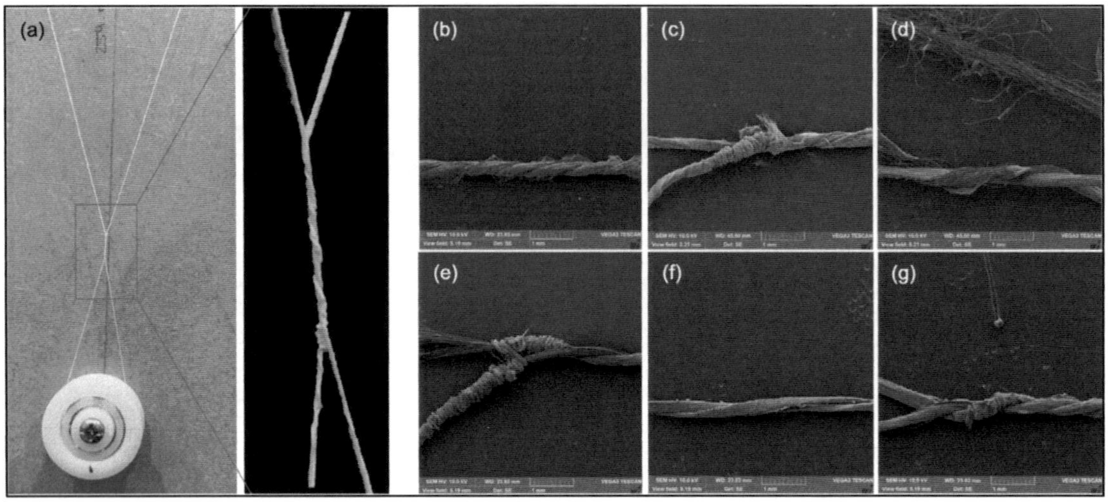

Figure 6 Yarn-on-yarn abrasion testing (a) abrasion zone; SEM Images of surface conditions of PCL-PLA nanofibrous core-sheath yarn abrasion resistance test (b) untreated after 25 cycles and (e) untreated yarn after 50 cycles; (c) E18/CMS60, (d) ElvanolTM, (f) ElvanolTM / Amitrolit 8169 and (g) ElvanolTM / Solamyl 9560 all after 25 cycles.

Conclusion

The aim of the study was to show proof of concept that textile sizing is a potential approach to improve the mechanical properties, particularly abrasion resistance, of nanofiber yarns for further textile processing, such as weaving or knitting. In this study, a PCL/PLA nanofiber core-sheath yarn was fabricated, and three different sizing processes were designed to determine the most efficient and controllable process. Subsequently, four different sizing recipes were prepared and piloted. The yarn is coated with different sizes utilizing the developed size box process with controlled yarn take-off and built-in temperature control. It was found that the type 3 sizing formulation containing the ElvanolTM / Amitrolit 8169 sizing agents had an acceptable overall performance. SEM images show the PVA starch mixture forms a polymeric film impregnating the nanofiber sheath and forming an outer layer. In addition, the nanofibrous sheath shows no visible damage after 25 cycles and therefore improves the abrasion resistance compared to untreated yarn.

In summary, this is an initial attempt to improve the performance characteristics of nanofiber yarns for use in the development of tissue engineering scaffolds for biomedical applications. Further research, especially on the development of suitable sizing recipes for PCL nanofibers, and a suitable sizing process is suggested. Furthermore, a desizing process should be explored to recover the nanofiber structure after manufacturing the textile scaffold. The biocompatibility, degradability and cell cytotoxicity of these sizing agents should be considered for biological applications in case of residuals of the sizing agents.

Therefore, in this study, mostly sizing agents derived from natural starch sources and with good biodegradability were selected. The nanofiber surface structure is essential for textile-based tissue engineering scaffolds. For this reason, it is recommended to test the biocompatibility after size removal.

References

ANAND, S. 2010. Medical and healthcare textiles.

CAI, J., XIE, X., LI, D., WANG, L., JIANG, J., MO, X. & ZHAO, J. 2020. A novel knitted scaffold made of microfiber/nanofiber core–sheath yarns for tendon tissue engineering. *Biomaterials Science,* 8, 4413-4425.

DADRAS CHOMACHAYI, M., JALALI-ARANI, A., BELTRÁN, F. R., DE LA ORDEN, M. U. & MARTÍNEZ URREAGA, J. 2020. Biodegradable Nanocomposites Developed from PLA/PCL Blends and Silk Fibroin Nanoparticles: Study on the Microstructure, Thermal Behavior, Crystallinity and Performance. *Journal of Polymers and the Environment,* 28, 1252-1264.

DOERSAM, A., TSIGKOU, O. & JONES, C. 2022. A Review: Textile Technologies for Single and Multi-Layer Tubular Soft Tissue Engineering. *Advanced Materials Technologies,* 7, 2101720.

FAKHRALI, A., EBADI, S. V. & GHAREHAGHAJI, A. A. 2014. Production of core-sheath nanofiber yarn using two opposite asymmetric nozzles. *Fibers and Polymers,* 15, 2535-2540.

GANDHI, K. L. 2020. 4 - Yarn preparation for weaving: Sizing. *In:* GANDHI, K. L. (ed.) *Woven Textiles (Second Edition).* Woodhead Publishing.

GORGIEVA, S., ZEMLJIĆ, L. F., STRNAD, S. & KOKOL, V. 2018. 9 - Textile-based biomaterials for surgical applications. *In:* THOMAS, S., BALAKRISHNAN, P. & SREEKALA, M. S. (eds.) *Fundamental Biomaterials: Polymers.* Woodhead Publishing.

IBRAHIM, N. A., SHOSHA, M. H. A., FAHMY, H. M. & HEBEISH, A. 1997. Effect of Size Formulation on Sizability and Desizability of Some Soluble Sizes. *Polymer-Plastics Technology and Engineering,* 36, 105-121.

JI, N., QIN, Y., LI, M., XIONG, L., QIU, L., BIAN, X. & SUN, Q. 2018. Fabrication and Characterization of Starch Nanohydrogels via Reverse Emulsification and Internal Gelation. *Journal of Agricultural and Food Chemistry,* 66, 9326-9334.

JIANG, C., WANG, K., LIU, Y., ZHANG, C. & WANG, B. 2021. Application of textile technology in tissue engineering: A review. *Acta biomaterialia,* 128, 60-76.

JOSEPH, J., KRISHNAN, A. G., CHERIAN, A. M., RAJAGOPALAN, B., JOSE, R., VARMA, P., MANIYAL, V., BALAKRISHNAN, S., NAIR, S. V. & MENON, D. 2018. Transforming Nanofibers into Woven Nanotextiles for Vascular Application. *ACS applied materials & interfaces,* 10, 19449-19458.

KHORSHIDI, S., SOLOUK, A., MIRZADEH, H., MAZINANI, S., LAGARON, J. M., SHARIFI, S. & RAMAKRISHNA, S. 2016. A review of key challenges of electrospun scaffolds for tissue-engineering applications. *Journal of Tissue Engineering and Regenerative Medicine,* 10, 715-738.

KOO, S., LEE, S. H., KIM, J. D., HONG, J. G., BAAC, H. W., KWAK, M. K. & OK, J. G. 2016. Controlled airbrush coating of polymer resists in Roll-to-Roll nanoimprinting with regimented residual layer thickness. *International Journal of Precision Engineering and Manufacturing,* 17, 943-947.

KOVAČEVIĆ, S., SCHWARZ, I., ĐORĐEVIĆ, S. & ĐORĐEVIĆ, D. 2020. Synthesis of Corn Starch Derivatives and Their Application in Yarn Sizing. *Polymers,* 12, 1251.

KUN, M., CHAN, C., RAMAKRISHNA, S., KULKARNI, A. & VADODARIA, K. 2019. 12 - Textile-based scaffolds for tissue engineering. *In:* RAJENDRAN, S. (ed.) *Advanced Textiles for Wound Care (Second Edition).* Woodhead Publishing.

LI, B., LIU, C., ZHOU, F., MAO, X. & SUN, R. 2018. Preparation of electrospun core–sheath yarn with enhanced bioproperties for biomedical materials. *Biotechnology Letters,* 40**,** 279-284.

LIMENEH, D. & TESEMA, A. 2019. Valuation of Polyvinyl alcohol and Maize Starch as Sizing Agent for Textile processing.

WANG, X., DING, B. & LI, B. 2013. Biomimetic electrospun nanofibrous structures for tissue engineering. *Materials Today,* 16**,** 229-241.

WU, S., DONG, T., LI, Y., SUN, M., QI, Y., LIU, J., KUSS, M. A., CHEN, S. & DUAN, B. 2022. State-of-the-art review of advanced electrospun nanofiber yarn-based textiles for biomedical applications. *Applied Materials Today,* 27**,** 101473.

YANG, G., LI, X., HE, Y., MA, J., NI, G. & ZHOU, S. 2018. From nano to micro to macro: Electrospun hierarchically structured polymeric fibers for biomedical applications. *Progress in polymer science,* 81**,** 80-113.

ZHOU, F.-L., GONG, R.-H. & PORAT, I. 2010. Nano-coated hybrid yarns using electrospinning. *Surface & coatings technology,* 204**,** 3459-3463.

A NEURAL NETWORK APPROACH FOR PREDICTING THE OPTIMUM DESIGNING PARAMETERS FOR AN ACTIVE COMPRESSION TEXTILE

D P Hedigalla[1], M Ehelagasthenna[2], G K Nandasiri[1], I D Nissanka[3], Y W R Amarasinghe[3]

[1]University of Moratuwa, Faculty of Engineering, Department of Textile and Apparel Engineering, Katubedda, Sri Lanka
[2] Advanced Textiles Research Group, Nottingham Trent University, Nottingham, United Kingdom
[3]University of Moratuwa, Faculty of Engineering, Department of Mechanical Engineering, Katubedda, Sri Lanka

Abstract

Compression therapy is an essential treatment in the management of all types of Chronic Venous Disease (CVD) which is the most common medical condition affecting to the veins in the lower extremities. However, majority of the existing compression modalities including Medical Compression bandages, Medical Compression Stockings and Intermittent Pneumatics Cuffs are constrained by their inability to provide consistent pressure over the lower limb circumference. A compression sleeve was designed with miniaturised air bladders to overcome this limitation by applying a radial force owing to the pressure exerted by a volume of air contained inside the miniaturised bladders. Thus, this study analysed the design parameters of the miniaturised bladders using numerical modelling and laboratory experiments. The optimum design parameters were established using machine learning techniques for the development of the active compression textile. It was observed that the percentage of pressure transmission had a positive correlation with thickness of the bladder membrane and a negative correlation with its side length. Moreover, the percentage of the area where uniform pressure is applied, is increased as the side length of the miniaturised bladder is increased while it was decreased as the thickness of the miniaturised bladder is increased.

Keywords

Chronic venous disease, Compression therapy, Active compression, Finite element analysis, Miniaturised air bladders, Neural network

Introduction

Chronic venous disease (CVD) is a commonly occurring medical condition in adults, which includes varicose veins and chronic venous insufficiency (CVI) (Bergan et al., 2006; Davies, 2019). While CVD affects people worldwide, its prevalence is highest in developed countries (Davies, 2019; Vuylsteke et al., 2018; Pappas et al., 2018). Among the various stages of CVD, venous leg ulcers represent the most severe form (Bergan et al., 2006). Both individuals and society experience several consequences associated with CVD. Symptoms such as pain, heaviness, itching, and reduced mobility significantly impact an individual's quality of life (QoL) (Bozkurt and Balkanay, 2017). Furthermore, the diagnosis and treatment of CVD, along with wound care and hospitalization for related complications, impose a substantial financial burden on the health care systems of the country (Rabe et al., 2012; Kahn et al., 2004). The prevalence of CVD varies significantly across studies due to the differences in research approaches, criteria for diagnosis, geographical locations, participant demographics, and ethnicities (Bergan et al., 2006; Davies, 2019; Vuylsteke et al., 2018; Pappas et al., 2018; Rabe et al., 2012; Wittens et al.,2017, Zolotukhin et al.,2017). There exists a scarcity of epidemiological investigations on elderly populations even though the research was conducted on the younger individuals (Vuylsteke et al., 2018; Kahn et al., 2004; Evans et al., 1999; Brand et al., 1988). However, a recent global survey named The Vein Consult Program, which involved more than 90 000 participants from 23 countries, with a mean age of 51.8 years, discovered that roughly 70% of the participants had clinically significant CVD (Vuylsteke et al., 2018). The survey further revealed a rise in the likelihood of CVD with increasing age, with the highest incidence observed in individuals aged 65 years or older (Vuylsteke et al., 2018). The population-based Bonn Vein Study (which included 3072 individuals aged 18 to 79 years) also observed a similar trend, with the highest prevalence of CVD among subjects aged 60 to 80 years (Maurins et al., 2008).

Chronic venous disease (CVD) is significantly influenced by age, which is considered one of the primary risk factors. The Edinburgh Vein Study provided evidence showing that approximately one in three patients who were initially diagnosed with varicose veins (VV) developed chronic venous insufficiency (CVI) during a 13-year follow-up period. Importantly, the incidence of CVI increased progressively with advancing age (Lee et al., 2015). Furthermore, CVD is more commonly observed in women due to various hormonal and physiological factors that contribute to their increased susceptibility to develop this condition (Lohr and Bush, 2013). Among women, pregnant individuals have a higher likelihood of experiencing CVD, with approximately 40% of them being affected (Ortega et al., 2021). Moreover, the number of pregnancies a woman has had may also be a noteworthy factor associated with the development of CVD (Cornu-Thenard and Boivin, 2014). In addition to being female and advancing in age, there are numerous other risk factors associated with the development of CVD. Overweight and obesity are well-established factors that are strongly associated with the various stages of chronic venous disease, regardless of other risk factors (Vlajinac et al., 2013). The correlation between overweight/obesity and CVD can be attributed to various plausible and interconnected factors. One factor is the presence of a pro-inflammatory state commonly observed in individuals with higher levels of body fat. Additionally, heightened intraabdominal pressure can contribute to increased reflux, dilation of veins, and elevated venous pressure (Patel, Shah and Gandhi, 2021). On the other hand, physical activity is not associated with an increased risk of developing CVD; in fact, it is considered an essential measure for preventing disease progression (Jones et al., 2018). However, prolonged periods of sitting or standing, particularly in occupational settings, are recognized as risk factors for CVD (Sudoł-Szopińska et al., 2011). The presence of a family history of CVD is recognized as a risk factor for the development of this condition (Ortega et al., 2021). Furthermore, a previous occurrence of blood clots in the veins, especially deep venous thrombosis (DVT), has been recognized as a possible risk factor for CVD (Vlajinac et al., 2012), along with regular cigarette smoking (Kondo et al., 2019).

The CEAP classification identifies two possible causes of chronic venous disease (CVD): primary CVD and secondary CVD (Lurie et al., 2020). Primary CVD refers to a gradual condition that develops within the venous wall or venous valve, causing the vein to become abnormally dilated and weak. Over a period of time, this leads to the presence of pathological reflux, or backward flow of blood in the veins, which can be clinically observed. This condition is often observed in the saphenous veins and is known as truncal insufficiency (Meissner et al., 2007). However, primary CVD can also manifest in other superficial veins (Labropoulos et al., 1999; Labropoulos et al., 2001). The development of this condition is attributed to a mix of genetic and environmental factors, with genetic factors playing a role in approximately 17% of CVD cases (Fiebig et al., 2010). The remaining cases are primarily influenced by environmental factors, including lifestyle choices and other risk factors (Jawien, 2003). Secondary CVD refers to cases where the development of the disease is a result of a previous event that can be further categorized into two types: secondary intravenous and secondary extravenous (Lurie et al., 2020). Secondary intravenous CVD occurs when the vein wall and valves are adversely affected by an underlying condition. On the other hand, secondary extravenous CVD is characterized by impaired venous hemodynamics either locally or systemically, without any evidence of vein damage (Lurie et al., 2020). This can be caused by various external factors such as central venous hypertension, extrinsic compression due to a tumoral mass, compression by arteries such as in May-Thurner syndrome, or dysfunction of the diaphragmatic or limb muscle pump (Lurie et al., 2020).

Compression therapy is a widely accepted approach for the management of chronic venous disease (CVD) that aims to reduce the effects of venous hypertension (Nandasiri, 2019; Nandasiri, Ianakiev and Dias, 2020; Hedigalla et al., 2022; Santler and Goerge, 2017). It involves the application of pressure to the main veins, leading to a decrease in the diameter of the veins and an increase in the velocity of blood flow (Hedigalla et al., 2022; Sarı and Oğlakcıoğlu, 2018). Medical compression stockings (MCSs), medical compression bandages (MCBs), and intermittent pneumatic compressions (IPCs) are frequently employed compression devices in the treatment of chronic venous disease (CVD) all of which can be categorized as either passive or active compression (Nandasiri, 2019; Nandasiri, Ianakiev and Dias, 2020; Hedigalla et al., 2022; Sarı and Oğlakcıoğlu, 2018). Passive compression therapies, such as MCBs and MCSs, apply pressure on the lower limb by taking into account the radius of the limb and the tangential tension component produced by stretching the elastic materials (Nandasiri, 2019; Nandasiri, Ianakiev and Dias, 2020; Hedigalla et al., 2022; Davies, 2019; Khaburi and Jawad, 2010). Nevertheless, passive compression devices have limitations in achieving consistent pressure distribution around the entire circumference of the limb due to the elastic nature of the materials used and the varying shape of the leg (Nandasiri, 2019; Nandasiri, Ianakiev and Dias, 2020; Hedigalla et al., 2022). Furthermore, the efficacy of passive compression devices may diminish over time due to factors such as material properties, stitching methods, and manufacturing techniques (Nandasiri, 2019; Nandasiri, Ianakiev and Dias, 2020; Hedigalla et al., 2022; Liu et al., 2017).

Intermittent pneumatic compression devices (IPCs) are frequently utilized in the treatment of chronic venous disease (CVD) and can be found in the form of single bladder cuffs or multi-chamber cuffs. Single bladder cuffs inflate to a consistent pressure level but lack the ability to establish a pressure gradient from the ankle to the knee (Nandasiri, 2019; Nandasiri, Ianakiev and Dias, 2020; Hedigalla et al., 2022). In contrast, multi-chamber cuffs have multiple compartments that can be inflated individually, allowing for a desired pressure gradient along the leg. However, both types of IPCs, due to their larger bladder size, are more challenging to wear and expensive compared to other compression devices like MCBs and MCSs (Nandasiri, 2019; Nandasiri, Ianakiev and Dias, 2020; Hedigalla et al., 2022). To address the limitations of existing devices, researchers have explored the use of active materials such as shape memory polymers, shape memory alloys, and dielectric elastomer actuators in compression devices (Pourazadi, Ahmadi and Menon, 2014; Kumar, Hu and Pan, 2016; Holschuh and Newman, 2015).

However, these active material-based devices also face challenges in achieving uniform pressure distribution around the limb's circumference, as they exert pressure based on the tension created by the deformation of the active materials and the limb's curvature.

A new compression sleeve has been developed, featuring miniaturized air bladders in a hexagonal shape, to provide even pressure distribution around the lower limb (Nandasiri, 2019; Nandasiri, Ianakiev and Dias, 2020). Unlike other compression methods such as MCSs, MCBs, and IPCs, this sleeve utilizes trapped air within the bladders to generate a radial force, resulting in uniform pressure applied to the limb circumference (Nandasiri, 2019; Nandasiri, Ianakiev and Dias, 2020). Previous studies have shown that these air bladders have the potential to distribute pressure uniformly in both the longitudinal and transverse directions across the skin, fat, and muscle layers (Hedigalla et al., 2022). However, the specific pressure transmission characteristics of the compression sleeve with miniaturized air bladders, considering different design parameters like side lengths and thickness, have not yet been investigated. Therefore, this study aims to explore how the compression sleeve, incorporating miniaturized air bladders, transmits pressure on curved surfaces and determine the percentage of the area where uniform pressure is applied using numerical modeling and laboratory experiments. Additionally, machine learning techniques were employed to identify the optimal design parameters for the design and development of the active compression textile.

Methodology

Design and Development of miniaturized air bladder arrays

Custom moulds were used to manufacture two sets of miniaturized air bladder arrays, each containing three hexagonal-shaped miniaturized air bladders. These mini bladders were designed with three layers: a top layer, a fabric layer, and a bottom layer. The main difference between the two sets of mini bladder units developed was the type of fabric layer utilized. In one of the arrays, the fabric layer was constructed using polyamide (PA) textile material of 225.8 GSM. In contrast, the other array used a fabric layer consisting of high modulus PA textile material bonded with a layer of carbon fabric of 443.8 GSM. Figure 1 illustrates the miniaturized air bladder array with the fabric layer composed of high modulus polyamide (PA) textile material bonded with a layer of carbon fabric.

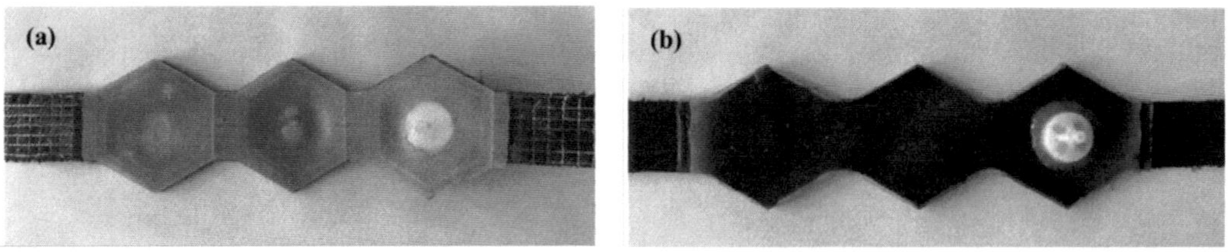

Fig. 1. a miniaturized air bladder array composed of a high modulus polyamide (PA) based textile layer bonded with carbon fabric (a) Top view (b) Bottom view

Experimental Setup for Evaluating Pressure Transmission Characteristics of Miniaturized Air Bladder Arrays

The experimental setup, as depicted in Figure 2, was designed to assess the pressure transmission characteristics of the hexagonal-shaped miniaturized air bladder units. It consisted of a perspex tube with a diameter of 60 mm, three AMI air pack sensors with a main unit, a hand pump equipped with a non-return valve, a pressure sensor (MPS20N0040D), and a Y connector. To conduct the evaluation, the

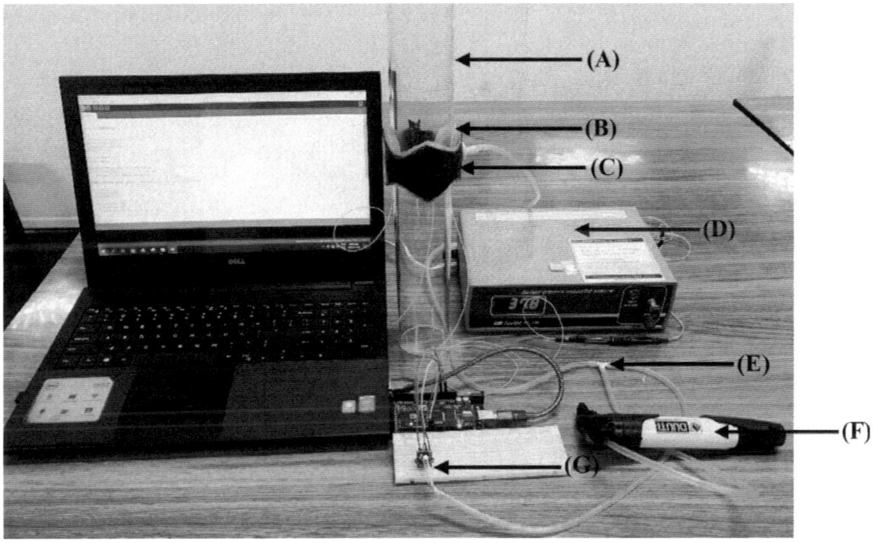

Fig. 2 Experimental setup (A) a transparent acrylic tube (B) AMI air pack sensor placed on to the acrylic tube (C) Mini-bladder array with high modulus PA fabric bonded with carbon fabric as the textile layer (D) Main unit of AMI sensors (E) Y connector (F) Hand pump with non-return valve (G) Pressure sensor

three AMI air pack sensors were affixed to the perspex tube using tape. The manufactured mini bladder array was then carefully wrapped around the cylinder, ensuring that each mini bladder aligned perfectly with one of the AMI air pack sensors. This arrangement allowed for accurate pressure measurements from each mini bladder during the experimental evaluation. To achieve various pressure levels, the hand pump equipped with a non-return valve was employed to inflate the mini-bladder unit. Using a Y connector, one end of the connector was linked to the mini-bladder array, while the other end was connected to a back pressure sensor responsible for measuring the inflated pressure. The inflation of the mini bladder array was carried out gradually, introducing 1 ml of air at a time. This inflation process was repeated multiple times to encompass a range of pressures for evaluation purposes. The measurements from both the AMI air pack sensors and the back pressure sensor were recorded manually. This recording process was repeated five times for each bladder to ensure accuracy and repeatability. Subsequently, the average interface pressure values were obtained and plotted against the corresponding inflated pressure values for each bladder.

Design and Development of a Numerical Simulation Setup for Comparative Analysis with Experimental Values

A numerical simulation setup was developed to accurately replicate the experimental setup. This involved creating a geometrical model of the miniaturized air bladder array and a transparent acrylic hollow cylinder using Solidworks (version). The mini bladder array consisted of three individual mini bladders of hexagonal shape with three different layers: a top elastomer layer of varying thickness (2.5 mm height), a bottom elastomer layer (0.3 mm thickness), and a fabric layer (1.2 mm thickness) positioned between the top and bottom elastomer layers. The hollow cylinder represents the Perspex tube, of a diameter of 60 mm, a thickness of 2 mm, and a height of 100 mm.

The material properties of each layer in the mini bladder array were defined using various experimental findings and available material models. The top and bottom layers made of Platsil® Gel OO30 (MouldLife, Suffolk, UK) were defined using a third order Yeoh model with specific material parameters ($C_{10} = 2.58\times10^{-2}$ MPa, $C_{20} = 1.464\times10^{-3}$ MPa, and $C_{30} = -9\times10^{-6}$ MPa and a density of 1100 kgm^{-3}). The PA warp knitted fabric was defined as a linear elastic material (density of 368.13 kgm^{-3}, young's modulus of 16.9315 MPa and a Poisson's ratio of 0.3.). The hollow cylinder was also defined as a linear elastic material (density of 1190 kgm^{-3}, Young's modulus of 2800 MPa, and Poisson's ratio of 0.4). A mesh sensitivity analysis was conducted to determine the appropriate mesh element size. All layers, including the top, bottom, and fabric layers, as well as the hollow cylinder, were meshed using first order tetrahedron solid elements with a mesh element size of 1.0 mm.

To achieve a more precise replication of the experimental setup, several boundaries and interfaces were defined between the geometries. These included bonded contacts between the top layer and fabric layer, as well as between the fabric layer and bottom layer. A frictional contact was established between the outer surface of the hollow cylinder and the top layer, with a coefficient of friction of 0.5. The interior surface of the hollow cylinder and the bottom surface of the mini bladder array were set as fixed surfaces. Furthermore, all the interior surfaces of the miniaturized air chamber and the air channels connecting the two mini bladders were designated as surfaces where a range of pressures would be applied.

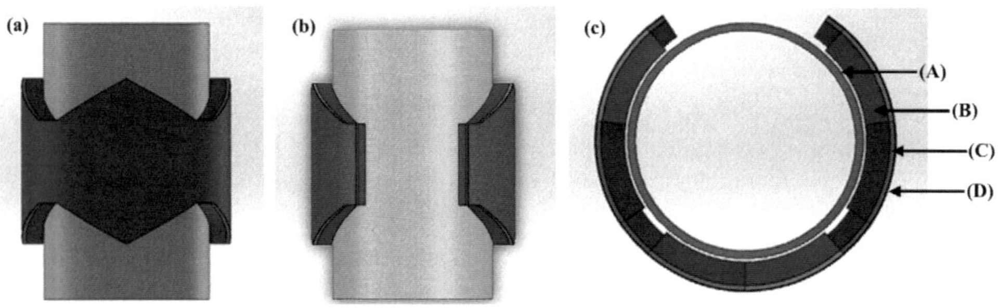

Fig. 3 SolidWorks® design of the experimental setup (a) Front view (b) Back view (c) Top Bottom view. (A) Transparent Acrylic tube of 60 mm diameter (B) Top layer (C) Fabric layer (D) Bottom layer

Determine the optimum parameters for the miniaturised air bladders utilized in the development of the active compression textile

Prior to develop neural network architectures to predict the pressure transmission percentage and the uniform are percentage, the correlation between predictor variables (side length, thickness, inflated pressure) and target variables (uniform area percentage and pressure transmission percentage) was investigated. The side length and the thickness were chosen as the input variables for both artificial neural networks utilized to predict the pressure transmission percentage (ANN1) and the uniform area percentage (ANN2). The trial-and-error approach was employed to identify the optimal parameters for both neural network architectures, which include the number of hidden layers and neurons, weight and bias initialization, activation functions, optimization functions, and the number of epochs.

A three-layered ANN architecture, consisting of an input layer, a hidden layer, and an output layer, was employed to predict both pressure transmission percentage and the percentage of area where the uniform pressure is applied. In both neural network architectures, input layer had two neurons and output layer had one neuron. The number of neurons in the hidden layer of the ANN1 and ANN2 are four and five respectively. Moreover, the rectified linear unit (ReLU) activation function was set for the hidden layer while the linear activation function was set for the output layer. Adam optimization algorithm with a learning rate of 0.001 and a batch size of 2 was employed. However, visualizing a 4D plot to determine the optimum parameters while maximizing both pressure transmission percentage and uniform area percentage was impractical. Hence, a new parameter, named pressure force, was defined as the product of pressure transmission percentage and the corresponding uniform area percentage. A 3D plot of pressure force as a function of side length and thickness was generated, enabling the identification of the optimum parameters for side length and thickness that maximize pressure force.

Results and Discussion

Pressure transmission characteristics of the manufactured mini bladder array

Figures 4 (a) and 4 (b) depict the percentage of pressure that the manufactured mini bladder arrays transmit to the Perspex tube across a range of inflation pressures. Each figure contains three curves that display the pressure transmission characteristics of three mini bladders, which are hexagonally shaped and located within a single array. In Fig. 4 (a), it can be observed that the percentage of pressure transmission rises until an inflation pressure of 25 mmHg is reached, beyond which it starts to decrease gradually. This decrease may be attributed to the inflation of the lower layer of the mini

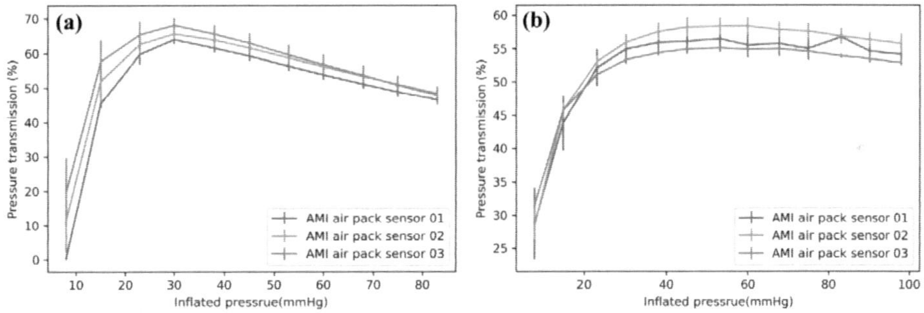

Fig. 4 Pressure transmission characteristics of the mini-bladder (a) made with only PA warp knitted fabric as the textile layer (b) made with PA based warp knitted fabric bonded with carbon fabric as the textile layer

bladder when subjected to inflation pressures exceeding 25 mmHg, instead of the surface that is in contact with the acrylic tube. In contrast, Figure 4 (b) demonstrates that the pressure transmission percentage of the mini bladder array, formed by combining a PA warp knitted fabric with carbon fabric, initially increases and then levels off at a plateau representing 50%-60% of the inflation pressure. Thus, the mini bladder array fabricated by bonding PA warp knitted fabric with carbon fabric was selected for integration into the final design of the active compression textiles.

Validation of simulation results with experimental results

Figure 5 (a) displays a graph that shows the average experimental interface pressure of the mini-bladder array across a range of 0-100 mmHg inflation pressure. It exhibits three curves representing the average interface pressure of individual units of hexagonal-shaped mini-bladders. The pressure values for all three curves are nearly identical, except for a slight deviation in one curve when the inflation pressure surpasses 30 mmHg. The plots in Figures 5 (b), (c), and (d) illustrate the comparison between the average simulated and experimental interface pressure as a function of the inflated pressure for the mini-bladder array. The high average R^2 value, exceeding 95%, obtained for all three plots indicates that the numerical model is capable of accurately determining the interface pressure for mini-bladders with various parameters, including thickness and size.

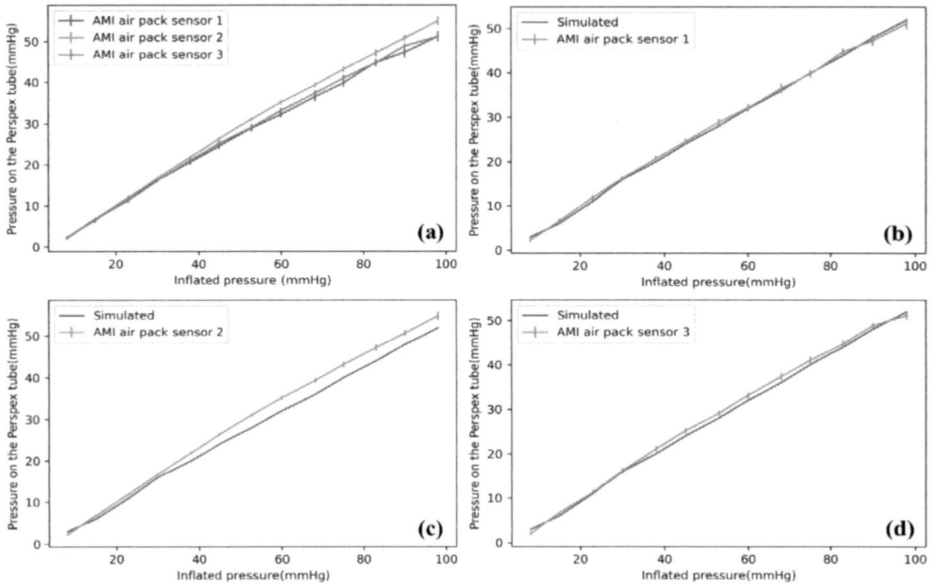

Fig. 5 (a) Interface pressure between the transparent acrylic cylinder and the three hexagonal shaped mini-bladders against inflated pressure (b), (c) and (d) comparison of experimentally measured values with the numerical simulated values.

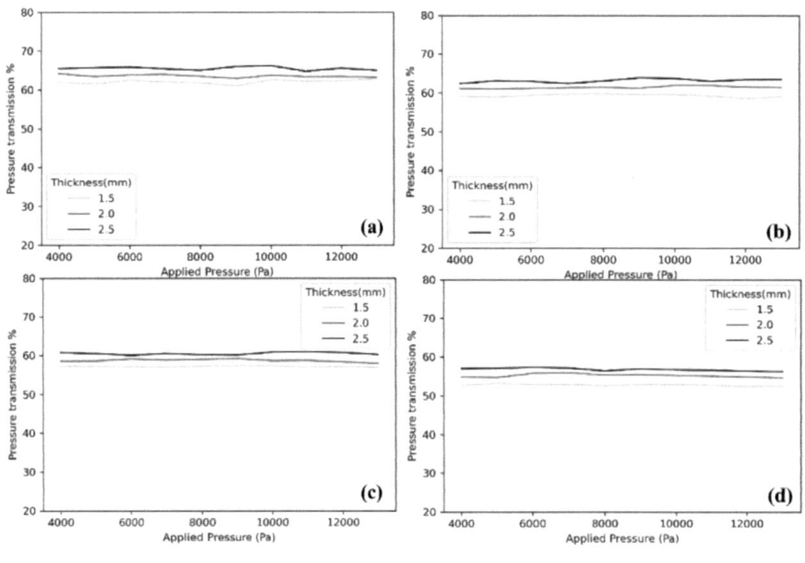

Fig. 6 Pressure transmission characteristics against inflated pressure at four different side lengths: (a) 1.0cm (b)1.5cm (c)2.0cm (d)2.5cm

Pressure transmission characteristics at different side lengths and thickness

The four plots depicted in Figure 6 showcase the variations in pressure transmission characteristics of miniaturized air bladders. These characteristics are examined across four different side lengths, namely 1.0 cm, 1.5 cm, 2.0cm, and 2.5 cm, in response to inflation pressures ranging from 4 kPa to 13 kPa (equivalent to 30 mmHg to 97.5 mmHg). Each of the four plots in Figure 6 shows three curves representing the pressure transmission characteristics of the miniaturised air bladders at three distinct thicknesses, namely 1.5 mm, 2.0 mm, and 2.5 mm. Among the mini bladder arrays, the one with a side length of 1.0 cm and a thickness of 2.5 mm exhibited the highest-pressure transmission percentage, approximately 65%. Conversely, the mini bladder array with a side length of 2.5 cm and a thickness of 1.5 mm displayed the lowest pressure transmission percentage, around 53%. Additionally, the plots in Figure 6 indicate that the influence of inflation pressure on the pressure transmission percentage of miniaturized air bladders with different thicknesses and side lengths seems to be insignificant.

Uniform area percentage at various side lengths and thicknesses

Figure 7 presents four plots showcasing the percentage of the area where uniform pressure is applied for mini bladders with different side lengths (1.0 cm, 1.5 cm, 2.0 cm, and 2.5 cm) as the inflated pressure varies from 4 kPa to 13 kPa. Each plot includes three curves that represent the change in the percentage of area with uniform pressure as the top layer's thickness is increased in 0.5 mm increments from 1.5 mm to 2.5 mm. The analysis of the plots revealed that there is a positive correlation between the side length of the mini bladders and the percentage of area with uniform pressure, while a negative correlation exists between the top layer's thickness and the uniform area percentage. Additionally, the impact of inflation pressure on the uniform pressure area percentage is minimal across all thicknesses and side lengths. The mini-bladder array with a side length of 2.5 cm and a thickness of 1.5 mm exhibited the highest percentage of area with uniform pressure, approximately 65% of the total area. Conversely, the mini-bladder array with a side length of 1.0 cm and a thickness of 2.5 mm displayed the lowest percentage, approximately 26% of the total area.

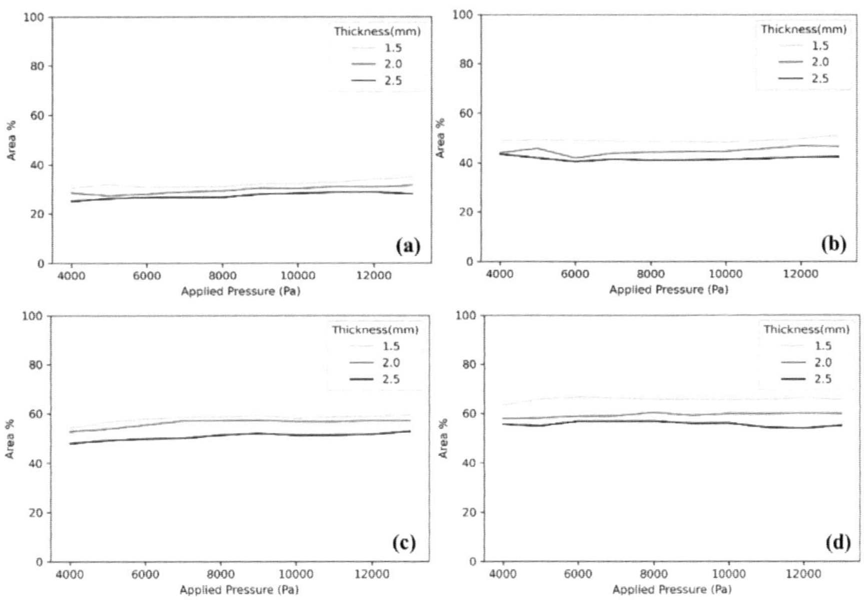

Fig. 7 Uniform area percentage against inflated pressure at four different side lengths: (a) 1.0cm (b)1.5cm (c)2.0cm (d)2.5cm

Optimizing Pressure Force through Side Length and Thickness Parameters

Figure 8 presents a 3D plot showcasing the variations in pressure force as the side length and thickness values are changed. The plot demonstrates a strong positive correlation between the pressure force and the side length, while a negative correlation exists with the thickness. Notably, the slope of the pressure force with respect to side length is steeper at lower side lengths compared to the slope with respect to the thickness at any given (thickness) value. This suggests that small adjustments in side length at lower lengths have much significant impact on the pressure force compared to the smaller changes in thickness. However, at higher side lengths, larger changes of side length are required to affect the pressure force, as indicated by the shallower slope. From Figure 8, it can be observed that the highest-pressure force is attained at a higher side length of 2.5 cm and a lower thickness level of 1.0

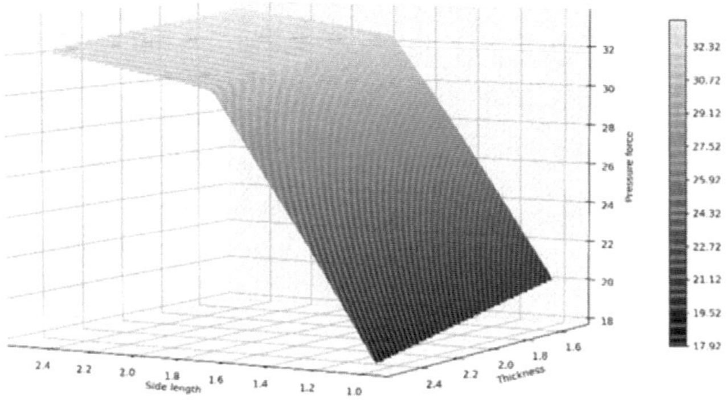

Fig. 8 3D surface plot of pressure force as a function of side length and thickness of the miniaturized air bladders

mm. Consequently, the optimal parameters for the miniaturized air bladders used in the development of the active compression device were determined as of side length of 2.5 cm and a thickness of 1.0 mm.

Conclusions

Chronic venous disease is a prevalent medical condition that hampers the blood flow from the legs to the heart due to venous dysfunction. This condition significantly impacts patients' quality of life and imposes a substantial economic burden, encompassing both productivity losses and treatment expenses. The objective of this study is to determine the ideal design parameters for miniaturized air bladders used in developing compression textiles that deliver uniform pressure around the lower limb circumference, leveraging the radial force generated by trapped air inside the bladder. Through a combination of numerical simulations and experimental validation, this study examines the pressure transmission percentage and the percentage of the area where uniform pressure is applied by the miniaturized air bladders at varying side lengths and thicknesses. Two artificial neural networks (ANNs) are developed to predict pressure transmission and uniform area percentage, and their performance is evaluated. The aim is to identify the optimal parameters that maximize pressure force.

The investigation reveals that the side length exhibits a strong positive correlation with the uniform area percentage, while displaying a significant negative correlation with the pressure transmission percentage. Furthermore, the top layer's thickness in the miniaturized air bladder demonstrates a positive correlation with the pressure transmission percentage but a negative correlation with the uniform area percentage. The ANN models effectively predict both the pressure transmission percentage and the uniform area percentage, with low and comparable average mean squared error (MSE) on the training and validation data. Based on the findings, the recommended design parameters for the miniaturized air bladders in active compression textiles, aiming to achieve effective pressure transmission while maintaining a higher uniform pressure area in compression therapy, are determined as 2.5 cm in side length and 1.0 mm in thickness level.

References

Bergan, J.J., Schmid-Schönbein, G.W., Smith, P.D.C., Nicolaides, A.N., Boisseau, M.R. and Eklof, B., 2006. Chronic venous disease. *New England Journal of Medicine, 355*(5), pp.488-498.

Davies, A.H., 2019. The seriousness of chronic venous disease: a review of real-world evidence. *Advances in therapy, 36*(Suppl 1), pp.5-12.

Vuylsteke, M.E., Colman, R., Thomis, S., Guillaume, G., Van Quickenborne, D. and Staelens, I., 2018. An epidemiological survey of venous disease among general practitioner attendees in different geographical regions on the globe: the final results of the Vein Consult Program. *Angiology, 69*(9), pp.779-785.

Pappas, P.J., Lakhanpal, S., Nguyen, K.Q. and Vanjara, R., 2018. The Center for Vein Restoration Study on presenting symptoms, treatment modalities, and outcomes in Medicare-eligible patients with chronic venous disorders. *Journal of Vascular Surgery: Venous and Lymphatic Disorders, 6*(1), pp.13-24.

Bozkurt, A.K. and Balkanay, O.O., 2017. Approach to venous diseases in the elderly. *Turk Kardiyoloji Dernegi Arsivi: Turk Kardiyoloji Derneginin Yayin Organidir, 45*(Suppl 5), pp.102-107.

Rabe, E., Guex, J.J., Puskas, A. and Scuderi, A., 2012. Epidemiology of chronic venous disorders in geographically diverse populations: results from the Vein Consult Program. *International angiology: a journal of the International Union of Angiology, 31*(2), pp.105-115.

Kahn, S.R., M'lan, C.E., Lamping, D.L., Kurz, X., Bérard, A., Abenhaim, L.A. and VEINES Study Group, 2004. Relationship between clinical classification of chronic venous disease and patient-reported quality of life: results from an international cohort study. *Journal of vascular surgery, 39*(4), pp.823-828.

Wittens, C.D.A.H., Davies, A.H., Bækgaard, N., Broholm, R., Cavezzi, A., Chastanet, S., de Wolf, M., Eggen, C., Giannoukas, A., Gohel, M. and Kakkos, S., 2015. Editor's choice–management of chronic venous disease: clinical practice guidelines of the European Society for Vascular Surgery (ESVS). *European Journal of Vascular and Endovascular Surgery, 49*(6), pp.678-737.

Zolotukhin, I.A., Seliverstov, E.I., Shevtsov, Y.N., Avakiants, I.P., Nikishkov, A.S., Tatarintsev, A.M. and Kirienko, A.I., 2017. Prevalence and risk factors for chronic venous disease in the general Russian population. *European Journal of Vascular and Endovascular Surgery, 54*(6), pp.752-758.

Evans, C.J., Fowkes, F.G.R., Ruckley, C.V. and Lee, A.J., 1999. Prevalence of varicose veins and chronic venous insufficiency in men and women in the general population: Edinburgh Vein Study. *Journal of Epidemiology & Community Health, 53*(3), pp.149-153.

Brand, F.N., Dannenberg, A.L., Abbott, R.D. and Kannel, W.B., 1988. The epidemiology of varicose veins: the Framingham Study. *American journal of preventive medicine, 4*(2), pp.96-101.

Maurins, U., Hoffmann, B.H., Lösch, C., Jöckel, K.H., Rabe, E. and Pannier, F., 2008. Distribution and prevalence of reflux in the superficial and deep venous system in the general population–results from the Bonn Vein Study, Germany. *Journal of vascular surgery, 48*(3), pp.680-687.

Lurie, F., Passman, M., Meisner, M., Dalsing, M., Masuda, E., Welch, H., Bush, R.L., Blebea, J., Carpentier, P.H., De Maeseneer, M. and Gasparis, A., 2020. The 2020 update of the CEAP classification system and reporting standards. *Journal of Vascular Surgery: Venous and Lymphatic Disorders, 8*(3), pp.342-352.

Meissner, M.H., Gloviczki, P., Bergan, J., Kistner, R.L., Morrison, N., Pannier, F., Pappas, P.J., Rabe, E., Raju, S. and Villavicencio, J.L., 2007. Primary chronic venous disorders. *Journal of vascular surgery, 46*(6), pp.S54-S67.

Labropoulos, N., Kang, S.S., Mansour, M.A., Giannoukas, A.D., Buckman, J. and Baker, W.H., 1999. Primary superficial vein reflux with competent saphenous trunk. *European journal of vascular and endovascular surgery, 18*(3), pp.201-206.

Labropoulos, N., Tiongson, J., Pryor, L., Tassiopoulos, A.K., Kang, S.S., Mansour, M.A. and Baker, W.H., 2001. Nonsaphenous superficial vein reflux. *Journal of vascular surgery, 34*(5), pp.872-877.

Fiebig, A., Krusche, P., Wolf, A., Krawczak, M., Timm, B., Nikolaus, S., Frings, N. and Schreiber, S., 2010. Heritability of chronic venous disease. *Human genetics, 127*, pp.669-674.

Jawien, A., 2003. The influence of environmental factors in chronic venous insufficiency. *Angiology, 54*(1_suppl), pp.S19-S31.

Lee, A.J., Robertson, L.A., Boghossian, S.M., Allan, P.L., Ruckley, C.V., Fowkes, F.G.R. and Evans, C.J., 2015. Progression of varicose veins and chronic venous insufficiency in the general population in the Edinburgh Vein Study. *Journal of Vascular Surgery: Venous and Lymphatic Disorders, 3*(1), pp.18-26.

Lohr, J.M. and Bush, R.L., 2013. Venous disease in women: epidemiology, manifestations, and treatment. *Journal of vascular surgery, 57*(4), pp.37S-45S.

Ortega, M.A., Sánchez-Trujillo, L., Bravo, C., Fraile-Martinez, O., García-Montero, C., Saez, M.A., Alvarez-Mon, M.A., Sainz, F., Alvarez-Mon, M., Bujan, J. and De Leon-Luis, J.A., 2021. Newborns of mothers with venous disease during pregnancy show increased levels of lipid peroxidation and markers of oxidative stress and hypoxia in the umbilical cord. *Antioxidants, 10*(6), p.980.

Cornu-Thenard, A. and Boivin, P., 2014. Chronic venous disease during pregnancy. *Phlebolymphology, 21*(3), pp.138-145.

Vlajinac, H.D., Marinkovic, J.M., Maksimovic, M.Z., Matic, P.A. and Radak, D.J., 2013. Body mass index and primary chronic venous disease–a cross-sectional study. *European Journal of Vascular and Endovascular Surgery, 45*(3), pp.293-298.

Patel, J., Shah, P. and Gandhi, F., 2021. A study of chronic venous insufficiency in relation with body mass index and diameter of saphenofemoral junction and great saphenous vein. *Indian Journal of Vascular and Endovascular Surgery, 8*(1), pp.58-62.

Jones, W.S., Vemulapalli, S., Parikh, K.S., Coeytaux, R.R., Crowley, M.J., Raitz, G., Johnston, A.L., Hasselblad, V., McBroom, A.J., Lallinger, K.R. and Sanders-Schmidler, G.D., 2018. Treatment Strategies for Patients with Lower Extremity Chronic Venous Disease (LECVD).

Sudoł-Szopińska, I., Bogdan, A., Szopiński, T., Panorska, A.K. and Kołodziejczak, M., 2011. Prevalence of chronic venous disorders among employees working in prolonged sitting and standing postures. *International journal of occupational safety and ergonomics, 17*(2), pp.165-173.

Ortega, M.A., Fraile-Martínez, O., García-Montero, C., Álvarez-Mon, M.A., Chaowen, C., Ruiz-Grande, F., Pekarek, L., Monserrat, J., Asúnsolo, A., García-Honduvilla, N. and Álvarez-Mon, M., 2021. Understanding chronic venous disease: A critical overview of its pathophysiology and medical management. *Journal of Clinical Medicine, 10*(15), p.3239.

Vlajinac, H.D., Radak, Đ.J., Marinković, J.M. and Maksimović, M.Ž., 2012. Risk factors for chronic venous disease. *Phlebology, 27*(8), pp.416-422.

Kondo, T., Nakano, Y., Adachi, S. and Murohara, T., 2019. Effects of tobacco smoking on cardiovascular disease. *Circulation Journal, 83*(10), pp.1980-1985.

Nandasiri, H.G.K., 2019. A study of use of mini-bladders in active compression as a treatment for venous disease and lymphoedema. Nottingham Trent University (United Kingdom).

Nandasiri, G.K., Ianakiev, A. and Dias, T., 2020. Hyperelastic properties of platinum cured silicones and its applications in active compression. *Polymers, 12*(1), p.148.

Hedigalla, D., Ehelagasthenna, M., Nissanka, I.D., Amarasinghe, R. and Nandasiri, G.K., 2022, July. Numerical Study to Investigate the Pressure Propagation Patterns by a Compression Sleeve with Miniaturised Air-Bladders. In *2022 Moratuwa Engineering Research Conference (MERCon)* (pp. 1-5). IEEE.

Santler, B. and Goerge, T., 2017. Chronic venous insufficiency–a review of pathophysiology, diagnosis, and treatment. *JDDG: Journal der Deutschen Dermatologischen Gesellschaft, 15*(5), pp.538-556.

Sarı, B. and Oğlakcıoğlu, N., 2018. Analysis of the parameters affecting pressure characteristics of medical stockings. *Journal of Industrial Textiles, 47*(6), pp.1083-1096.

Davies, Alun H. "The seriousness of chronic venous disease: a review of real-world evidence." *Advances in therapy 36*, no. Suppl 1 (2019): 5-12.

Khaburi, A. and Jawad, J.A., 2010. Pressure mapping of medical compression bandages used for venous leg ulcer treatment. University of Leeds.

Liu, R., Guo, X., Lao, T.T. and Little, T., 2017. A critical review on compression textiles for compression therapy: Textile-based compression interventions for chronic venous insufficiency. *Textile research journal, 87*(9), pp.1121-1141.

Pourazadi, S., Ahmadi, S. and Menon, C., 2014. Towards the development of active compression bandages using dielectric elastomer actuators. *Smart Materials and Structures, 23*(6), p.065007.

Holschuh, B. and Newman, D., 2015. Two-spring model for active compression textiles with integrated NiTi coil actuators. *Smart Materials and Structures, 24*(3), p.035011.

Kumar, B., Hu, J. and Pan, N., 2016. Smart medical stocking using memory polymer for chronic venous disorders. *Biomaterials, 75*, pp.174-181.

MEDICAL TEXTILES IN THE GREENER OPERATING THEATRE

Holly Morris[1] *and Richard Murray*[2]

[1] *Orthopaedic Surgeon, Derby, DE22 3NU, United Kingdom*
[2] *Emeritus Professor, Manchester Metropolitan University, United Kingdom*

Abstract

By generating almost 5% of the world's carbon emissions, healthcare, if it were a country, would be the world's fifth biggest polluter. In fact, in 2017 the health sector was responsible for 4.4% of net global greenhouse gas emissions and 6.3% of the UK's carbon footprint. In 2020, the UK National Health Service became the first health service to announce its intention to achieve net zero emissions. Between 20% and 33% of health care waste is thought to originate from a hospital's operating rooms and up to 90% of this is sent for costly and unnecessary hazardous-waste disposal. The textile components can account for up to 30% of the waste from operating theatres and regionally collected data from trauma and orthopaedic operations suggests that 6-8kg of the waste generated per procedure originates from gowns and drapes. Current practice allows the use of disposable or re-usable items or a combination of the two.

This paper, by the authors of the textile guidance given in the UK Green Surgery Report, is intended to alert textile researchers to the need for improved evidence not only on infection prevention whilst using re-usable textiles but also on the implementation of effective laundry processes for its cleaning and microbiological decontamination. The steps which still need to be taken to achieve a circular economy for the medical textiles used in operating theatres will be discussed.

Keywords

medical textiles, sustainability, laundering, decontamination, circular economy

Introduction

Climate change has been labelled as "the biggest global health threat of the 21st century" ('Editorial. A Commission on climate change', 2019) and in 2018, the Intergovernmental Panel on Climate Change (IPCC) announced that to limit global warming to 1.5°C, greenhouse gas emissions must decrease 45% by 2030 (compared with 2010) and reach Net Zero by 2050 (Intergovernmental Panel on Climate Change, 2018). In the United Kingdom, Public Health England and the National Health Service estimated the health and social care climate footprint within England in 2017 to be around 6.3% of the country's climate footprint (Public Health England, 2018) and in 2020, it was reported that the NHS in the UK generates approximately 4-5% of the country's greenhouse gases and a quarter of all public-sector waste **(NHS England Press Release, 2020; Rizan C *et al.*, 2020; Watts N *et al.*, 2021)**.

In 2020, the NHS launched its campaign for a "Greener NHS" and commissioned an Expert Panel to set out a practical, evidence-based, and quantified path to a 'Net Zero NHS'. Their report sets out a strategy and two clear targets to respond to this challenge:

- Net Zero by 2040 for the emissions the NHS controls directly
- Net Zero by 2045 for the emissions the NHS can influence (NHS England, 2022).

Key areas identified in this report include waste management and procurement of lower carbon options. The main components of healthcare waste are plastic (39.3-50%), textile (14-31%), paper (11.2-25%), glass (0.3-22.7%), woodware (3.2-20%), rubber (3.4-6.6%), metal (0.3-5%) and other waste (1.4-18.6%) **(Zlaugotne B *et al.*, 2022)**. It has been suggested that between 20% to 33% of health-care waste originates from a hospital's operating room **(Rizan C *et al.*, 2020)**; contaminated or hazardous waste includes waste that has been directly used in patient care. Up to 90% of operating room waste is improperly sorted and sent for costly and unnecessary hazardous waste processing (Lee RJ and Mears SC, 2012) and there is often significant variation in the way in which health care waste is disposed of between different NHS Trusts.

A wide range of medical textiles are important to medicine and healthcare, and that range continues to grow. Items described as 'linen,' as defined by the UK Department of Health, include bed linen (blankets, counterpanes, cot sheets and blankets, duvets, duvet covers, pillowcases, and sheets including those which are woven, knitted, half sheets, draw and slide sheets); bibs; blankets; canvases; curtains; hoist slings; patient clothing (gowns, nightdresses and shirts, pyjama tops and bottoms); staff clothing (coats, scrub suits, tabards, uniforms); and towels (Department of Health, 2016).

Design and Use of Medical Textile Items

There are several ways of reducing clinical waste and particular attention has been paid to implementing a circular healthcare-economy model, through reducing, reusing, and recycling:

- Reducing what is consumed can be achieved by identifying whether the item needs to be opened for the set. Do we need to open multi-pack options of surgical tools and drapes when single-packed items may suffice? (Department for Business, 2021).
- Reusing items, by selecting items that can be laundered reduces the production and subsequent consumption of items.
- Recycling allows for a reduction in waste going forward to incineration and landfill.
- If waste must be disposed, utilise methods that allow for energy harvesting and prioritise reuse over landfill.

Clinicians sit at the end of the medical device life cycle and will therefore need to play their part in achieving the circular economy, but medical-device designers, manufacturers and procurement have key roles to play in waste reduction, because over 80% of the environmental impact of a product is determined at the design stage (European Commission, 2012). Using resources efficiently by considering the use of renewable and sustainable raw materials and reducing the amount of resource consumed in a device's manufacture, delivery, operation, and waste minimization all impact upon the environmental impact. Medical devices should be designed for circularity (a product suitable for several life cycles), durability, disassembly, and reuse. Finally, the product life should be able to be extended to keep the product in use for the highest number of safe-use cycles rather than just designed for disposal after a single use.

The textile industry has not yet developed the ways and means to achieve the overarching aim of a circular economy for itself, and this will have implications for those working towards sustainable healthcare procurement (Claxton S and Kent A, 2020; United Nations Development Programme, 2022). A further issue which poses challenges to transparency is the highly globalized, complex and extremely fragmented supply chain (Ki CW, Chong SM and Ha-Brookshire JE, 2020). Many medical textile manufacturing firms are located outside the Western consumer markets; corporate governance is an issue hence it can be challenging to access sufficient reliable information either for full traceability or to determine a procurement footprint. It would be easy to dismiss fashion-clothing retail supply chains as unrelated and irrelevant, but important lessons can be learned from fashion retailers about the intensity of supervision necessary to ensure that product designs are followed through and that items meet the required specifications. Every significant operation is checked through the meticulous application of over 300 specified supply-chain-management supervisory steps (Chen Z, Murray R and Jones RM, 2007). Retailer representatives visit the suppliers' design, production, and warehousing facilities to run checks on a weekly basis. Nothing operates only on the basis of trust.

Within a medical-textile circular economy, alongside satisfying the existing performance and reliability criteria, the design and procurement of items must also focus on

- Resource efficiency: using renewable and sustainable raw materials, reducing the amount of resource consumed, using sustainable manufacturing processes, and minimising waste
- Circular design: designing for circularity, disassembly, and reuse
- Product life extension: increasing the safe uses of a textile product, overhauling current standards to ensure they are relevant to the technology of today

Reuse of Textile Items

Sun recommended when considering disposable and reusable medical textiles that individuals should seek improved protective performance, low cost, and minimised environmental impact from medical textiles which could be disposable or reusable (Sun G, 2011). Whilst single-use disposables do have some attractive features and are the sensible choice in certain settings where the numbers of particular types of medical textile items being used is small, in others, they have the potential to incur high costs and to make a substantial contribution to pollution. How to make the decision about which type to use was well-addressed by Walton, who recommended that the use of disposables should be reserved for low-volume applications, where it was not going to be economical to equip to be able to decontaminate and sterilise them (Walton JR, 1986).

Waste minimisation in a circular economy can be achieved by a variety of methods including

- Reuse of products

- Product life extension (discussed in previous section)

Reuse of products in healthcare requires effective sterilization and repurposing such that the product remains fit for purpose and a further life cycle. In Europe, according to Regulation (EU) 2017/745 of 5 April 2017 on medical devices (European Union, 2017), remanufactured 'single-use' medical devices must meet manufacturer standards and be able to receive a CE mark to indicate that they are as safe and effective as the new equipment, and are required to be sterilised to the level of the original device. Healthcare units keen to save on resources and address the growing need to generate less waste, but concerned about the expense of decontamination and sterilisation, are now able to find systems and products to help them to achieve those aims. Practice Greenhealth™ (Reston, VA, USA) has developed the business case and step-by-step guidance documents to give practical help in developing a sound basis from which to reach such a decision about what choice to make for the operating room (Practice GreenHealth, 2022). Adaption of their work should prove of significant assistance in the development of processes, procedures, and equipment to suit other regions and other areas of healthcare.

With reusable items, laundering requires attention. A recent review concluded that a number of studies had indicated that microorganisms survive on textiles for extended periods of time and can transfer onto skin and other surfaces hence raising the possibility that they might be transferred through contact between staff or patients and contaminated textiles. A number of case studies linked small outbreaks with inadequate laundering or poor infection-control processes for healthcare laundry. Others demonstrated the survival of potential pathogens during the laundering of healthcare textiles, which might possibly increase the risk of infection. Although there are no large-scale epidemiological studies demonstrating a direct link between healthcare-acquired infections and contaminated textiles, evidence of outbreaks given in the case studies needs further attention (Owen L and Laird K, 2020).

The United Kingdom National Health Service has produced Best Practice guidance for the laundering of linens (Department of Health, 2016). Traditionally, disinfection has been by thermal methods although chemical methods at a lower temperature may also be employed. For thermal disinfection methods, the washing process should have a cycle in which the temperature of the load is either maintained at 65°C for not less than ten minutes or 71°C for not less than three minutes. Alternative time/temperature relationships may be used provided the efficacy of the process chosen is equal to the 65°C or 71°C processes. Mixing time should be added to ensure heat penetration and assure disinfection. Those laundering workwear uniforms at home in domestic facilities should be washing for ten minutes at 60°C (Department of Health, 2016).

For heat-labile items, chemical disinfection is preferred, and the Best Practice suggests that the exact process should be chosen in discussion and agreement with the infection control team, such that the entire process, should be capable of passing the required microbiological tests (Department of Health, 2016).

Heating water to 65°C or 71°C requires significant energy. Attention should be turned towards more sustainable practice including scrutinising whether this temperature could be lowered without undue effect on the cleaning of the textiles. Attention should also be paid to the chemicals used in chemical disinfection including how they are disposed of. A system for decontamination and disinfection at low temperatures involving the generation of ozone within the washing machine which automatically provides the data demonstrating its efficacy required for use in audit trails has been shown to be cost-effective within both hospitals and care-home settings and deserves serious attention and implementation on a wider scale. Details about such a system approved in accordance with the Biocidal Products Regulations (EU) 528/2012 (European Chemicals Agency, 2012) are available from one UK-based provider (JLA, 2021). There is a European trade organisation (EUO_3TA) which provides useful information about other suppliers and the standards with which their equipment must comply (Ozone Industries, 2023).

A review undertaken at the behest of the Department of Health in 2007 found no robust evidence of a difference in efficacy of decontamination of uniforms/clothing between industrial and domestic laundry processes, or that the home laundering of uniforms provides inadequate decontamination. However, only a small number of relevant studies were found, and they provided only limited evidence directly related to the decontamination of uniforms. Studies concerning domestic laundry processes were small scale and largely observational whereas the large-scale practice current at the time together with the related guidance for laundering uniforms was extrapolated from studies of industrial hospital linen processing **(Wilson JA *et al.*, 2007)**. More-recent work, in 2015, demonstrated that 44% of healthcare workers laundering at home do not launder their uniforms at the 60°C stipulated by the national guidance (Riley K, Laird K and Williams J, 2015). There is scope now for improvement in the laundering of hospital textiles, both in the hospital and domestic setting given advances in laundering abilities.

The laundering of synthetic textiles is a significant cause of microplastic release to the environment (Palacios-Marín AV and Tausif M, 2021). Natural cellulosic microparticles, alongside microplastics, have also been identified in the environment and stomachs of other species **(Parton KJ *et al.*, 2020)**. Washing at higher temperatures increases the rate of release of microparticles, another reason to reconsider the laundering process **(Cotton L *et al.*, 2020)**. Filtration of effluent water used in the laundering process can reduce the escape and leaching of microplastics into the environment by up to 80% (Napper IE, Barrett AC and Thompson RC, 2020).

Disposal of Textile Items

Rather than the disposal of items, consideration should be given to the circular economy and the appropriateness of reuse, recycling or remanufacture either for the item itself or its constituent panels or by harvesting the constituent fibres for recycling and reuse elsewhere. Reuse of single-use textile items via schemes to process, clean and test the item to the relevant standard for use are already in use in countries such as the USA and New Zealand (MedSalv, 2021). Reuse of the textile item may not be as was originally intended, but may have value in a different setting; for example, reusing a surgical gown as protective wear in a non-sterile environment (MacNeill AJ *et al.*, 2020).

Remanufacturing which allows disassembly of the item, recovery of useful component parts, redesigning and reassembling them to make new items (to be used within healthcare or elsewhere) should be considered. Whilst this is often a cost- and resource-effective treatment method on its own, compared to recycling (which often requires energy and chemical intensive treatment processes), the design process can further facilitate easy disassembly by incorporating suitable bonding and stitch types (Dissanayake DG and Weerasinghe D, 2021).

Whilst recycling facilities in hospitals are highly variable, good recycling systems are usually environmentally and financially preferable (Campion N *et al.*, 2015). A potential problem of recycling in the hospital environment is the mixing of textile wastes e.g., polyester/cotton blends with 100% cotton or with other fibre types. Near-infrared spectroscopy is a fast, non-destructive, and efficient method of determining fibre type for textile waste separation (Wei B *et al.*, 2015). Chemical processes are already available to enable the separation of the fibre components from fabrics made with fibre blends such as the polyester/cotton blend used in surgical gowns and drapes. Already used in the fashion sectors, there have been several studies reporting promising results, which suggest a recycling rate of more than 93-96%, a carbon footprint reduction of 1440-1534kg of CO_2 and an economic return of US$1466-1629 per tonne (Zamani B, 2014; Yousef S *et al.*, 2020).

If a garment or fabric is not in a condition to be reused, consideration should be given to the recycling of fibres wherever possible. Recycling technologies are divided broadly into mechanical and chemical recycling. Mechanical recycling pulls at the fabric reduce it to short fibres which can be separated physically, whereas thermal recycling applied to thermoplastic waste materials involves melting of the fabric then extrusion of the molten polymer to obtain new filaments then be cut into fibres and respun into yarns. In chemical recycling, synthetic-polymer materials are first depolymerized into their constituent monomers and then re-polymerized so that the polymer can be spun and converted back into fibres. Recycling does not only apply to synthetic fibres; for example, although cotton can decompose during composting, recycling improves sustainability because the use of water, pesticides and insecticides employed in growing virgin cotton does not have to be repeated (Hvass KK, 2016). One tonne of reused cotton has been estimated to conserve 765m^3 of water of which 98% would otherwise have been used in the production of virgin cotton (Deguine JP, Ferron P and Russell D, 2008).

Rather than reusing the fibre as fibre, fibres from the medical textile can also be converted to yield other products which may be used in other industry sectors; for example, cellulose can be converted to glucose or nanocellulose (Ling C *et al.*, 2019; Li X *et al.*, 2019). This may offer possible routes for reuse that avoid the need for incineration or landfill disposal.

Textiles Within Surgery

High-volume textiles used within the surgical directorate include surgical gowns and drapes and the workwear of the operating personnel. There was an attitude initiated in the mid-1950s that was labelled at that time by LIFE magazine as 'Throwaway Living', and this became adopted in the hospital and healthcare settings by the 1970s and more widely ever since (Strasser BJ and Schlich T, 2020). The advantages in terms of the convenience of disposable items had been strongly marketed to the healthcare sector with attention drawn to similarities in cost and the possibility of moving away completely from the management responsibilities involved with in-house or outsourced institutional laundering and sterilisation.

Overcash, in 2012, completed a useful analysis that combined the findings from six independent lifecycle assessments for gowns and drapes, carried out in terms of cost-effectiveness analyses and life-cycle testing. The analysis showed there was little cost difference to the hospital/institutional user between using disposable gowns and drapes compared with using reusables, but there were significant differences in terms of environmental impact (Overcash M, 2012). The environmental impact differences arise because disposables require proportionately much greater amounts of energy and water per single use than the reusables which require the energy and water for only a single manufacturing stage then much less when they are subsequently cleaned 50 times. Reusables demonstrate a 28% reduction in energy consumption, a 30% reduction in greenhouse gas emissions, a 41% reduction in blue-water consumption and a 93% reduction in solid-waste generation when compared with disposables (Vozzola E, Overcash H and Griffing E, 2018). Even considering just the water and energy requirements of the laundering of reusable gowns, the manufacturing of disposables creates significantly more impact to the environment than using a reusable. That said, it would be presumptuous to assume that a reusable gown is appropriate in every setting.

Surgical Gowns

Surgical gowns were introduced primarily as a protective barrier to reduce surgical site infection and protect the patient from the surgical team. They are intended to accomplish this by one of two means:

i. by controlling the skin and bacteria being shed by the user, and
ii. by preventing the direct transmission of bacteria.

In current practice, the gown now also offers protection to the surgical team from the patient. A surgical gown needs to be able to demonstrate good wearing comfort whilst keeping the surgeon and team comfortable, particularly when worn for lengthier periods of time (Aibibu D, Lehmann B and Offermann P, 2003; Aslan S, Kaplan S and Cetin C, 2013).

There have been significant advances in both the design and construction of surgical gowns since the time of the frock coat donned in the mid-19th Century. Recognition of the importance of pore size between woven fibres in terms of enabling the transmission of microbes, the difference in transmission of bacteria through wet and dry fabric, the effect of pressure on the fabric for microbial transmission, the introduction of synthetic fibres and the development of effective finishing treatments have seen a variety of changes in the construction, durability, impenetrability, and soil repellency of the fabrics used over the decades (Leonas KK and Jinkins RS, 1997).

Some well-intentioned changes to the design of the gown, such as the introduction of total body exhaust suits by Charnley, have since been shown to increase wound contamination and surgical site infection in arthroplasty (Hooper GJ *et al.*, 2011). The glove-gown interface also remains a source of potential contamination (Fraser JF *et al.*, 2015). There is room for further research into the improved surgical gown.

Despite great enthusiasm being shown for investigating the best types of fabric for a surgical gown, replication of results in vivo and in vitro often differ (Gulihar A, Taub NA and Taylor GJS, 2009). A recent review of surgical-site infection in orthopaedic and spine surgery found no available evidence to support a difference between reusable or disposable gowns and drapes, and called for further research (Kieser DC *et al.*, 2018). WHO Global Guidelines for the Prevention of Surgical Site Infection recommend use of reusable or disposable gowns and drapes and acknowledge the paucity of evidence in this domain (World Health Organization, 2016).

Clinicians should be critically evaluating how to drape patients and to consider whether the quantity of drapes routinely used for operations can be reduced. Analyses of the economic and environmental benefits have already been undertaken across a range of surgical specialities (van Demark Jr RE, Smith VJ and Fiegen A, 2018; Tan E and Lim D, 2021) but again, however, there remains a paucity of evidence; no randomised controlled trials assessing surgical site infection in clean surgery with reusable and disposable drapes have been undertaken (Kieser DC *et al.*, 2018).

Surgical Masks

The terms *Surgical Mask* or *Facemask* when used in the United States, refer to Food and Drug Administration (FDA)-cleared surgical-, laser-, isolation-, dental-, and medical-procedure masks with or without a face shield. They are approved based on test results for particle filtration efficiency, bacterial filtration efficiency, fluid resistance, differential pressure, and flammability. In addition, there are also European Standards for masks; EN 14683:2019+AC:2019 (Morris H and Murray R, 2021).

The need for Filtering Facepiece Respirators (FFRs) with Surgical Mask capabilities such as fluid resistance and flammability, with the capacity to act as one form of personal protective equipment (PPE), was first addressed by the FDA in 1996 through the introduction of the Surgical N95 Respirator. In the European Union, the recommended FFR must meet the requirements of the more demanding

European standard for filtering half masks, EN 149:2001+A1:2009, wherein there is also a need to be able to withstand additional challenge by an oily substance and to demonstrate higher efficiency levels than the N95 respirator (although the difference in efficiency between the N95 and the widely used FFP2 is only slight). High filtration efficiency is essential, and to achieve this it has proved essential to incorporate an electret to attract and capture passing particles, but typically, with higher efficiency comes increased difficulty in drawing the air through the filter as the wearer breathes in, pointing to the need for great care to ensure completeness of fit to ensure that unfiltered air is not drawn in from the sides. Guidance on the use of respirators and facial protection equipment has been developed and published by the Healthcare Infection Society Working Group on Respiratory and Facial Protection (Coia JE *et al.*, 2013).

The use of surgical N95 FFRs in both surgical and non-surgical environments increases markedly during outbreaks involving a known or suspected respiratory pathogen. To combat transmission of diseases, respirators need to be able to filter out fine particles. Such particles are classified by BS EN ISO 16890 into three categories PM10 (less than 10μm), PM2.5 (less than 2.5μm), and PM1(less than 1μm). BS EN 149:2001 + A1:2009 is the British Standard to which respiratory protective devices are tested using sodium chloride and paraffin oil at 95 l/min.

With the advent of a growing number of pandemic threats, there has been interest in finding the 'best mask' to prevent infection amongst healthcare workers. The 2009 H1N1 influenza pandemic and 2020 COVID-19 pandemic saw contradictory recommendations given by leading authorities depending upon whether the authority wrote guidance based upon a risk or a hazard analysis. It would seem reasonable to suggest that, rather than pedantically abiding by just one or the other, organisations such as The Health & Safety Executive in the UK, and The Occupational Safety and Health Administration in the US, should recommend using both risk and hazard analysis.

Research to produce a 'better' mask has taken several different avenues including the fit of the mask and the style of the mask. There has been work assessing the effectiveness of finishing the fabrics with antimicrobial and antiviral compounds to improve filtration (Shen H and Leonas KK, 2005; Li Y *et al.*, 2006) and attention has also been paid to improving filtration not just by the physical properties of the fabric but by the inclusion of electrets to attract particles and aerosols and nanofibre webs (Wang N *et al.*, 2018; Suen LK *et al.*, 2020). It has been shown for example that N95 masks (which do contain an electret layer) retain their performance through storage for periods of at least ten years, which is something to be considered when planning the stocks for future pandemics (Viscusi DJ, Bergman M, *et al.*, 2009).

The limited supply of masks in the COVID-19 pandemic also saw guidance released upon reuse and decontamination of masks. There is still, however, not only a lack of research around how long a mask remains effective in actual use but also a problem with acceptance of possible treatments for the decontamination of respirators and a lack of official certification rather than emergency permission to enable their re-use after exposure to viruses. This remains the current situation despite useful baseline research conducted ten years ago by Viscusi, Bergman, and co-workers (Viscusi DJ, Bergman MS, *et al.*, 2009; Bergman MS *et al.*, 2010). Their work was undertaken following concerns in the USA by health officials about a potential shortage of masks following the H1N1 outbreak in 2009. Ten years later, when COVID-19 had been spreading for six months, it was clear that preparations for such a pandemic were seriously inadequate. Responding to the worsening situation, Stanford University established a rapidly growing group called N95DECON to create a set of best practices on the decontamination and reuse of N95 respirators (N95DECON, 2020). Several decontamination methods were proposed, some of which were based on the earlier work published by Viscusi et al. and Bergmann et al. Whilst other papers have emerged, there is still no review assessing the best method of decontamination of a facemask with consideration given to number of cycles of decontamination, environmental impact of repeated decontamination or the ease to by which decontamination could be upscaled in a future pandemic.

Healthcare Workers' Clothing

Healthcare workers' clothing requires separate attention because operating-theatre attire differs from the sterile gown. The objective of wearing an operating theatre scrub suit is to reduce contamination of the air by healthcare provider skin squames. One observational study found that single use polypropylene scrub suits reduced the median quantity of bacterial CFU/m^3 compared to two different reusable scrub suits (Kasina P *et al.*, 2016). However, there is no study assessing the impact of scrub suit on surgical site infection rate. That said, if WHO guidance on surgical site infection with reusable and disposable drapes and gowns is extrapolated, then the impact is likely to be minimal (World Health Organization, 2016).

A life-cycle assessment published in 2022 of reusable and disposable scrub suits has demonstrated that the reusable solution had a 62% lower impact on climate change compared to the disposable (Burguburu A *et al.*, 2022). This supports previous publications and would support recommending using reusable scrub suits where practical. Further work is needed to determine whether there is a more appropriate fibre that could replace the traditional polyester and cotton blend used in scrub suits currently.

There has been much attention turned towards the production of antimicrobial and antiviral fabrics, driven most recently by the COVID-19 pandemic. Properly designed trials undertaken in realistic situations are few and far between, but the trials which have been carried out show that existing antimicrobial treatments are ineffective in the practical setting. However, there will continue to be a need for fabrics able not only to act as barriers, but also to deactivate pathogens that may have ended up on the fabrics and thereby keep staff and patients safe despite the possible transfer of contamination built up during the treatment. As a consequence, robust research is required into the role that antimicrobial textiles may play in the workplace. The groundwork was established in the 'Three-Tier' protocol published by the OECD in 2008 in their Guidance Document on the Evaluation of the Efficacy of Antimicrobial Treated Articles with Claims for External Effects (OECD, 2008).

i. Tier 1 concerns 'Proof of Principle'
ii. Tier 2 'Simulated Use'
iii. Tier 3 'In-Use Evaluation'

It is at Tier 2 where current test methods currently fall seriously short. Realistic, reliable, laboratory-based Tier 2 test methods now need to be developed for medical textiles and validated for their alignment with practice before they could be considered for acceptance as suitable standards. The OECD does give guidance on the development of appropriate Tier 2 test methods in their 2019 document "Guidance Document on Use and Development of Tier-2 Laboratory Based Tests Used to Substantiate Claims for Efficacy of Biocide Treated Articles", (OECD, 2019) and this is what more researchers now need to build on to develop effective antimicrobial PPE.

References

Aibibu D, Lehmann B and Offermann P (2003) 'Barrier effect of woven fabrics used for surgical gowns', *AUTEX Research Journal*, 3(4), pp. 186–193.

Aslan S, Kaplan S and Cetin C (2013) 'An investigation about comfort and protection performances of disposable and reusable surgical gowns by objective and subjective measurements', *Journal of the Textile Institute*, 104(8), pp. 870–882.

Bergman MS *et al.* (2010) 'Evaluation of multiple (3-cycle) decontamination processing for filtering facepiece respirators', *Journal of Engineered Fibers and Fabrics*, 5(4), p. 155892501000500400.

Burguburu A *et al.* (2022) 'Comparative life cycle assessment of reusable and disposable scrub suits used in hospital operating rooms', *Cleaner Environmental Systems*, 4, p. 100068.

Campion N *et al.* (2015) 'Sustainable healthcare and environmental life-cycle impacts of disposable supplies: a focus on disposable custom packs.', *Journal of Cleaner Production*, 94, pp. 46–55.

Chen Z, Murray R and Jones RM (2007) 'Fashion supply chain organisation and management between the UK and China', *Journal of Fashion Marketing and Management: An International Journal* [Preprint].

Claxton S and Kent A (2020) 'The management of sustainable fashion design strategies: An analysis of the designer's role', *Journal of Cleaner Production*, 268(122112).

Coia JE *et al.* (2013) 'Guidance on the use of respiratory and facial protection equipment', *Journal of Hospital Infection*, 85(3), pp. 170–182.

Cotton L *et al.* (2020) 'Improved garment longevity and reduced microfibre release are important sustainability benefits of laundering in colder and quicker washing machine cycles', *Dyes and Pigments*, 177(108120).

Deguine JP, Ferron P and Russell D (2008) 'Sustainable pest management for cotton production. A review.', *Agronomy for Sustainable Development*, 28(1), pp. 113–137.

van Demark Jr RE, Smith VJ and Fiegen A (2018) 'Lean and green hand surgery', *The Journal of Hand Surgery*, 43(2), pp. 179–181.

Department for Business, E.& I.S. (2021) *Greenhouse Gas Reporting: Conversion Factors 2021*. Available at: https://www.gov.uk/government/publications/greenhouse-gas-reporting-conversion-factors-2021 (Accessed: 22 May 2022).

Department of Health (2016) Health Technical Memorandum 01-04: Decontamination of linen for health and social care. Management and provision.

Dissanayake DG and Weerasinghe D (2021) 'Towards circular economy in fashion: review of strategies, barriers and enablers', *Circular Economy and Sustainability*, pp. 1–21.

'Editorial. A Commission on climate change' (2019) *The Lancet*, 373(9676), p. 1659.

European Chemicals Agency (2012) *Understanding BPR*. Available at: https://echa.europa.eu/regulations/biocidal-products-regulation/understanding-bpr (Accessed: 2 August 2022).

European Commission (2012) Ecodesign your future: how ecodesign can help the environment by making products smarter.

European Union (2017) Regulation (EU) 2017/745 of the European Parliament and of the Council of 5 April 2017 on medical devices, amending Directive 2001/83/EC, Regulation (EC) No 178/2002 and Regulation (EC) No 1223/2009 and repealing Council Directives 90/385/EEC and 93/42/EEC (Text with EEA relevance. Available at: https://eur-lex.europa.eu/legal-content/EN/TXT/?uri=CELEX%3A32017R0745 (Accessed: 27 June 2022).

Fraser JF *et al.* (2015) 'The gown-glove interface is a source of contamination: a comparative study', *Clinical Orthopaedics and Related Research*, 473(7), pp. 2291–2297.

Gulihar A, Taub NA and Taylor GJS (2009) 'A randomised prospective comparison of Rotecno versus new Gore occlusive surgical gowns using bacterial air counts in ultraclean air', *Journal of Hospital Infection*, 73(1), pp. 54–57.

Hooper GJ *et al.* (2011) 'Does the use of laminar flow and space suits reduce early deep infection after total hip and knee replacement?', *Bone & Joint Journal*, 93(1), pp. 85–90.

Hvass KK (2016) Weaving a path from waste to value: exploring fashion industry business models and the circular economy.

Intergovernmental Panel on Climate Change (2018) *Global Warming of 1.5 °C*. Available at: https://www.ipcc.ch/sr15/ (Accessed: 4 May 2022).

JLA (2021) *Ozone Laundry Disinfection*. Available at: https://jla.com/otex/laundry/ (Accessed: 17 April 2022).

Kasina P *et al.* (2016) 'Comparison of three distinct clean air suits to decrease the bacterial load in the operating room: an observational study', *Patient Safety in Surgery*, 10(1), pp. 1–6.

Ki CW, Chong SM and Ha-Brookshire JE (2020) 'How fashion can achieve sustainable development through a circular economy and stakeholder engagement: A systematic literature review', *Corporate Social Responsibility and Environmental Management*, 27(6), pp. 2401–2424.

Kieser DC *et al.* (2018) 'Does the type of surgical drape (disposable versus non-disposable) affect the risk of subsequent surgical site infection?', *Journal of Orthopaedics*, 15(2), pp. 566–570.

Lee RJ and Mears SC (2012) 'Greening of orthopedic surgery', *Orthopedics*, 35(6), pp. e940–e944.

Leonas KK and Jinkins RS (1997) 'The relationship of selected fabric characteristics and the barrier effectiveness of surgical gown fabrics', *American Journal of Infection Control*, 25(1), pp. 16–23.

Li X *et al.* (2019) 'Recovery of glucose and polyester from textile waste by enzymatic hydrolysis.', *Waste and Biomass Valorization*, 10(12), pp. 3763–3772.

Li Y *et al.* (2006) 'Antimicrobial effect of surgical masks coated with nanoparticles', *Journal of Hospital Infection*, 62(1), pp. 58–63.

Ling C *et al.* (2019) 'Separation of waste polyester/cotton blended fabrics by phosphotungstic acid and preparation of terephthalic acid.', *Polymer Degradation and Stability*, 161, pp. 157–165.

MacNeill AJ *et al.* (2020) 'Transforming the medical device industry: Road map to a circular economy: Study examines a medical device industry transformation', *Health Affairs*, 39(12), pp. 2088–2097.

MedSalv (2021) *MedSalv*. Available at: https://www.medsalv.com (Accessed: 13 March 2022).

Morris H and Murray R (2021) *Medical Textiles*. CRC Press.

N95DECON (2020) *N95DECON*. Available at: https://www.n95decon.org (Accessed: 1 August 2022).

Napper IE, Barrett AC and Thompson RC (2020) 'The efficiency of devices intended to reduce microfibre release during clothes washing', *Science of The Total Environment*, 738(140412).

NHS England (2022) *Delivering a net zero NHS*. Available at: https://www.england.nhs.uk/greenernhs/a-net-zero-nhs/ (Accessed: 10 August 2022).

NHS England Press Release (2020) *Greener NHS campaign to tackle climate 'health emergency'*. Available at: https://www.england.nhs.uk/2020/01/greener-nhs-campaign-to-tackle-climate-health-emergency/ (Accessed: 21 December 2022).

OECD (2008) Guidance Document on the Evaluation of the Efficacy of Antimicrobial Treated Articles with Claims for External Effects. ENV/JM/MONO(2008)27. Available at: https://www.oecd.org/officialdocuments/publicdisplaydocumentpdf/?cote=env/jm/mono(2008)27&doclanguage=en (Accessed: 4 August 2022).

OECD (2019) Guidance Document on Use and Development of Tier-2 Laboratory-base Tests Used to Substantiate Claims for Efficacy of Biocide-treated Articles. ENV/JM/MONO(2018)20. Available at: https://www.oecd.org/officialdocuments/publicdisplaydocumentpdf/?cote=env/jm/mono(2018)20&doclanguage=en (Accessed: 4 August 2022).

Overcash M (2012) 'A comparison of reusable and disposable perioperative textiles: sustainability state-of-the-art', *Anaesthesia & Analgesia*, 114(5), pp. 1055–1066.

Owen L and Laird K (2020) 'The role of textiles as fomites in the healthcare environment: a review of the infection control risk', *PeerJ*, 8, p. e9790.

Ozone Industries (2023) *Compliance EUTOA*.

Palacios-Marín AV and Tausif M (2021) 'Fragmented fibre (including microplastic) pollution from textiles', *Textile Progress*, 53(3), pp. 123–182.

Parton KJ *et al.* (2020) 'Investigating the presence of microplastics in demersal sharks of the North-East Atlantic', *Nature. Scientific Reports.*, 10(1), pp. 1–11.

Practice GreenHealth (2022) *Greening the OR*. Available at: https://practicegreenhealth.org/topics/greening-operating-room/greening-or (Accessed: 27 June 2022).

Public Health England (2018) Reducing the use of natural resources in health and social care.

Riley K, Laird K and Williams J (2015) 'Washing uniforms at home: adherence to hospital policy', *Nursing Standard (2014+)*, 29(25), p. 37.

Rizan C *et al.* (2020) 'The carbon footprint of surgical operations: a systematic review', *Annals of Surgery*, 272(6), pp. 986–995.

Shen H and Leonas KK (2005) 'Study of repellent finish of filtration ability of surgical face masks', *International Nonwovens Journal*, 1558925005os-400403.

Strasser BJ and Schlich T (2020) 'A history of the medical mask and the rise of throwaway culture', *The Lancet*, 396(10243), pp. 19–20.

Suen LK *et al.* (2020) 'Comparing mask fit and usability of traditional and nanofibre N95 filtering facepiece respirators before and after nursing procedures', *Journal of Hospital Infection*, 104(3), pp. 336–343.

Sun G (2011) 'Disposable and reusable medical textiles', in *Textiles for hygiene and infection control.* Woodhead Publishing, pp. 125–135.

Tan E and Lim D (2021) 'Carbon footprint of dermatologic surgery', *Australasian Journal of Dermatology*, 62(2), pp. e170–e177.

United Nations Development Programme (2022) *Saving Lives Sustainably*. Available at: https://savinglivesustainably.org (Accessed: 26 June 2022).

Viscusi DJ, Bergman MS, *et al.* (2009) 'Evaluation of five decontamination methods for filtering facepiece respirators', *Annals of Occupational Hygiene*, 53(8), pp. 815–827.

Viscusi DJ, Bergman M, *et al.* (2009) 'Evaluation of the filtration performance of 21 N95 filtering face piece respirators after prolonged storage', *American Journal of Infection Control*, 37(5), pp. 381–386.

Vozzola E, Overcash H and Griffing E (2018) 'Environmental considerations in the selection of isolation gowns: A life cycle assessment of reusable and disposable alternatives', *American Journal of Infection Control*, 46(8), pp. 881–886.

Walton JR (1986) 'A new controversy in respiratory equipment management: reusables versus disposed disposables versus reused disposables', *Respiratory Care*, 31(3), pp. 213–217.

Wang N *et al.* (2018) 'Electret nanofibrous membrane with enhanced filtration performance and wearing comfortability for face mask', *Journal of Colloid and Interface Science*, 530, pp. 695–703.

Watts N *et al.* (2021) 'The 2020 report of the Lancet Countdown on health and climate change: responding to converging crises', *The Lancet*, 397(10269), pp. 129–170.

Wei B *et al.* (2015) 'Fabrication and property of discarded denim fabric/polypropylene composites.', *Journal of Industrial Textiles*, 44(5), pp. 798–812.

Wilson JA *et al.* (2007) 'Uniform: an evidence review of the microbiological significance of uniforms and uniform policy in the prevention and control of healthcare-associated infections', *Journal of Hospital Infection*, 66(4), pp. 301–307.

World Health Organization (2016) Global guidelines for the prevention of surgical site infection.

Yousef S *et al.* (2020) 'Sustainable green technology for recovery of cotton fibers and polyester from textile waste.', *Journal of Cleaner Production*, 254, p. 120078.

Zamani B (2014) Towards understanding sustainable textile waste management: Environmental impacts and social indicators. Sweden.

Zlaugotne B *et al.* (2022) 'Quantitative and Qualitative Assessment of Healthcare Waste and Resource Potential Assessment', *Environmental and Climate Technologies*, 26(1), pp. 64–74.

A COMPARATIVE STUDY ON PHOTO-DEGRADATION AND PHOTO-YELLOWING CHARACTER OF JUTE FABRICS TREATED WITH SELECTED UV-ABSORBERS AND ANTIOXIDANTS USING EXHAUST PROCESS (IN BLEACH-BATH) AND PAD-BATCH-DRY PROCESS

A K Samanta[1] and Deepali Singhee[2]

[1]*Department of Jute and Allied Fibre Technology, Institute of Jute Technology, University of Calcutta, Kolkata, West Bengal, India;*
[2]*J.D. Birla Institute, Affiliated to Jadavpur University, Kolkata, West Bengal, India*

Abstract

Raw bleached jute textiles have been treated with selective UV-absorbers, such as MEK, benzophenone, and benzotriazole, as well as antioxidants, such as sodium azide and lauryl-thio-di-propionate (LTDP), using both exhaust (in H_2O_2 bleach-bath) and the pad-batch-dry approach (after H_2O_2 bleaching) with an objective to minimize the fading effect of UV-light exposure and loss of tensile strength and other properties of the jute fabrics. Both raw (untreated) and treated/bleached jute fabrics (treated with UV-absorbers and antioxidants) were exposed to light in a weather-o-meter for varying time periods (0-320 h), and the resulting changes in tensile as well as rates of whiteness index reduction and yellowness index increase were assessed. According to the observed outcomes, sodium azide and LTDP render better protection, but these treatments are not wash-fast. On the other hand, treatment with 1% benzotriazole results in medium to good level of photo-yellowing resistance with acceptable wash fastness, but it is associated with some reduction in tensile strength. 2% benzotriazole applied to jute fabric using pad-batch-dry technique in acidic medium (with acetic acid) exhibits higher stabilization of whiteness retention (i.e. less extent of photo-yellowing) under UV-light exposure than other UV-absorbers/antioxidants. Though the pad-batch-dry technique renders better stabilization of whiteness, but it is associated with higher loss in tensile strength. Exhaust process of application of UV-absorbers/antioxidants in bleach-bath under alkaline medium renders less strength loss and is associated with ease of application during bleaching in a jigger in mills. Reaction mechanism was studied using another comparable model chemical (benzaldehyde in place of jute-hemicellulose-CHO) for reaction with benzotriazole followed by and UV-VIS spectrophotometric and FTIR analysis.

Keywords

Antioxidant, jute fabric, photo-degradation, photo-fading, UV-light, UV-stabilizer.

Introduction

Exposure of textile products to ultra-violet radiation is one of the most common and pervasive causes of its UV-light induced degradation. Although some earlier discrete reports are also available for jute (Mukhapadhyay, 1976; Abdullah et al., 1984; Abdullah et al., 1985), there are many scientific reports related to this in the literature also for different textile materials other than jute (Shamey et al., 2003; Egerton, 1949; Jones et al., 1998; Slater, 1991). The action of light on jute has been examined and discussed by several authors (Callow et al., 1949; Sengupta, 1967; Macmillan et al., 1954).

The extent of photo-degradation of raw and bleached jute in terms of change in surface whiteness/yellowness and tensile strength against different times of photo-fading has also been explored and reported in a similar study (Samanta and Deepali, 2022) using pad-batch-dry process in acidic medium. But when benzotriazole is used to stabilize the whiteness of H_2O_2 bleached jute fabric, the pad-batch-dry method causes a higher strength loss. As a result, a study using an alkaline bleach-bath was conducted and the results of both the processes (acidic pad-batch-dry process and alkaline bleach-bath process) are compared in the following section of the paper.

Jute is prone to photo-yellowing on exposure to UV or sunlight and this effect is more predominant in bleached jute compared to raw jute. Lignin present in bleached jute has poor photo-stability against UV-light and the lignin-based phenolic –OH in the lignin moieties gets transformed to coloured carbonyl or quinone forms due to photo-oxidation as reported by spectroscopic studies on lignin (Callow et al., 1949; Sengupta, 1967; Macmillan et al., 1954) and this is responsible for early fading or discoloration and darkening of the surface colour.

Even though the entire ultraviolet region of the solar spectrum causes bleached jute to yellow, the most significant region that causes this lies between 350-355 nm in the UV-region (Mukhapadhyay, 1976; Abdullah et al., 1984a; Callow et al., 1949; Banerjee et al., 1985). The effect also extends up to 433 nm in the visible region. Several attempts have been made to minimize/reduce this photo-yellowing impact on bleached jute (Banerjee et al., 1985; Bag et al., 1984; Farouqui et al., 1990; Achwal et al., 1994; Sanyal et al., 1995; Sanyal et al., 1995) either by partial removal of lignin from the jute fibre surface through H_2O_2 bleaching called stabilized bleaching of jute which however is a costly process and exhibits high strength loss and weight loss as well.

There are several reports on how to lower the likelihood of jute's photo-yellowing propensity or increase its photo-stability and UV-light fastness by carefully controlled and selective removal of lignin (from the fabric surface, prior to H_2O_2 bleaching) through use of several mediums like weakly acidic NaOCl solution or chlorine gas to remove some lignin from the surface of jute (Sengupta et al., 1958; A monograph, Indian Jute Industries Research Association, Kolkata, 1977), acidified $KMnO_4$ (Slater, 2001; Sanyal et al., 1995), and treatment with some chlorinated compounds. The Huetex process of stabilized bleaching of jute (A monograph, Indian Jute Industries Research Association, Kolkata, 1977) offers some degree of improvement in this direction and stabilizes the whiteness of bleached jute. However, it is associated with the drawback of high weight and tensile strength loss. Moreover, it is long sequential process also causes undesirable air pollution through emission of chlorine gas during the treatment. In contrast the modified Huetex process known as 'Chemical dehairing and light fast bleaching (CDB process)" (Chatterjee et al., 1991) partially addresses this issue and with much dehaired appearance that is free from surface fuzziness, stabilizing UV-light fading to a certain extent but not sufficient. Few discrete reports on use of singlet oxygen (Abdullah et al., 1984), nickel chelates (Abdullah et al., 1985) and UV-absorbers/UV-absorbable dyes (Banerjee et al., 1985; Farouqi et al., 1990; Achwal et al., 1994) have been reported. None of these techniques, however, provide a wash fast solution to maintain certain level of whiteness without observable strength loss and most are expensive processes. So, identifying an economic wash fast reactive UV-absorber for jute might give a solution to this problem.

A significant class of chemicals, including 2-hydroxy-benzophenone, hydoxy-benzoyl-pyridines, benzotriazole or hydroxyl-benzotriazole, and O-hydroxy-ketone derivatives (such as methyl-ethyl-ketone) are known to act as effective UV-absorbers (Banerjee et al., 1985; Achwal et al., 1994, Katrizky et al., 2000). The use of 2,4,-dihydroxy-benzophenone (Farouqui et al., 1990) and Tinuvin-320 (Achwal et al., 1994) as UV-absorbers to prevent the photo-fading of jute has been studied. Jute's photo-fading can be stabilized by applying benzotriazole using the pad-batch-dry method (Samanta and Deepali, 2022) though it causes some strength loss. In order to compare the photo-fading behaviour of raw (unbleached) and bleached jute fabrics as well as those treated with 1,2,3 benzotriazole (UV-absorber) in an alkaline medium of the bleach-bath exhaust process.

Experimental

Raw ex-loom, plain weave fine hessian 100% jute fabric of decorative variety with 195 tex jute yarns as warp and 214 tex jute yarns as weft, having 63 ends/dm and 59 pick/dm, with fabric area density 260 g/m^2 and fabric thickness 0.90 mm, obtained form M/s Gloster Ltd., Bowreah, Howrah was used in the present work.

Commercial grade (a) 1,2,3-benzotriazole ($C_6H_5N_3$); (b) benzophenone ($C_6H_5CO.C_6H_5$); (c) sodium azide (NaN_3); (d) methyl-ethyl-ketone ($CH_3.CO.C_2H_5$); (e) lauryl-thio-di-propionate i.e LTDP ($C_{11}H_{23}.CS.O.CH_2.CH_2.CH_2$); (f) acetic acid and (g) sodium acetate obtained from E.Merck, India were used.

Conventional hydrogen peroxide bleaching (ordinary bleaching process)

Untreated (control ex-loom raw jute fabric) or suitably desized/scoured jute fabric were subjected to conventional hot H_2O_2 bleaching (Samanta et al., 2004) using 3% (owf) H_2O_2 (i.e 6% H_2O_2 of 50% vol strength), 8% (owf) sodium metasilicate, 0.7% (owf) NaOH, 0.05% (owf) EDTA, 0.5% (owf) non-ionic wetting agent (Lissapol-D) at 85°C to 90°C for 2 h at pH 10.5-11.0. In each of the above cases, the pH of the bleach-bath was maintained by adding small amounts of 1% aqueous solution of NaOH, drop wise, as and when the pH level went down. After bleaching, each fabric was washed thoroughly under running water, followed by a wash in 3 gpl (w/v) of acetic acid (30%) solution for neutralization, further followed by hot and cold water wash. Finally, the washed fabric samples were dried in air.

Application of UV-absorber in H_2O_2 bleach-bath

Jute fabric samples were bleached using 3% H_2O_2 following conventional method in exhaust-bath process (as mentioned above) at alkaline pH (10.5-11.0) with simultaneous addition of varying concentrations (1-4%) of selected UV-absorbers (benzophenone and benztriazole) and antioxidants (sodium azide, lauryl-thio-di-propionate and methyl-ethyl-ketone) and further varying concentrations (0.1%, 0.5%, 1%, 2%, 3% and 4%, owf) of benzotriazole in the H_2O_2 bleach-bath. The different UV-absorbers and antioxidants were pre-dissolved in hot water or in organic solvent or as an emulsion form as required which are mentioned below.

- Sodium azide: in water at room temperature
- 1,2,3 Benztriazole: in hot water at 60°C
- BenzophenoneL in butanol (vigorous stirring)
- Laurylthiodi-propionate (LTDP): in warm (40°C) propane-2ol
- Methyl-ethyl-ketone (MEK): in emulsion of 50% water and 50% butanol with 2% emulsifier

After bleaching, the fabric was rinsed and washed under running water followed by neutralization treatment with 1% acetic acid (30%) and then further washed with water. Finally, the washed fabric samples were finally squeezed and dried in air.

The chemical structure/formula of 1,2,3 benzotriazole (UV-Absorber) used is shown in Figure 1. Although benzotriazole solution is naturally acidic, it is made alkaline for application on bleached jute fabric by adding a small amount of NaOH drop by drop. This raises the pH to 10.5-11.0, which creates an active species of benzotriazole that reacts with the jute-aldehyde present in jute hemicellulose.

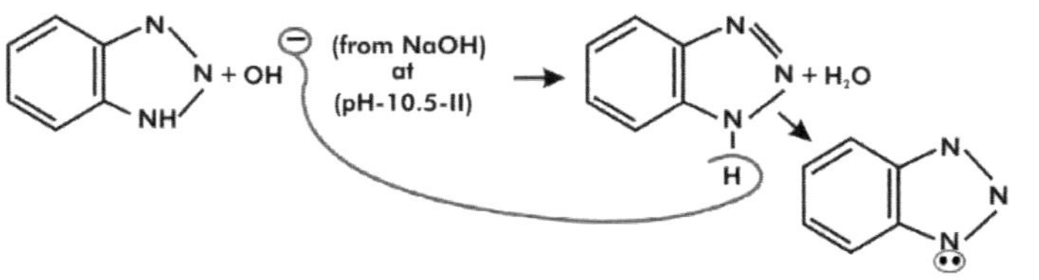

Figure 1. *Chemical structure of the 1,2,3 Benzotriazole (UV-Absorber) and its active species in alkaline medium*

Method of washing to test wash stability

Jute fabric samples treated with selected dosages of the selective UV-absorber (i.e benzotriazole) in the H_2O_2 bleach-bath by the exhaust process were subjected to different number of wash cycles (1, 5 and 10 wash cycles) following the ISO-II method (ASTM D 435-42) using 5 gpl solution of non-ionic detergent, MLR of 1:50 at 50 ±2**°C** for 45 min in a launder-o-meter with beakers rotating at 40 rpm. The treated samples were subsequently washed, dried and exposed to photo-exposure under in a weather-o-meter for different time periods, to assess their photo-stability/whiteness stability.

Testing

Weight loss or gain: Weight loss or gain in the modified jute fabrics after the selective treatments were determined by the standard gravimetric procedure (ASTM 1974) by measuring the oven/bone dry weight of the jute before and after the treatments and expressing the results as a percentage of the initial bone dry weight of untreated jute taken.

Breaking strength and breaking extension: The breaking tenacity (cN/tex) and breaking extension (%) in both the warp and weft directions of selective fabric samples were determined by the ravelled strip method (ISI 1982) using an Instron (model-1445) CRE universal tensile tester with a traverse speed of 100 mm/min and a pretension of 0.5 N.

Determination of –COOH and –CHO group content: Following the standard procedures (ISO-1560, 1974; Braidy and Nevell, 1963), changes in copper number (Braidy and Nevell, 1963) as a measure of the –CHO group concentration) for treated and untreated jute fibres were evaluated.

Determination of lignin content: The TAPPI standard method (T 222-05-74, 1974) was used to determine the lignin content of the treated and untreated jute cloth samples.

Determination of alpha-cellulose and hemicelluloses: The TAPPI standard method (Doree, 1947; T 222-05-74, 1974) was used to determine the alpha-cellulose and hemicellulose content of the treated and untreated jute fabric samples.

Surface colour strength and other colour parameters: Surface colour strength in terms of K/S values (Shah and Gandhi, 1990) and total colour difference (ΔE) using the CIE-Lab equations (Shah and Gandhi, 1990) of raw and differently treated jute fabric samples was assessed at λ_{max} using a Macbeth 2020-plus reflectance spectrophotometer and the corresponding Colour-Lab plus software.

Measurement of whiteness, yellowness and brightness indices: Whiteness index (Hunter Lab-Scale) (Shah and Gandhi, 1990), yellowness index (ASTM 1974) and brightness index (ISO 1977) of the selected fabric samples were evaluated from the measured reflectance value and corresponding tri-stimulus values of the samples, which were pre-determined using a computer aided Macbeth 2020 plus reflectance spectrophotometer and associated Colour-Lab Plus software using D_{65} standard illuminant and 10^0 standard observer setting using the following relationships:

Whiteness Index (WI) (Hunter Lab-Scale)	$= L - 3b = 10\sqrt{Y} - [21(Y - Z\ \%)] / \sqrt{Y}$
Yellowness Index (YI) (ASTM-E313/1973)	$= 100\,[(1 - (B/G)] = 100 \left[1 - \dfrac{0.8477}{Y}\right]$
Brightness Index (BI) [ISO-2470-(1977)]	$\dfrac{\text{Reflectance value of the substrate}}{\text{Reflectance value of the standard white diffuser / white tile at 457 nm}} \times 100$

Where, X, Y and Z are the CIE tri-stimulus values of the sample, L^* is the lightness/darkness indicator, b^* is the blueness/yellowness indicator and $B = Z/1.181 = 0.847\ Z$, $G = Y = L^2/100$.

Test for photo-fading behaviour: The photo-fading behaviour of the raw, bleached and UV-absorber (benzotriazole) treated jute fabric samples were evaluated as per IS: 2454-1967 method (ISI 1989) like light fastness testing where the fabric samples (10 cm × 1 cm) were exposed to standard UV-light in a

Q-Lab, USA-make Q-Sun model weather-o-meter along with the eight blue wool standards (BS 1006: BOI: 1978) for different hours of UV-light/photo-exposure.

By contrasting the observed fading of the fabric specimen with the fading of the respective (1-8) blue wool standards after exposure to UV-light in a weather-o-meter for different hours, the rate of reduction/retention in whiteness of bleached jute and benzotriazole treated jute on photo-exposure, and the corresponding light fastness rating (appearance of first fading of samples corresponding to 1 to 8 scale of the blue wool standards) were assessed. The raw, bleached, and benzotriazole-treated jute specimen's light fastness grade was recorded and matched with standard number of the blue wool standards that displayed a comparable rating of photo-fading/colour changes for the same number of fading hours.

Assessment of photo-yellowing rate/photo-stability character: Selected raw-untreated and bleached/treated jute fabric samples mounted on an appropriate sample holder were placed inside USA-make Q-Sun model weather-o-meter for exposure to acetate accelerated UV-light (xenon arc) for different time periods (20, 40, 80, 160 & 320 h). One half portion of each fabric sample was appropriately covered with a piece of black paper (to provide a control section for visual as well as instrumental assessment of extent of photo-fading) before placing the fabric samples in the weather-o-meter. Hunter lab scale whiteness index, ASTM E-313-1973 (1974) yellowness index and ISO-2470 (1977) brightness index values for the photo-exposed part of the samples were measured after different durations of UV-light exposure for each sample.

Average rate of reduction in whiteness index (RRWI) per hour of UV-light exposure: The average rate of reduction in whiteness index per hour of UV-light exposure for each sample was calculated using the following relationships (Samanta et al., 2004) described earlier:

RRWI: Average Rate of Reduction in Whiteness Index (RRWI) per hour of UV-light exposure =

$$10^3 \times \left[\text{Initial whiteness index of the unexposed sample} - \text{Whiteness index of the photo-exposed sample after specified hour of photo-exposure} \right] \div \text{Total hours of photo-exposure}$$

The factor of 10^3 is multiplied in the above relationship, in order to avoid small fractional data, which is difficult to handle for the purpose of comparing.

a. *Average rate of increase in yellowness index (RIYI) per hour of UV-light exposure:* The average rate of increase in yellowness index per hour of UV-light exposure was calculated using the following relationships:

Average rate of increase in yellowness index (RIYI) per hour of UV-light exposure =

$$10^3 \times \left[\text{Initial yellowness index of the unexposed sample} - \text{Yellowness index of the photo-exposed sample after specified hour of photo-exposure} \right] \div \text{Total hours of photo-exposure}$$

The factor of 10^3 is multiplied in the above relationship, in order to avoid fractional data.

b. *Combined effect of both whiteness and strength retention by an index, Whiteness and Strength Retention Index (WSRI):* The combined effect of whiteness and strength retention was measured by an index, WSRI for a fixed hour of photo-fading of a sample. This indicates the effect of overall photo-stability/photo-degradation behavior, which can be assessed at a glance by the following empirical equation as an indicator:

$$\text{WSRI} = \frac{\text{Tensile strength value of the sample after photo-fading for fixed hour}}{\text{Initial tensile strength value before photo-fading}} \times \frac{\text{Whiteness index value of the sample after photo-fading for a pre-fixed hour of photo-exposure}}{\text{Initial whiteness Index value of a bleached and unexposed sample (before photo-fading)}}$$

UV-VIS spectral analysis: UV-VIS spectra of (a) 1% aqueous solution of benzaldehyde (as a model compound) at pH of 10.5-11.0 with addition of NaOH, (b) 1% aqueous solution of benzotriazole (as an UV-absorber) at pH of 10.5-11.0 with addition of NaOH and (c) mixture of (a) and (b) heated at boil (100°C) for 30 min at alkaline pH (10.5-11.0), were obtained using Hitachi-U-200 double beam UV-VIS spectroscope to study the said reaction mechanism. Benzaldehyde was used as a model compound instead of jute-hemicellulose based aldehyde, taking into consideration that if this benztriazole can react with benzaldehyde, then it can also react with jute-hemicellulose and aldehyde.

Fourier transform infrared spectroscopy (FTIR): The chemical composition of selected jute samples was assessed by examining 3 mg of finely crushed fibre samples taken out from raw (untreated) and treated jute fabrics after bleaching followed by washing with distilled water and methanol in sequence on a double beam Spectrophotometer (Perkin Elmer Pvt. Ltd., Singapore make SPECTRUM-II model) using KBr disc technique (O'Connor et al., 1957).

Assessment of ultra-violet protective factor (UPF): The Ultraviolet protective factor that is expressed as the ratio of extent of time required for the detectable browning of the treated and untreated jute fibre with and without protection of the UV-absorber under continuous exposure to radiant energy was calculated using the following equation (Das, 2010):

$$\text{UPF} = \frac{\text{MED treated fabric}}{\text{MED treated fabric}}$$

Where, MED is the minimal erythemal dose or quantity of radiant energy needed to produce the first detectable browning of fibre after 22 ± 2 hours of continuous exposure.

Results and Discussions

Effect of photo-exposure (UV-light) on tensile and surface appearance properties of raw (untreated) and bleached jute fabrics

Table 1 shows the changes in whiteness and yellowness indices, average rate of reduction in the whiteness index (per hour of photo-exposure), average rate of increase in yellowness index per hour of photo-exposure as well as changes in warp-way and weft-way tenacities, warp-way and weft-way breaking extensions for different hours of accelerated photo-exposure to UV-light in a standard weather-o-meter for raw (untreated) and bleached jute.

In general, the breaking tenacity of jute decreases in both warp and weft directions when exposed to UV-light due to photo-oxidation of polymer chains of jute constituents. This loss is approximately 15% for the first 80 hours of accelerated UV-fading and is approximately 33.5% after 320 hours of fading. There is also a gradual reduction in both warp and weft breaking extension for longer hours of photo-exposure due to possible chain scission in jute celluloses and hemi-celluloses. One of the primary cause of the reduction in both breaking strength and breaking extension in jute fabric (fibre and yarn assembly) for various hours of photo-exposure may be the more photo-oxidative chain scission of long polymeric chains of major jute constituents which leads to loss of long fibre entity. There is a gradual reduction in whiteness index and a consecutive increase in the yellowness index with increase in time of photo-exposure. These changes mostly become more pronounced after 80 hours of photo-fading. As a result of photo-oxidation of phenolic –OH group of lignin component of jute forming brownish quinone structure (Callow and Speakman, 1949; Macmillan and Bhattacharya, 1954; Sengupta 1967), the whiteness index of jute fabric gradually reduces and corresponding yellowness index and surface colour strength gradually increases for different hours of photo-exposure under UV-light.

Comparatively conventionally H_2O_2 bleached jute cloth fades and degrades more quickly than untreated (control) jute fabric when exposed to UV-radiation, as seen by some loss in tensile strength. The larger amount of voids, pores and cracks that were created during oxidative bleaching may be the cause of the faster rate of photo-oxidation/photo-fading of bleached jute when exposed to UV-radiation. Additionally, compared to raw (control) jute, bleached jute contains greater level of aldehyde and carboxylic acid group functionality, which fosters a favourable environment for a considerably higher order of oxidative chain scission during photo-fading under UV-light. Further, reduction in whiteness index (hunter lab scale) and increase in yellowness index in bleached jute is higher than in raw jute sample.

The effect of photo-exposure is less aggressive in case of raw jute fabric due to preferential absorption of UV-light by the coloured matter present in raw jute causing less action of UV-light on lignin, a major jute constituent. While the increased vulnerability and proneness of H_2O_2 bleached jute fabric to photo-fading and photo-degradation under to UV-light/sunlight exposure due to the absence of colouring matter and openness of the fibre structure which induces higher conversion rate of the phenolic –OH content of jute lignin to coloured quinone structure.

The strength loss of raw and bleached jute on UV-light exposure is due to chain session of jute cellulose, jute hemicelluloses and lignin by photo-oxidation reaction. Conventionally H_2O_2 bleached jute fabric is more susceptible to textile strength loss due to its more open structure that causes greater oxidative degradation of jute cellulose/hemicelluloses and lignin chains along with oxidation of phenolic –OH of lignin to quinone structure resulting in accelerating photo-fading compared to raw jute. As a result of the less availability of more accessible lignin molecules for photo-oxidation after the more accessible lignin molecules have already been photo-oxidized within shorter period of photo-exposure, the severity of action on rendering whiteness and increasing yellowness is less for longer period of UV-light exposure.

Table 1. *Effect of UV-light exposure on raw and H_2O_2 bleached jute fabric.*

Sl. No.	Properties	Raw (Untreated Jute)	Conventionally H_2O_2 Bleached Jute
1	Retention of warp-way and weft-way tenacity (cN/tex) after 320 hours of photo-fading	56-60%	65-68%
2	Reduction in breaking extension (%) after 320 hours of photo-fading	42-48%	53-61%
3	Reduction in whiteness index (Hunter Lab-scale) after 320 hours of photo-fading	10.9 from 48.5	16.4 from 70.2
4	Increase in yellowness index (ASTM E-313)after 320 hours of photo-fading	65.70 from 40.00	58.40 from 22.40
5	Average rate of reduction in whiteness index per hour (RRWI)	Increasing trend with increase in time periods of photo-exposure	Much higher for lower time of photo-exposure; it reduces with increase in time of photo-exposure
6	Average rate of increase in yellowness index per hour (RIYI)	Increasing trend with increase in time periods of photo-exposure	Much higher for lower time of photo-exposure; it reduces with increase in time of photo-exposure

Hence, it may be possible to partially prevent or delay the photo-degradative action of jute fibre (jute celluloses, hemicelluloses, and lignin as the major jute constituents) by treating it with various UV-absorbers and antioxidants, which preferentially absorb UV-light than its action on jute constituents including lignin.

Therefore, by treating jute fibre with various UV-absorbers and antioxidants, which preferentially absorb UV-light rather preventing its action on jute constituents including lignin, it may be possible to partially prevent or delay the photo-degradative action of jute fibre (with celluloses, hemicelluloses, and lignin as the major jute constituents).

Effect of treatment with different types of UV-absorbers and antioxidants in H_2O_2 bleach-bath on whiteness retention and yellowing propensity of on raw (untreated) jute fabric

Table 2 shows the data for change in whiteness and yellowness indices for different hours of exposure (20 h to 160 h) of the corresponding jute samples after application of different concentration (1-4%) of each of sodium azide, methyl-ethyl-ketone, lauryl-thio-di-propionate, benzophenone and benzotriazole as UV-absorbers/antioxidants in the bleach-bath and subsequently exposed to UV-light in a standard weather-o-meter. Corresponding data indicates that amongst the five types of UV-absorbers and antioxidants used, sodium azide, methyl-ethyl-ketone, lauryl-thio-di-propionate show the same level of performance, while benzophenone and benzotraizole offer better retention of the whiteness index probably due to their higher ability of preferential UV-absorption than others. There is however no significant difference in the corresponding increase in yellowness index after different hours of photo-exposure for these five UV-absorbers and antioxidants when applied as UV-absorbers/antioxidants in the bleach-bath.

Table 2. Surface appearance properties of raw (untreated) jute fabric treated with different concentrations of different UV-absorbers/antioxidants by exhaust process in H_2O_2 bleach-bath after different hours of photo-exposure.

Expt. No.	Treatments	% Application (owf)	Whiteness Index (Hunter Lab-scale)					Yellowness Index (ASTM E-313)				
			0 h	20 h	40 h	80 h	160h	0 h	20 h	40 h	80 h	160h
1	**Nil (ordinary H_2O_2 bleached jute fabrics)**	Nil	70.2	64.5	58.2	57.5	38.1	22.4	26.1	28.2	32.6	45.0
2	**H_2O_2 bleached jute fabrics treated with sodium azide in H_2O_2 bleach-bath ***	1%	67.6	55.9	55.1	53.0	44.0	25.5	28.6	29.1	31.1	39.2
		2%	67.0	58.4	56.3	51.2	44.1	25.6	29.1	29.9	32.2	42.0
		3%	67.3	58.2	57.4	50.7	44.5	25.4	28.6	29.0	32.9	48.7
		4%	67.1	58.1	57.2	51.6	44.9	25.4	27.7	27.0	30.0	42.5
3	**H_2O_2 bleached jute fabrics treated with methyl-ethyl -ketone (MEK) in H_2O_2 bleach-bath ***	1%	61.1	54.3	55.8	54.8	43.5	26.0	27.0	28.5	27.3	38.8
		2%	63.1	56.2	56.4	53.7	42.8	25.5	29.7	30.0	31.2	37.6
		3%	64.5	55.4	56.2	52.3	43.6	25.0	28.7	29.3	28.9	36.6
		4%	65.9	56.1	56.4	52.0	44.2	25.2	28.8	28.2	28.5	35.6
4	**H_2O_2 bleached jute fabrics treated with lauryl-thio-di-propionate (LTDP) in H_2O_2 bleach-bath ***	1%	62.2	58.7	58.1	52.3	46.2	27.5	28.3	28.3	29.0	36.3
		2%	63.5	59.3	57.8	52.0	45.6	27.3	29.1	28.0	30.4	37.6
		3%	64.0	59.0	58.0	51.6	44.2	27.3	28.9	28.6	29.2	36.0
		4%	65.9	59.7	57.2	50.2	44.6	26.5	27.1	26.5	28.1	32.5
5	**H_2O_2 bleached jute fabrics treated with bezophenone in H_2O_2 bleach-bath ***	1%	66.8	59.1	58.8	55.6	50.2	30.7	32.6	34.7	36.5	38.0
		2%	66.4	58.1	57.2	53.0	46.2	27.1	27.0	28.9	30.2	37.1
		3%	66.7	58.6	56.4	52.0	46.1	27.6	28.0	28.3	29.3	35.4
		4%	66.1	58.3	58.29	51.3	46.0	26.6	27.5	28.4	28.5	32.1
6	**H_2O_2 bleached jute fabrics treated with benzotriazole in H_2O_2 bleach-bath ***	1%	68.91	57.8	54.9	45.2	44.0	29.0	26.6	27.6	29.0	33.9
		2%	68.04	67.2	674.1	56.0	53.1	32.7	31.1	32.2	33.3	34.4
		3%	68.15	67.6	66.1	56.0	50.8	29.7	28.8	28.0	29.2	33.1
		4%	67.20	66.8	65.3	54.3	48.4	29.6	28.6	28.3	29.3	32.1

*** H_2O_2 bleach-bath: 6% H_2O_2 owf , 8% owf sodium metasilicate, 0.7% owf NaOH, 0.5% owf non-ionic detergent, 0.05% owf EDTA, MLR 1:20, p^H 11, for 2 h, at 80- 85°C.**

However amongst these two, benzopheneone and benzotriazole, 2-4% benzotriazole treatments show somewhat better whiteness retention and less increase in yellowness. Hence, the study has further been carried out using only benzotriazole as promising UV-absorber for jute by conventional exhaust process of bleach-bath technique.

Table 3. Surface appearance and tensile properties of raw (untreated) jute fabric treated with different concentrations of benzotriazole (as UV-absorber) by exhaust process in H_2O_2 bleach-bath after different hours of photo-exposure.

Treatments	Properties	Duration of Exposure to UV-Light					
		Zero h	20 h	40 h	80 h	160 h	320 h
Raw untreated jute fabric treated with benzotriazole (0.1%) in H_2O_2 bleach-bath *	*K/S* (at 420 nm)	0.9	2.2	2.2	4.2	6.2	5.7
	dE	--	4.2	5.9	6.7	7.6	8.3
	Whiteness Index (Hunter Lab-scale)	65.4	54.9	54.4	46.7	40.3	43.1
	RRWI	--	52.1	27.4	23.4	15.7	7.0
	Yellowness Index (ASTM E-313)	27.5	45.3	50.0	55.7	61.3	61.0
	RIYI	--	88.9	56.2	35.2	21.1	10.5
	Warp-way Breaking Tenacity (cN/tex)	6.6	6.0	5.5	4.7	4.1	3.8
	Weft-way Breaking Tenacity (cN/tex)	5.3	4.8	4.6	4.1	3.5	3.0
	WSRI (warp-way)	--	0.8	0.7	0.5	0.4	0.4
	WSRI (weft-way)	--	0.8	0.7	0.6	0.4	0.4
	Warp-way Breaking Extension (%)	7.7	7.1	6.7	6.3	6.1	5.4
	Weft-way Breaking Extension (%)	7.8	7.3	7.0	6.5	6.0	5.7
Raw untreated jute fabric treated with benzotriazole (0.5%) in H_2O_2 bleach-bath *	*K/S* (at 420 nm)	0.7	1.8	2.7	4.3	5.1	5.6
	dE	--	3.6	5.2	6.0	7.0	7.8
	Whiteness Index (Hunter Lab-scale)	67.4	57.7	54.0	46.2	40.4	39.0
	RRWI	--	48.5	33.5	26.5	16.8	8.9
	Yellowness Index (ASTM E-313)	25.2	42.5	50.0	56.6	59.8	63.2
	RIYI	--	86.6	62.1	39.2	21.7	11.9
	Warp-way Breaking Tenacity (cN/tex)	6.7	6.2	5.9	5.4	4.7	4.2
	Weft-way Breaking Tenacity (cN/tex)	5.4	5.0	4.6	4.2	3.7	3.1
	WSRI (warp-way)	--	0.8	0.7	0.6	0.4	0.4
	WSRI (weft-way)	--	0.8	0.7	0.5	0.4	0.3
	Warp-way Breaking Extension (%)	7.3	7.0	6.6	6.0	5.5	5.1
	Weft-way Breaking Extension (%)	7.6	7.2	6.9	6.4	5.9	5.3
Raw untreated jute fabric treated with benzotriazole (1%) in H_2O_2 bleach-bath *	*K/S* (at 420 nm)	0.7	2.0	2.6	4.7	5.8	5.5
	dE	--	3.5	4.7	5.7	5.9	6.4
	Whiteness Index (Hunter Lab-scale)	68.9	57.6	54.3	45.2	41.6	39.8
	RRWI	--	56.8	36.6	29.7	17.1	9.1
	Yellowness Index (ASTM E-313)	25.1	44.6	49.6	58.0	60.7	63.4
	RIYI	--	97.7	61.4	41.1	22.3	12.0
	Warp-way Breaking Tenacity (cN/tex)	6.8	6.4	6.1	5.8	5.2	4.6
	Weft-way Breaking Tenacity (cN/tex)	5.6	5.1	4.8	4.5	4.1	3.6
	WSRI (warp-way)	--	0.8	0.7	0.6	0.5	0.4
	WSRI (weft-way)	--	0.8	0.7	0.5	0.5	0.4
	Warp-way Breaking Extension (%)	7.2	7.0	6.9	6.4	5.9	5.3
	Weft-way Breaking Extension (%)	7.5	7.2	6.9	6.6	6.1	5.7
Raw untreated jute fabric treated with benzotriazole (2%) in H_2O_2 bleach-bath *	*K/S* (at 420 nm)	0.7	1.8	2.4	3.8	4.3	5.2
	dE	--	3.6	3.9	4.5	5.1	5.9
	Whiteness Index (Hunter Lab-scale)	68.0	67.9	67.1	66.0	65.0	48.4
	RRWI	--	0.6	2.5	2.5	1.9	6.1
	Yellowness Index (ASTM E-313)	21.3	22.9	23.8	25.2	28.8	38.4
	RIYI	--	8.0	6.2	4.9	4.7	5.3
	Warp-way Breaking Tenacity (cN/tex)	7.0	6.7	6.3	6.1	5.6	4.9
	Weft-way Breaking Tenacity (cN/tex)	5.7	5.3	4.9	4.6	4.4	3.9
	WSRI (warp-way)	--	1.0	0.9	0.9	0.8	0.5
	WSRI (weft-way)	--	1.0	0.9	0.8	0.7	0.5
	Warp-way Breaking Extension (%)	7.4	7.1	6.9	6.5	6.2	5.7
	Weft-way Breaking Extension (%)	7.8	7.3	7.0	6.8	6.4	5.9

* H_2O_2 bleach-bath: 6% H_2O_2 owf , 8% owf sodium metasilicate, 0.7% owf NaOH, 0.5% owf non-ionic detergen7.30t, 0.05% owf EDTA, MLR 1:20, pH 11, for 2 h, at 80 – 85°C.

RRWI: Average rate of reduction in whiteness index; RIYI: Average Rate of increase in Yellowness Index; WSRI: Whiteness and Strength Retention Index

Descending order of properties of raw untreated jute fabric treated with varying amounts (1-2%) of benzotriazole (1%) in H_2O_2 bleach-bath (alkaline medium) after 320 hours of photo exposure under UV-light

RRWI: **9.1** (for 1%) **>** **8.9** (for 0.5%) **>** **7.0** (for 0.2%) **>** **6.1** (for 2%)

RIYI: **12.0** (for 1%) **>** **11.9** (for 0.5%) **>** **10.5** (for 0.2%) **>** **5.3** (for 2%)

WSRI (Warp-way): **0.4** (for 1%) **>** **0.4** (for 0.5%) **>** **0.4** (for 0.2%) **>** **0.5** (for 2%)

WSRI (Warp-way): **0.4** (for 1%) **>** **0.3** (for 0.5%) **>** **0.4** (for 0.2%) **>** **0.5** (for 2%)

The above analysis indicate that both the average rate of reduction in whiteness index (RRWI) and the average rate of increase in yellowness index (RIYI) is lowest for 2% application of benzotrizole.

Test of wash-stability of 2% benzotriazole application

The wash-stability of jute fabric treated with 2% (owf) benzotriazole by the exhaust technique in bleach-bath has also been studied and the results obtained are shown in Table 4. Wash stability of application of 2% benzotriazole treated jute fabric renders whiteness stability at the nearly the same level, or sometimes at marginally reduced level when the fabric is subsequently subjected to 1, 5 and 10 wash cycles (under the ISO-II conditions of washing, i.e. 5 gpl non-ionic soap solution, 1:50 MLR, at 50 ±2°C for 45 min.) thereby indicating reasonably good wash resistance. This indicates that benzotriazole possibly causes some reactive interaction with jute which offers good wash stability. Hence, before proceeding further, it is considered appropriate to study the possible reaction mechanism or interactions between benzotriazole and jute and the same is discussed below.

Table 4. Surface appearance properties of H2O2 bleached jute fabric treated with benzotriazole (2%) in bleach-bath and subsequently washed (IOS-II) under different number of wash cycles.

Treatment	No. of Wash Cycles	Properties	Duration of Exposure to UV-Light					
			Zero h	20 h	40 h	80 h	160 h	320 h
Raw untreated jute fabric treated with benzotriazole (2%) in H_2O_2 bleach-bath ** and subsequently washed once (1 wash cycle) using ISO-II method *	1	*K/S* (at 420 nm)	0.6	2.0	3.2	4.1	4.8	5.1
		dE	--	1.3	2.4	3.6	4.0	4.5
		Whiteness Index (Hunter Lab-scale)	71.2	66.7	62.8	54.6	50.8	45.8
		RWRI	--	22.5	21.0	20.8	12.8	8.0
		Yellowness Index (ASTM E-313)	22.0	25.4	34.7	42.3	50.3	55.5
		RIYI	--	10.8	31.6	25.2	17.7	10.5
Raw untreated jute fabric treated with benzotriazole (2%) in H_2O_2 bleach-bath ** and subsequently washed 5 times (5 wash cycles) using ISO-II method *	5	*K/S* (at 420 nm)	0.5	1.8	2.2	3.5	4.1	5.4
		dE	--	1.4	3.0	4.1	4.5	4.8
		Whiteness Index (Hunter Lab-scale)	73.1	67.7	60.8	59.1	50.1	44.9
		RWRI	--	26.9	30.8	17.6	14.4	8.8
		Yellowness Index (ASTM E-313)	21.8	24.1	33.0	41.9	50.5	57.5
		RIYI	--	12.1	28.0	25.2	18.0	11.2
Raw untreated jute fabric treated with benzotriazole (2%) in H_2O_2 bleach-bath ** and subsequently washed 10 times (10 wash cycles) using ISO-II method *	10	*K/S* (at 420 nm)	0.4	1.0	2.1	3.2	4.4	5.6
		dE	--	1.9	3.0	4.1	4.7	5.0
		Whiteness Index (Hunter Lab-scale)	74.3	68.5	61.6	56.6	51.8	44.1
		RWRI	--	29.2	31.7	22.1	14.1	9.4
		Yellowness Index (ASTM E-313)	20.4	26.3	33.5	42.3	51.2	58.4
		RIYI	--	29.6	32.8	27.4	19.3	11.9

* ISO II Method: 5 gpl soap, MLR of 1:50, at temperature of 50 ± 2°C C for 45 min, with rotation of beakers at 40 ± 2 rev/min.
* * For jute fabric: H_2O_2 bleach-bath: 6% H_2O_2 owf, 8% owf sodium metasilicate, 0.7% owf NaOH, 0.5% owf non-ionic detergent, 0.05% owf EDTA, MLR 1:20, pH 11, for 2 h, at 80-85°C.

Study of the reaction mechanism between benzotriazole and jute

It has been reported in the literature (Katrizky et al. 2000) that benzotriazole may react with aliphatic aldehydes to give isodisable adducts which are usually crystalline in nature in both acidic media and also in alkaline media. The isomerization of benzotriazole to amino-diazo-ol structure in acid medium has been reported earlier (Samanta and Singhee, 2022). The possible reaction mechanism between the small amount of –CHO group (an aromatic aldehyde with large nucleophilicity character) present in jute hemicellulose and benzotriazole under alkaline medium (addition of NaOH) is shown in Reaction Scheme A, Figure 2. The reaction shows the rearrangement reaction forming amino-diazo-ol structure under alkaline medium.

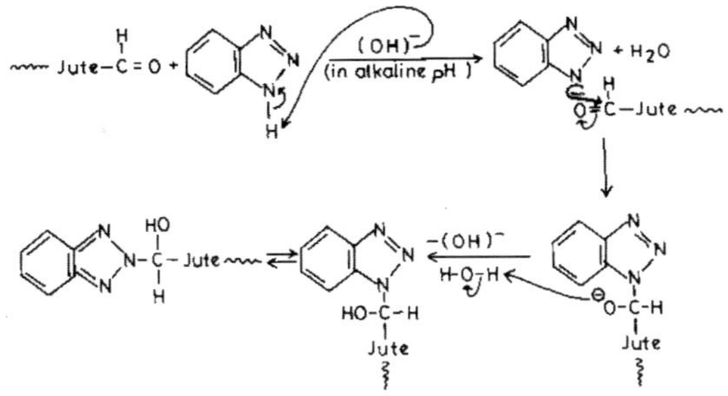

Figure 2. Reaction Scheme A between –CHO group of jute hemi-cellulose and benzotrizole under alkaline medium in bleach-bath

Though such an attempt r is said to be effective with benzilydene (benzal) chloride, it depends on the nucleophilicity of aromatic aldehydes, according to conflicting reports on the creation of such adducts with benzaldehyde (an aromatic aldehyde) (Katrizky et al., 2000). Since jute hemicellulose contains aromatic aldehydes with high nucleophilicity, it is therefore impossible to rule out the potential of a reaction between these groups, leading to the suggestion of Reaction Scheme A (Figure 2).

To check the possibility of such reaction in practice, a model compound like benzaldehyde has been taken as aldehyde model compound instead of jute hemi-cellulose. 1% solution of aqueous benzotriazole was heated (boiled for 1 h) separately with 1% aqueous solution of benzaldehyde under pH 10.5-11.0 to examine the reactive interaction of benzaldehyde (as a model compound) with benzotriazole and the three solutions (a) only benzaldehyde alkaline solution, (b) only benzotriazole alkaline solution and (c) mixture of (a) and (b) boiled and cooled in alkaline medium examined. Figure 3 show the UV-VIS scans of the (a), (b), and (c) solutions, all in an alkaline medium (to simulate an H_2O_2 bleach-bath). The UV-VIS-Spectra were analyzed to verify the possible reaction/changes taking place and whether there is any formation of adduct between benzaldehyde (as a model compound instead of hemicellulose) and benzotriazole in an alkaline medium (at pH 10.5-11.0).

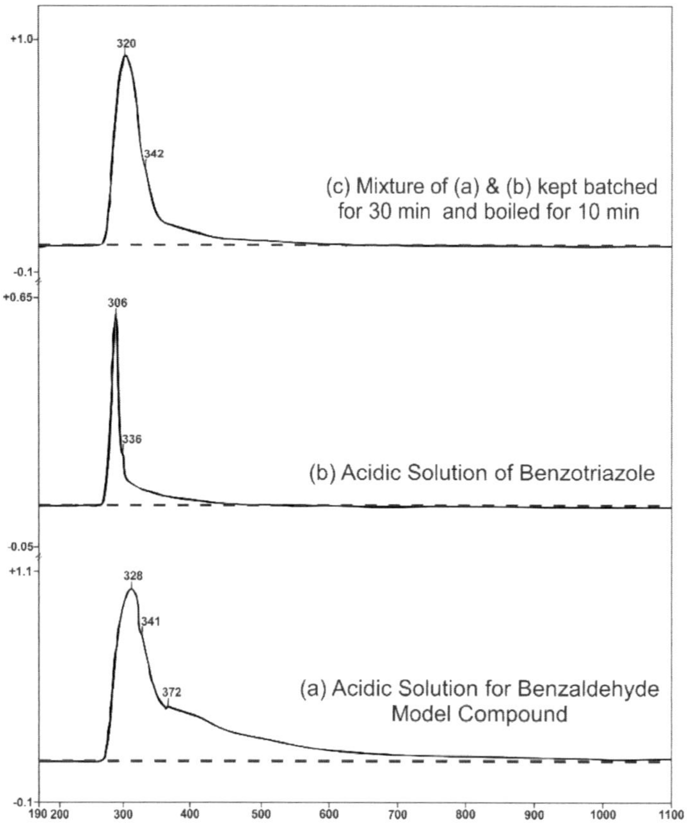

Figure 3. *UV-VIS Spectra of (a) 1% alkaline solution of benzaldehyde at pH 10.5-11.0, (b) new 1% alkaline solution of benzotriazole at pH 10.5-11.0 with addition of sodium hydroxide and (c) mixture of (a) & (b) kept batched for 30 min and boiled at 100°C for 30 min, and cooled at pH 10.5-11.0 (alkaline)*

In Figure 3, the UV-VIS Spectrum (a) for pure 1% aqueous solution of benzaldehyde shows a sharp peak at 328 nm and a small notch at 341 nm and 372 nm, which are the known characteristic peak for aldehyde group of benzaldehyde absorbing near visible UV-light component. The UV-VIS spectra (b) for alkaline solution (pH 10-11) of benztriazole shows major peaks at 306 nm and knotch at 338 nm in the UV-zone showing strong UV-absorption. However, when an aqueous solution of 1% benzaldehyde is mixed with 1% aqueous solution of benzotraizole (both having alkaline pH of 10.5-11.0) and batched for 30 min followed by boiling for 30 min i.e. following equivalent alkaline pH conditions, the corresponding UV-VIS spectra [Spectrum (c) in Figure 3] shows a significant widening in the peak width showing peak at 320 nm and disappearance of peak at 328 nm and 306 nm for (a) & (b) thus indicating a new adduct with new peak at 320 nm. This result thus indicates partial consumption of aldehyde group of benzaldehyde evidencing some sort of reactive interactions between benzaldehyde and benzotriazole under specific treatment conditions of alkaline pH. Therefore, this study with benzaldehyde model compound suggests that there is a possibility for a reaction between the aldehyde group of compounds and benztriazole in an alkaline pH, supporting Reaction Scheme A.

This result indicates that the benzotriazole added to the benzaldehyde solution is fully consumed/transformed forming some adducts in association with benzaldehyde when both are batched and boiled/heated (for 30 min) in an alkaline media (pH 10.5-11.0), Spectra (c). This result thus confirms the reactive interaction between benzotriaozle and –CHO group of benzaldehyde. In order to support Reaction Scheme A, Figure 2, a comparable reaction is also anticipated between the CHO group of jute hemicelluloses and CHO in cellulose produced in bleach-baths as well as in aldehyde of lignin. Hence, the similar reaction can also be expected between –CHO group of jute hemi-celluloses and –CHO in cellulose in bleach-bath and with aldehyde of lignin also reacting with benzotriazole supporting Reaction Scheme A, Figure 2.

Moreover, Table 5 gives data on compositional analysis of major jute components (lignin, hemicelluloses and cellulose) and also copper number and –COOH group content of raw jute, bleached jute and benzotriazole treated jute. Data in Table 5 indicate that raw (control) jute fabric samples contain 13.4% lignin, 60.9% α-cellulose, 23.8% hemicellulose, which after bleaching has the final composition as 10.9% lignin, 59.40% α-cellulose and 21.6% hemicelluloses. This reduction is due to partial weight loss of each component totaling approx 6.02% on 3% H_2O_2 bleaching. On subsequent 2% benzotriazole treatment by pad-batch-dry process, it is observed that there is gain in weight of 1.42% (0.4% of cellulose + 0.22% of lignin + 0.8% of hemicelluloses), which may be due to addition of benzotriazole (approx 0.8% to hemicellulose and approx 0.4% addition to α-cellulose and rest probably to lignin).

The higher weight gain of 1.70% (0.5% of cellulose + 0.30% of lignin + 0.9% of hemicelluloses) for application of 2% benzotriazole by the exhaust process in the bleach-bath compared to the pad-batch-dry process (weight gain 1.42%) may be due to higher addition of benzotriazole in alkaline medium (bleach-bath)

Reduction in –COOH group content (0.9%) after 2% benzotriazole treated jute confirms that some of the benzotriazole is attached to the –COOH group of hemicellulose and hence –COOH group content of jute available is reduced accordingly by that proportion. Same trend is seen for bleach-bath treatment for overall weight loss is rather much higher and weight gain is somewhat less due to more bleaching action of jute constituents in bleach-bath process. Similarly, copper number gives the measure of –CHO group content in jute; higher the –CHO group content, higher is the copper number. After bleaching, the copper number is increased by a small amount. Further, after 2% benzotriazole treatment, copper number is reduced indicating consumption of some –CHO group of jute hemicellulose to form adduct with benzotriazole. Copper number for jute treated with 2% benzotriazole by the exhaust process in the bleach-bath (alkaline medium) is higher than when benzotriazole is applied by the pad-batch-dry-cure process (acidic medium) indicating that more under alkaline conditions of bleach-bath application, more aldehyde is generated causing higher reaction rate/weight gain in bleach-bath for availability of more accessible aldehyde. Similarly, reduction in –COOH group for jute constituents in jute treated with 2% benzotriazole by the exhaust process in the bleach-bath (alkaline medium) to 23.9 also indicate more aldehyde group formation and hence possibility of more addition of benzotriazole in jute is evident.

Table 5. Weight loss/gain, α-cellulose, hemicellulose, lignin content, copper number of jute before and after treatment.

Treatment	Weight loss(-) /gain(+) (%)	α-cellulose	Hemicellulose	Lignin content (%)	Copper Number	-COOH group content meq/100 gm
NIL (Control jute fabric)	--	60.90	23.80	13.40	2.06	4.80
Conventional 3% H_2O_2 bleaching at 85°C for 2 h at pH 11	-6.02	59.60	21.6	10.90	2.72	25.2
H_2O_2 bleaching along with 2% (owf) benzotriazole in pad-pad-batch-dry process	-4.6 (weight loss 6.02 and gain 1.42)	60.00 (60.00-59.60 =0.4%)	22.4 (22.4-21.6=0.8%)	11.12 (11.12-10.90 = 0.22%)	2.05	24.3
Bleach bath by alkaline exhaust process	-4.32 (weight loss 6.02 and gain 1.70)	60.10 (60.1-59.6 = 0.50%)	22.50 (22.5-21.6 = 0.90%)	11.20 (11.20-10.90 = 0.30%)	2.20	23.9

FTIR Spectral Analysis

FTIR spectra of (a) raw untreated (control) jute, (b) jute bleached by conventional process of bleaching using H_2O_2 with usual additives at 85°C-100°C for 2 h at pH 10.5, and (c) 1% aqueous solution of benzotriazole added in bleach-bath at pH 10.5-11.0 followed by H_2O_2 bleaching at 85°C-100°C for 60 min are shown in Figure 4. These FTIR spectra indicate that some common FTIR bands and peaks of raw jute (Ghosh et al., 1997) and bleached jute appear due to common functional groups present in them. Besides the common peaks for jute constituents, FTIR Spectrum (c) in Figure 4 show an additional small peak at 1550 cm^{-1} alongside another small peak of jute at 1540 cm^{-1} forming a duplet which may be ascribed to –C–N stretching vibrations (Ghosh et al., 1998) for reaction between jute aldehyde and benzotriazole in alkaline medium (bleach-bath) supporting Reaction Scheme A in Figure 2.

Also at 2347 cm^{-1} another singlet shoulder appears in FTIR Spectrum (c), Figure 4 by merging the duplet at 2340 cm^{-1} and 2365 cm^{-1} present in FTIR Spectra (a) and (b), Figure 4 which is ascribed to stretching vibration of –C–N– and –N–CH– groups (Ghosh et al., 2008) supporting Reaction Scheme A in Figure 2. Reaction Scheme A is further supported by evidence of appearance of a small additional peak at 1698 cm^{-1} in FTIR Spectrum (c), Figure 4, which is ascribed for –C–NH– stretching (Ghosh et al., 1998; Ghosh et al., 2008).

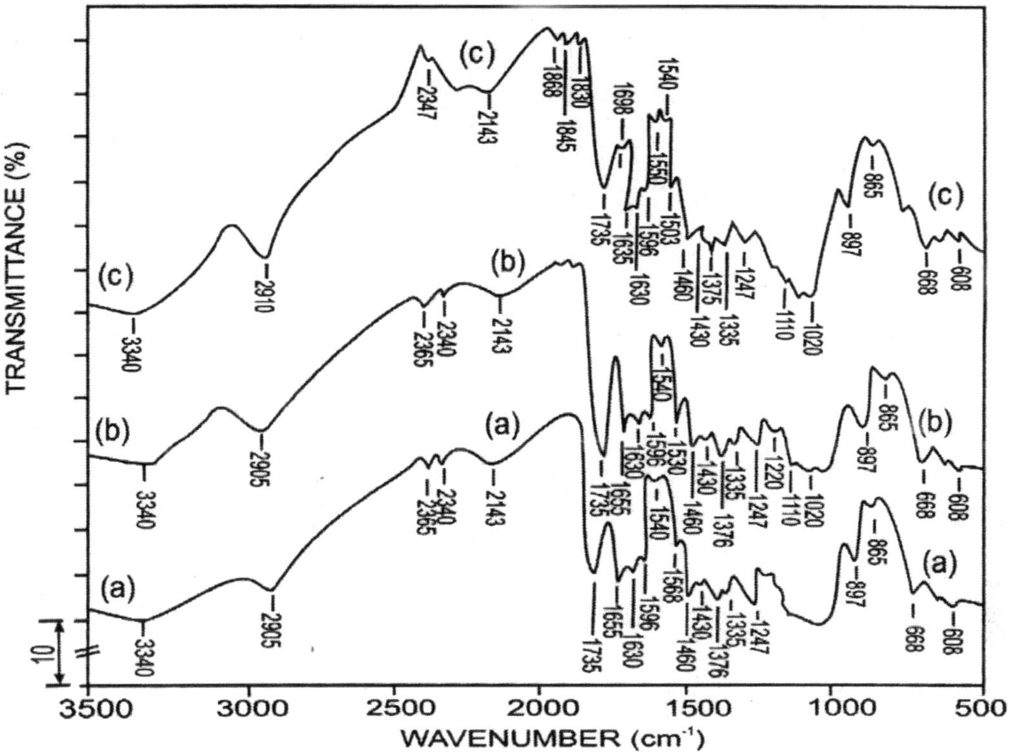

Figure 4. *FTIR Spectra of (a) raw untreated (control) jute, (b) jute bleached by conventional 3% H_2O_2 bleaching process with usual additives, (c) jute bleached with 3% H_2O_2 with usual additives in the presence of 1% benzotriazole at 85°C for 2 h at pH of 10.5-11.0*

Thus, the mechanisms postulated in Reaction Scheme A in Figure 2 can be further evidenced from the FTIR analysis which suggests chemical anchoring reaction between benzotriazole and jute in alkaline media producing an adduct with the presence of –C–N– or –C–N–H stretching vibration in benzotriazole and jute cellulose adduct.

Study of improvement in light fastness rating and UV-protection factor by application of benzotriazole on bleached jute fabric

Table 6 shows the results of light fastness and ultra violet protection factor (UPF) with absorption of UVA% and UVB% for raw and H_2O_2 bleached jute fabric and also raw jute treated with 2% benzotriazole (UV-absorber) in the conventional H_2O_2 bleach-bath by the exhaust process. Relevant results indicate that light fastness rating of raw jute is found to be grade 2, while bleached jute fading earlier at higher rate shows light fastness rating of grade 1-2 (i.e less than raw jute fabric). Both raw and bleached jute fabric shows same UPF values of 5 only (poor) with 12-15% absorption of UVA% and UVB%. Application of benzotrizaole on jute by the exhaust process in H_2O_2 bleach-bath gives higher UPF of 12 compared to its application on bleached jute by the pad-batch-dry process due to higher reactivity of benzotriazole in an ionized form (in alkaline pH) having two lone pair of electrons, of which one i.e. =N– is formed by reaction with alkaline –OH.

Table 6. *Results of light fastness rating and UPF for raw, bleached and UV-absorber (benzotriazole) treated jute fabric.*

Treatment	Light Fastness Rating	UVA%	UVB%	UPF
Untreated jute fabric (Control)	2	15.0	14.7	5
Ordinary 3% H_2O_2 bleached jute fabric	1-2	14.4	12.7	5
H_2O_2 bleached jute fabric treated with 2% benzotriazole by pad-batch-dry method	3-4	8.8	7.1	10
Raw jute fabric treated with 2% benzotriazole by in the conventional H_2O_2 bleach-bath by the exhaust process	3-4	7.8	6.9	12

Conclusion

It is therefore indicated from the present study that bleached jute fades at a faster rate than raw untreated (control) jute because of structural openness and generation of more number of functional groups in bleached jute. Further the average rate of reduction in WI and rate of increase in YI for raw jute is lower than that for H_2O_2 bleached jute specially for initial time of photo-exposure of 10-40 hrs, while, for higher time of UV light/sunlight exposure above 40 hrs, the rate of reduction in WI and rate of increase in YI gradually become lower, particularly for bleached jute than that for raw jute, with increase in time of photo-exposure. The higher rate of reduction of WI and rate of increase in YI for ordinary H_2O_2 bleached jute is probably due to more structural opening and more air and moisture content with generation of more number of functional groups in jute on oxidative bleaching.

Among all the UV-absorbers and antioxidant compounds applied, sodium azide and LTDP offers good stability against photo-fading, when applied in their optimum doses in bleach-bath, but these are not fast to washing. Application of 1% benzotriazole gives good resistance to photo-yellowing and photo-degradation with reasonable wash fastness, as compared to benzophenone and other UV-absorber/antioxidant compounds applied. Under specified reaction conditions i.e in an alkaline pH or (pH of 10.5-11.0) or in acidic conditions (pH of 5.5), benzotraizole may specifically react with the aldehyde groups present in jute hemi-cellulose during the boiling (100°C) or drying (90°C) condition of treatment to be reasonably wash-fast even after 1-0 number of wash cycles.

Among the two techniques of application of benzotriazole singly i.e. pad-batch-dry method and blench-bath methods, application of benzotraizole by the pad-batch-dry technique offers relatively higher whiteness stability although this process suffers from relatively higher degree of loss in tensile strength than that obtained by applying the same dose level of benzotriazole in the bleach-bath. On the other hand, the bleach-bath technique is simpler to adopt and gives better retention of tensile strength although it provides marginally less whiteness stability to photo-exposure as compared to that obtained by the pad-dry technique.

This work has high commercial significance as natural and bleached jute and jute/cotton union fabrics are widely used as window curtains, tent and furnishing fabrics for temporary structure under sunlight, where photo-fading of jute was a serious complaint. Earlier methods of stabilized bleaching of jute partially removing lignin was non-ecofriendly, cumbersome, costly and destructive process, whereas this process is much economical and ecofriendly, easy process for jute and jute/cotton union fabrics.

References

Abdullah, A. B. M., and L. W. C. Miles. 1984. Photostability of jute action on singlet oxygen. Textile Research Journal 54 (6): 415-417

Abdullah, A. B. M., and L. W. C. Miles. 1985. Photostability effect of some nickel chelates on bleached on bleached jute fabric. Journal of Textile Institute 76 (6): 415-418.

Achwal, W.B., and U. Sinkar. 1994. Modified processing of jute fabric to minimize photoyellowing: Part II - Use of UV-absorbers. Indian Journal of Textile and Fibre Research 19 (1): 30-33.

ASTM D 2974-87. 1993. Standard Test method for moisture, ash and organic matter of peat and other organic soils. Annual book of ASTM Standard. Philadelphia, Pennsylvania: American Society for Testing Materials.

ASTM-E-313-67/E-313-73 (Re-approved-1979). 1974. Determination of Yellowness. Annual Book of ASTM Standards E 12.02. Philadelphia, Pennsylvania: American Society for Testing Materials.

Bag, S. C., Banerjee S. K., Day A., and Sen, S. K. 1984. Chromophores of lignin responsible for yellowing of bleached jute. Cellulose Chemistry and Technology 18 (2): 149-153.

Banerjee, S. K., A. Day, S. C. Bag, P. Sengupta, and N. L. D. Sarkar. 1985. Textile Chemical finishing of jute yarn and fabric. Colourage 32 (1): 23-28.

Braidy M. H. and Nevell T. P. Methods in Carbohydrate Chemistry. R. L. Whistler (ed). Academic Press, Vol 3, New York, 1963, 43.

Callow, H. J., and J. Speakman. 1949. The action of light on jute. Journal of Society Dyers Colorists 65 (12): 758-763.

Chatterjee, S., B. Dasgupta, D. Adhikari and T. K. Guha Roy. 1991. Improvement in fastness properties of bleached and dyed jute products. Paper presented at the Proc. 15th Technological Conference, Indian Jute Industries Research Association, Kolkata, India.

Das, B. R. 2010. UV radiation protective clothing. Open Textile Journal 3: 14-21.

Doree, C. The Methods of Cellulose Chemistry, Chapman & Hall, London, (1947) 20-26.

Egerton, G.S. 1949. The Mechanism of the Photochemical Degradation of Textile Material. Journal of Society Dyers Colorist 65 (12): 764-780.

Farouqui, F. I., and I. Hossain. 1990. Prevention of photochemical degradation of undyed and dyed jute by UV-absorbers. Textile Dyer Printer 23 (1): 15-18.

Ghosh, P., A. K. Samanta, and D. Dev. 1997. Simultaneous free radical polymerization and acidic polycondensation of acrylamide-formaldehyde resin on jute fabric. Journal of Applied Polymer Science 64 (13): 2473-2489.

Ghosh, P., D. Dev, and A. K. Samanta. 1998. Effects of graft copolymerization of mixture of acrylamide and methyl-methacrylate on mechanical properties jute fibres of different compositions. Journal of Applied Polymer Science 68 (7): 1139-1147.

Ghosh, P., A. K. Samanta, and G. Basu. 2008. Enzyme and silicone treatment of jute fibers: Part I- Effect on textile related properties. Journal of Textile Institute 99 (4): 295-306.

Indian Jute Industries Research Association. 1977. Chemical processing of jute fabrics for decorative end-uses: Part-I. Bleaching treatment for jute: A monograph. Indian Jute Industries Research Association, Kolkata, March: 5-10.

IS:1560 (Part I)-1974. Handbook of Textile Testing" SP-15-1981 (Bureau of Indian Standards, New Delhi), 1982, 315.

IS:1670-1970. 1982. Determination of breaking load, elongation at break and tenacity of yarns." ISI Handbook of Textile Testing. New Delhi, India: Bureau of Indian Standards.

IS:2454-1985. 1989. Determination of colour fastness of textile materials to artificial light (xenon light). ISI Handbook of Textile Testing. New Delhi, India: Bureau of Indian Standards.

ISO-2470-1977 (E) and ISO-2469, 1977. Determination of Brightness. Switzerland: International Organization for Standardization (ISO)

Jones, D. C., C. M. Carr, W. D. Cooke, and D.M. Lewis. 1998. Investigating the Photo-Oxidation of Wool Using FT-Raman and FT-IR Spectroscopies. Textile Research Journal 68 (10): 739-748.

Katrizky, A.R., Y. Fang, A. Donker, and J. Xu. 2000. Imine acylation via benztriazole derivatives: The preparation of enaminones. Journal of Synthetic Organic Chemistry. 14 (6): 2029-2032.

Macmillan, W.G. and H.P. Bhattacharya.1954. The action of light on jute. Journal of Textile Institute 45(9): 700.

Mukhapadhya, U. 1976. Effect of different wavelength of the solar spectrum on discolouration of jute. Textile Research Journal 46 (2): 153-154.

O'connor, R. T., E. F. Dupre, and E. R. McCall. 1957. In-frared spectrophotometric procedure for analysis of cellulose and modified cellulose. Analytical Chemistry, 29 (7): 998-1005.

Samanta, A.K., and D. Singhee. 2022. Effect of application of selective UV-absorbers/antioxidants on raw and bleached jute fabrics by pad-batch-dry process for reduction of its photo-degradation and photo-yellowing character. Journal of Natural Fibers, Article 19(15): 12100-12118.

Samanta, A.K., K.K. Mahalanobis and D. Singhee. 2004. Effect on different methods of bleaching and resin finishing on physical properties and photoyellowing character of jute fabric. Journal of Institution of Engineers (India), Textile Engineering Division. 85 (2): 1-17.

Sanyal, A. K., and A. C .Chakraborty. 1995. Partial prevention of yellowing of bleached jute. Indian Patent No. 143 3441, Fibre Research Laboratory (FRL) for Bleached Jute, 1968. US Patent 3:427.

Sengupta, A.B. 1967. Action of light on jute. Jute Chronicle May-June: 1.

Shah, H. S., and R. S. Gandhi. 1990. Instrumental Measurements and Computer Aided Colour Matching for Textiles. Ahmedabad, India: Mahajan Book Distributers.

Shamey, R., and Sinha K. 2003. A review of degradation of nylon 6.6 as a result of exposure to environmental conditions. Review of Progress in Coloration and Related Topics. 33 (1): 93-107.

Slater, K. 1991. Textile Degradation. Textile Progress (The Textile Institute) 21(1-2): 1-150.

TAPPI Standard and Suggested Methods. T-222-05-74. Technical Association of the Pulp and Paper Industries, Atlanta, USA, 1974.

RECYCLED PARA-ARAMID YARN FOR PROTECTION AGAINST THERMAL HAZARDS

Shubham Srivastava[1], *Nandan Kumar*[2] *and Chandra Shekhar Malvi*[1]

[1]*Mechanical Engineering Department, Madhav Institute of Technology and Science, Gwalior, Madhya Pradesh, India-474005*
[2]*High Performance Textiles Pvt. Ltd. Panipat, Haryana, India-132103*

Abstract

In this study, a blend of recycled and virgin para-aramid fibres were used to develop both knitted and woven fabrics for a comparative study. The fibres were processed into roving which were further knitted (warp and weft) into fabric. Owing to lower strength of roving, a wrapping technique was used where Ne 30 ring spun yarn was wrapped with roving to increase strength required for knitting. The developed fabrics of 100% virgin para-aramid were compared with that of 70/30 virgin-recycled blended para-aramid for their thermal performance. It was observed that the thermal performance of virgin and virgin-recycled based fabrics were comparable in vertical flammability, contact, convective and radiant heat transmission testing as no significant difference was observed. Further, the knitted roving fabrics can also be designed as middle layer of multi-layered protective clothing such as firefighters suit, proximity suit to name a few. Also, warp knitted roving fabric can be knitted in any shape and could be used as 'three dimensional removable panels' for additional protection.

Keywords

Recycled para-aramid, roving, textile, thermal, aramid.

Introduction

Recycling is the process of collecting, sorting, processing, and transforming used materials into new products. It helps conserve natural resources, reduce energy consumption, and minimize the amount of waste sent to landfills or incinerators and burning (Chuang et al., 2019). Recycling is closely related to sustainability that refers to practices and actions that aim to meet present needs without compromising the ability of future generations to meet their own needs. It encompasses various aspects, including environmental, social, and economic considerations. Recycling comes in the 12th goal of sustainable development goals (SDGs) which precisely refers to responsible consumption and production. Excessive use of water and enormous energy consumption resulted textile industry in one of the most polluted manufacturing sector. The pollution is not only limited to the manufacturing process of textile products but also present in post-consumer side which worsen the environmental burden globally (Irfan et al., 2022). Moreover, rapid changes in fashion, fabric designs and increased consumer's buying capability in urban areas promote post-consumer waste in the form of clothing. These wastes are primarily generated from households and goes into landfill or sometimes even burnt due to unavailability of other prominent options available for disposal (Todor et al., 2019). Recent involvement of world into sustainability and recyclability promoted the culture of environment conservation by utilizing the maximum possible resources with minimum wastage. Extensive researches are being carried out to recycle traditional or conventional textiles (Bartl et al., 2005; Domina and Koch, 1999; Farrant et al., 2010; Hawley, 2000; Sandin and Peters, 2018), however, recycling of technical textiles presents unique challenges due to their specific composition and construction. The material complexities are involved in technical textiles as these are designed to serve special purposes and often incorporate high performance materials, coatings, or finishes. Further, the reinforcement of metals (stainless steel, tungsten etc. wires) in core along with high performance fibres present unique challenges in sorting and separation. Various methods are being developed to recycle high performance textiles with special interest to aramid based products owing to its versatility, strength, durability, resistance to mechanical, chemical and thermal hazards. Aramid fibres are most commonly being used in protective textiles such as gloves, socks, sleeves, workwear, undergarments to name a few. The fibre's strength upholds the modulus required to develop a new product from used protective textile of aramid because the properties of protective products may only be slightly damaged (Chuang et al., 2019). Therefore, an efficient recycling process and design can pave a way to add value to the material being used already. In this context, recycled aramid are already being used in thermal insulation and other commercial applications such as recycled para-aramid fibres have been used as insulation material in automotive (Bailey, 2021), aerospace, construction (Tilioua et al., 2018; Xiong et al., 2020), protective clothing (Irfan et al., 2022; Narayanan and Natick, 2015), climbing ropes, communication lines (Okajima et al., 2019) as well in composites (Amado et al., 2021) and reinforcement applications (Chuang et al., 2019). Due to high strength and various finishes present on the fibres, it's still challenging to recycle para-aramid fibres and most commonly shredding & grinding are being adopted for recycling the same. However, this method of mechanical recycling reduced the properties of fibres that ultimately results into a lower quality yarns and hence the products.

This paper aims to overcome these problems by utilising a new innovative 'roving' yarn technique where recycled para-aramid fibres are blended with virgin para-aramid fibres in different ratios (e.g. 30/70, 50/50, 70/30) to achieve a consistent and uniform knitted and woven fabrics which could replace 100% virgin para-aramid based products in certain applications. Further, same ratio was used to compare needle punch nonwovens which are commonly used as a thermal liner in multi-layered protective textiles.

Materials and Method

The 100% virgin para-aramid fibre and blended recycled & virgin para-aramid (ratio: 30/70, 50/50, 70/30) fibres were processed in blow room, carding and speed frame. However, the roving (recycled: virgin, 30/70, 50/50) were not used for further sample preparation on account of lower quality and difficulty in processing. Samples were prepared using 100% virgin and 70/30, virgin/recycled para-aramid roving. Three types of samples were developed by utilising warp, weft knitting and weaving. The roving obtained from speed frame (figure 1(a)) was directly used in weft knitting (Kumar et al., 2022) whereas roving wrapped with Ne 16 para-aramid yarn (figure 1(b)) was used in warp knitting. The woven samples were processed by Ne 16/2 para-aramid yarn (figure 1(c)) in warp with roving yarn in weft.

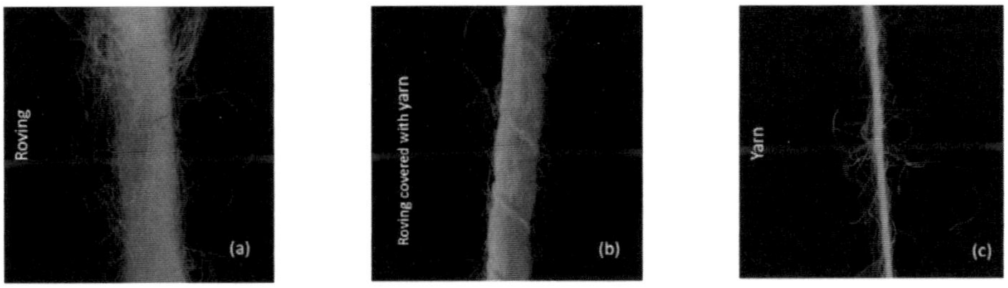

Figure 1 (a) Roving yarns (b) roving yarn wrapped with fine count of para-aramid yarn and (c) yarn; captured at same magnification.

The course per inch (CPI) and wales per inch (WPI) of knitted fabrics were 6 and 10 whereas, ends and picks per inch of prepared woven fabrics were 50 and 27 respectively. Later, for comparative study 70/30 ratio was utilized for the development of nonwoven fabric. Additionally, a blend of recycled and virgin (70/30) was used to develop woven fabrics. The Ne 16/2 yarn was used in warp and roving was used in weft. The ends and picks per inch for the developed woven fabric was 50 and 27 respectively. The developed samples are shown in figure 2.

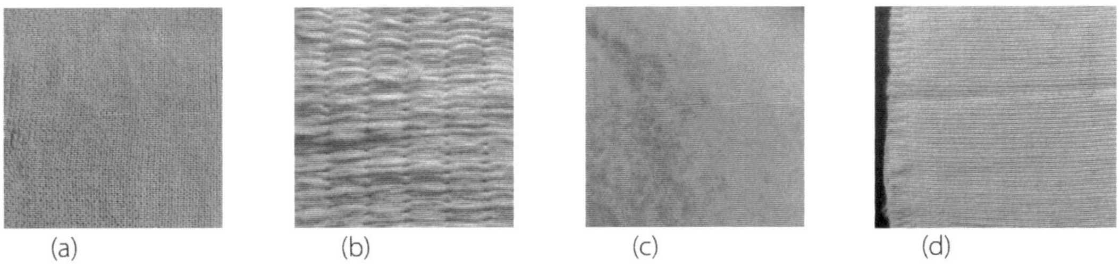

Figure 2 (a) Weft knitted sample using 70/30 virgin-recycled para-aramid roving (Sample 2), (c) Warp knitted sample using 70/30 virgin-recycled para-aramid roving (Sample 3), (c) Nonwoven sample using 70/30 virgin-recycled para-aramid (Sample 4) and (d) woven roving fabric

The performance of developed fabrics was tested against different thermal hazards such as flame spread, contact, convective and radiant heat transmission. All the samples were conditioned at 65% relative humidity(RH) and 27°C for 24 hours before testing. The physical properties such as air permeability, thickness, areal density were evaluated in accordance with ASTM D 737, ASTM D 1777, ASTM D 3776 and thermal conductivity was determined as per ASTM C 518 on ALAMBETA. These properties are given in table 1.

The flame spread test was conducted as per ISO 15025:2016 on *T419 Protective Clothing Flame Spread Tester*. The flame length was adjusted to 25 ± 2 mm in horizontal direction and 40 ± 2 mm in vertical direction before the start of the test. The bottom edge ignition method was used where the samples were placed vertically and burner was 30° from the vertical respectively. The flame was kept in contact with the samples for 10 seconds.

The contact heat transmission testing was conducted on *T424 Contact Heat Transmission Testing Machine* in accordance with ISO 12127-1:2015. The samples were placed on aluminium calorimeter and made to contact with the heated cylinder at 100°C and 250°C and threshold time was evaluated for accessing the performance of different fabrics which is the time required to raise the temperature of the fabric by 10°C.

Table 3 Physical and thermal properties of developed fabrics

	Fabric samples	Areal density (gram/m^2)	Thickness (mm)	Air permeability (cm^3/cm^2/s)	Thermal conductivity (W/mK)
100% virgin para-Aramid Roving	Warp	523.38	2.20	29.8	72.7
	Weft	1534.4	4.72	62.7	50.7
	Woven	529.15	1.51	10.2	10.8
70/30 virgin recycled para-aramid roving	Warp	527.1	2.40	26	76.5
	Weft	1540.4	4.74	73	54.2
	Woven	530.15	1.52	13.5	13.8

The convective heat transmission testing was performed in accordance with ISO 9151: 2016 and the performance of the samples was evaluated by measuring the Heat Transfer Index (HTI), which is the mean time required for a test specimen to achieve a temperature rise of 24 ± 0.2 °C. The samples were placed horizontally on a specimen holder and exposed to the heat flux of 80 ± 2 kW/m^2 by adjusting the flow of gas through the Meker burner.

The performance of the samples against different radiant heat exposure was evaluated as per the test standard ISO 6942:2002. In this test the samples were exposed to the heat flux of 20 kW/m^2 and the Radiant Heat Transfer Index (RHTI) for 24°C temperature rise was evaluated to access the performance of different fabrics.

Results and Discussion

The thermal performance of newly developed samples is shown in figure 4 and 5 whereas the flammability against vertical flame (inclined at 30^0 from horizontal) is compared in table 2. There was no significant after flame and afterglow observed for any of the samples. The charred area of different samples was compared and given in figure 3. Due to the presence of inherent flame retardant para-aramid fibres (Jassal and Ghosh, 2002; Srivastava et al., 2023, 2022), shrinkage, melting or dripping were not observed in any sample under the exposure of flame for 10 s at the bottom edge.

Table 4 Vertical flammability result of developed fabrics

	Fabric Samples	Flammability	
		After flame time (s)	Afterglow time (s)
100% virgin para-Aramid Roving	Warp	0	1
	Weft	0	0
	Woven	0	0
70/30 virgin recycled para-aramid roving	Warp	0	1
	Weft	0	1
	Woven	0	0

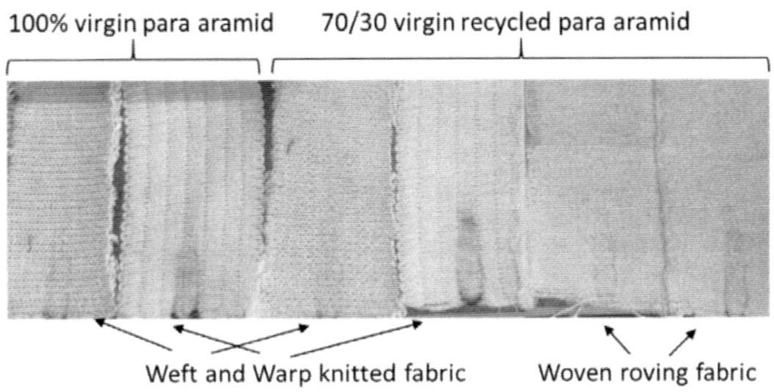

Figure 3 Char area of developed fabrics against 10 s of flame exposure

The threshold time for weft knitted samples were higher than both warp knitted and woven samples as per contact heat transmission test, mainly due to higher areal density (Mandal et al., 2020; Mandal and Song, 2016) and presence of wrapped roving yarn that ultimately resulted in the more compact structure. The pores present in fabrics were responsible for heat transfer through it at a faster rate. The performance of 100% virgin and 70/30 virgin-recycled fabrics were almost same as no considerable differences was observed (figure 4). Also, increase in the temperature of hot cylinder resulted in lower performance of fabric due to reduction in time required to raise the temperature of sample by 10^{0}C.

In order to determine the fabric resistance against convective heat, the developed fabrics were exposed to direct flame which were capable of transferring the heat flux of 80 ± 2 kW/m^2. It was observed that the time required to raise the sample temperature by 24^{0}C was 25s & 23s for warp knitted, 33s & 31s for weft knitted, 19s & 17s for woven fabrics in case of 100% virgin and 70/30 virgin recycled fabrics respectively. From the obtained data it can be anticipated that performance of 100% virgin and 70/30 virgin recycled fabrics were comparable (figure 5). The samples after being exposed to direct flame are shown in figure 6.

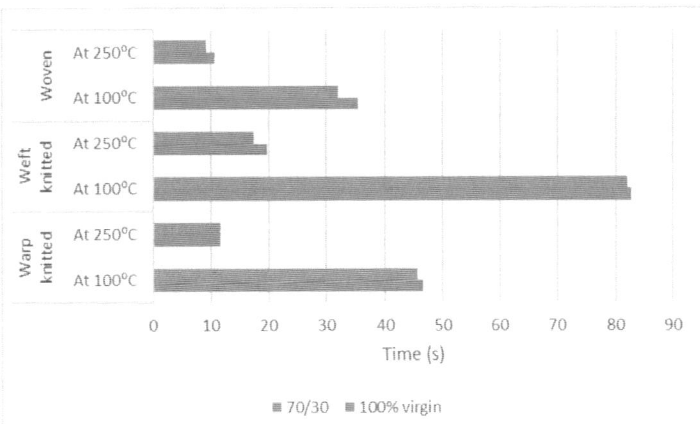

Figure 4 The performance of fabrics at 100^{0}C and 250^{0}C contact heat temperature

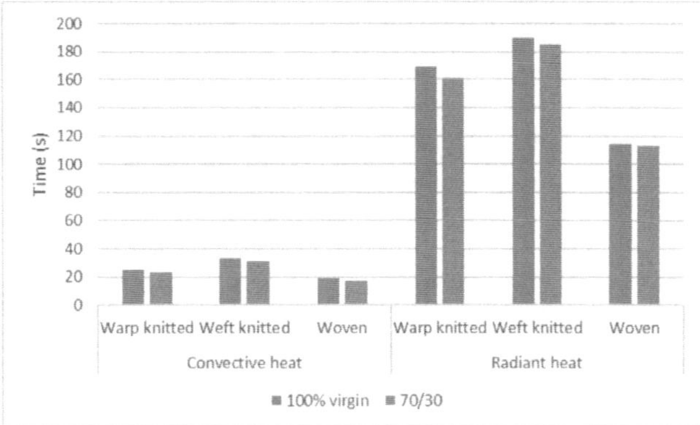

Figure 5 Performance of fabrics against convective and radiant heat

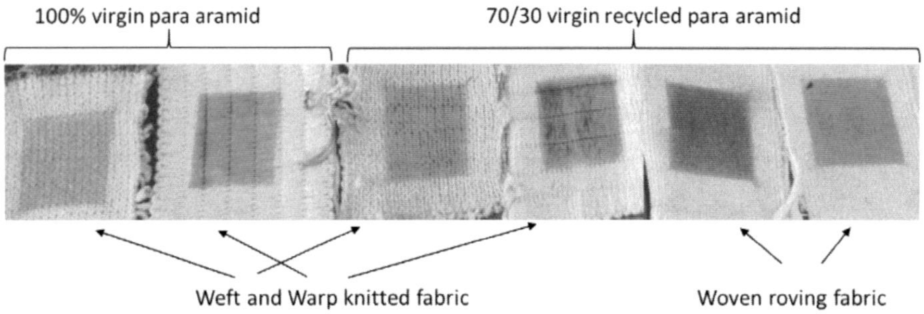

Figure 6 Fabric samples after exposing them to direct flame as per ISO 9151

There were no significant differences observed in the performance of 100% virgin and 70/30 virgin recycled fabric when tested against 20 kW/m^2 radiant heat flux. The radiant heat transfer index ($RHTI_{24}$) for warp knitted 100% virgin and 70/30 virgin recycled fabric was 169s and 161s respectively, whereas for weft knitted 100% virgin and 70/30 virgin recycled fabric was 190s and 185s respectively. Likewise, the $RHTI_{24}$ for woven 100% virgin and 70/30 virgin recycled fabric was 114s and 113s respectively. The comparative graph for the performance of fabrics can be seen in figure 5. It can also be anticipated from the test results of contact, convective and radiant heat that the performance of woven fabric was significant in comparison with knitted fabric even after having lower areal density and thickness due compact and less porous structure.

Further, the test results obtained in recycled warp knitted fabrics showed that it possesses good thermal properties and can be knitted in different shapes with less wastage. The woven fabric along with the knitted fabrics can also be used as middle layer of multi-layered thermal protective clothing. In order to compare newly developed fabrics with nonwoven structure, a 70/30 virgin-recycled para-aramid nonwoven fabric (VRPANW) was developed as shown in figure 2 (c). The physical the thermal properties of nonwoven fabrics are given in table 3.

Table 5 Physical and thermal properties of 70/30 virgin-recycled para-aramid nonwoven fabric

Fabric samples	Areal density (gram/m^2)	Thickness (mm)	Air permeability (cm^3/cm^2/s)	Thermal conductivity (W/mK)	Vertical flammability		Contact heat		Convective heat	Radiant heat
					After flame	After glow	@100^{0}C	@250^{0}C		
70/30 VRPANW	240	3.34	121.2	51.8	0	2	64.1	22	35	141

It can be observed from table 2 that the test results of 70/30 VRPANW fabric were comparable with that of warp knitted and woven fabric for both convective and radiant heat although both fabrics were thinner than nonwoven fabrics. However, the contact heat results were better for nonwoven fabric due to the more air trap and thickness. The favourable results of roving fabrics proved their applicability in certain application.

Conclusion

A feasible study was conducted to utilize the virgin-recycled blended para-aramid roving yarn in knitted (weft and warp) and woven fabrics. The thermal results for 100% virgin and blend of virgin-recycled (70/30) para-aramid fabrics were comparable. Also, a wrapping of fine count yarn over roving increased their strength and yarn developed through this innovative technique, found suitable for knitted fabrics having higher areal density that could be cut and sewn into different protective wear. In addition to this, weft knitted roving fabrics can be developed in different shape and sizes resulting in lower wastage during garmenting and faster production rate. Further, knitted and woven fabrics developed through roving yarn can also be used as an inner thermal layer in case of multi-layered thermal protective clothing. Further study is required to compare usage of roving fabric and nonwoven as thermal layer in multi-layered clothing assembly.

References

Amado, J.Q., Alvarez, H., Moreno, E., Sanchez, M., Escobar, G., 2021. Thermal conductionof elastomeric Polyurethanes with Titanium dioxide as mineral filler and its composites with recycled KEVLAR fabric. Fibre Chem. 53, 39–45. https://doi.org/10.1007/s10692-021-10236-8

Bailey, M., 2021. Teijin Aramid demonstrates fiber-recycling breakthrough.

Bartl, A., Hackl, A., Mihalyi, B., Wistuba, M., Marini, I., 2005. Recycling of fibre materials. Process Saf. Environ. Prot. 83, 351–358. https://doi.org/10.1205/psep.04392

Chuang, Y., Bao, L., Lou, C., Lin, J., 2019. Hybrid-Fiber-Reinforced Composite Boards Made of Recycled Aramid Fibers : Preparation and Puncture Properties 20, 398–405. https://doi.org/10.1007/s12221-019-8868-1

Domina, T., Koch, K., 1999. Consumer reuse and recycling of post-consumer textile waste. J. Fash. Mark. Manag. 3, 346–359. https://doi.org/10.1108/eb022571

Farrant, L., Olsen, S.I., Wangel, A., 2010. Environmental benefits from reusing clothes. Int. J. Life Cycle Assess. 15, 726–736. https://doi.org/10.1007/s11367-010-0197-y

Ghosh, S., 2022. The Report on India's Textile Waste Needs a Closer Look and an Update.

Hawley, J.M., 2000. Textile recycling as a system: A micro/macro analysis. J. Fam. Consum. Sci. 92, 40–43.

Irfan, M., Afzal, A., Nazir, A., Qadir, M.B., Khaliq, Z., Hassan, M., Ali, U., 2022. Development and characterization of protective gloves using waste para aramid fibers 52, 1–15. https://doi.org/10.1177/15280837221113363

Jassal, M., Ghosh, S., 2002. Aramid fibres - An overview. Indian J. Fibre Text. Res. 27, 290–306.

Kumar, N., Srivastava, S., Malvi, chandra S., MITS, 2022. A THREE-DIMENSIONAL TEXTILE AND A PROCESS OF PREPARING THE SAME.

Mandal, S., Gaan, S., Camenzind, M., Annaheim, S., Membranes, B., Federal, S., 2020. Thermal characterization of fire-protective fabrics. https://doi.org/10.1016/B978-0-08-100572-9.00021-5

Mandal, S., Song, G., 2016. Characterizing thermal protective fabrics of firefighters 'clothing in hot surface contact. https://doi.org/10.1177/1528083716667258

Narayanan, V., Natick, U.S.A., 2015. FR / HR SEWING THREAD.

Okajima, I., Okamoto, H., Sako, T., 2019. Recycling of aramid fiber using subcritical and supercritical water. Polym. Degrad. Stab. 162, 22–28. https://doi.org/10.1016/j.polymdegradstab.2019.01.034

Pensupa, N., Leu, S.Y., Hu, Y., Du, C., Liu, H., Jing, H., Wang, H., Lin, C.S.K., 2017. Recent Trends in Sustainable Textile Waste Recycling Methods: Current Situation and Future Prospects. Top. Curr. Chem. 375, 1–40. https://doi.org/10.1007/s41061-017-0165-0

Sandin, G., Peters, G.M., 2018. Environmental impact of textile reuse and recycling – A review. J. Clean. Prod. 184, 353–365. https://doi.org/10.1016/j.jclepro.2018.02.266

Srivastava, S., Kumar, N., Malvi, C.S., 2023. Determining the performance of thermal protective gloves against the exposure of flame as per ISO 9151:2016. Asian Tech. Text. J. 17, 58–63.

Srivastava, S., Kumar, N., Malvi, C.S., 2022. Study of Multi-layered inherent flame retardant fabrics for protection against contact heat transmission as per ISO 12127-1, in: Fire Engineer. pp. 63–70.

Tilioua, A., Libessart, L., Lassue, S., 2018. Characterization of the thermal properties of fibrous insulation materials made from recycled textile fibers for building applications: Theoretical and experimental analyses. Appl. Therm. Eng. 142, 56–67. https://doi.org/10.1016/j.applthermaleng.2018.06.071

Todor, M.P., Bulei, C., Kiss, I., Alexa, V., 2019. Recycling of textile wastes into textile composites based on natural fibres: The valorisation potential. IOP Conf. Ser. Mater. Sci. Eng. 477. https://doi.org/10.1088/1757-899X/477/1/012004

Xiong, C., Lan, T., Li, Q., Li, H., Long, W., 2020. Study of Mechanical Properties of an Eco-Friendly Concrete Containing Recycled Carbon Fiber. Materials (Basel). 13.

Business
Models
& Supply Chain

AN INVESTIGATION INTO THE IMPACTS OF BREXIT, COVID-19 AND CLIMATE CHANGE ON THE UK FASHION AND TEXTILE INDUSTRY

Anja Connor-Crabb[1], Emmanuel Sirimal Silva[3], Caroline Hemingray[1], Steven Toms[2], Alessandra Vecchi[3], Francesca Bonetti[3], Sophie Bulman[1], Yue Guo[1], Laura Solomon[3], Jane Harris[3], Stephen Russell[1]*

[1]School of Design, University of Leeds, Leeds LS2 9JT
[2]Leeds University Business School, University of Leeds, Leeds LS2 9JT
[3]Fashion Business School, London College of Fashion, University of the Arts London, 20 John Prince's Street, London W1G 0BG

Abstract

The aim for this mixed methods study is to examine the current state of affairs and the attitudes towards the future of UK businesses and professionals in the fashion, textiles and technology sectors, against the current political, social and economic background, characterized by the impacts of COVID-19, Brexit, and the climate emergency. To date, detailed aspects of these factors and their interplay are poorly understood by the sector, hence the need for this research. Participants were recruited through industry databases and professional networks (April - November 2022). The results show that the majority of respondents view sustainability as a high priority concern, in response to consumers becoming increasingly environmentally aware, and despite increased market uncertainty and rising costs caused by Brexit and Covid-19. R&D emerged as the most popular area of investment to reduce carbon footprint. Reducing carbon emissions is also cited as a key factor for reshoring, post-Brexit. Furthermore, COVID-19 has led to consumers increasingly demanding operational transparency. The opportunities and challenges posed by external factors, such as COVID-19 and Brexit, point to future trends, including a focus on sustainability and an increase in technology, particularly mobile commerce and digital marketing. These insights have the potential to benefit industry stakeholders and influence policy.

Keywords

Fashion, Textiles, Covid-19, Brexit, Sustainability, Climate Change, UK, Reshoring, Digital Technologies

Introduction

In recent times, the UK fashion and textile industry has been impacted by significant political, economic, social, technological, legal and environmental factors, including Brexit and the Covid-19 pandemic. This research seeks to quantify the impact of these macro events and provide insight into the ways in which UK fashion and textile companies have responded to the challenges and opportunities that Brexit and Covid-19 have presented.

A survey of UK fashion and textile businesses (including micro, SME and large enterprise) was conducted. The survey results, presented in this paper, form part of a larger study based on a sequential explanatory research design leading to the collection of preliminary quantitative data followed by qualitative data collection through semi-structured interviews.

This paper will focus on the quantitative research findings alone and it is noteworthy that the purpose of the quantitative study was primarily to inform the design and direction of the qualitative data collection by identifying some key insights around the impact of Brexit and Covid-19 whilst also considering the impact of climate change and digital practices on business activities.

Literature Review

The textile and fashion industry in the UK produces £9 billion worth of product for export (UKFT, 2020). It contributes more than £35 billion to the British economy and there are 890,000 people employed by the British fashion industry (British Fashion Council, 2020). The sector is formed mainly from micro firms and small medium sized enterprises (Harris et al., 2021) and represents an essential part of the UK economy, crucial for job creation, innovation and productivity growth (NESTA, 2017). The impacts of the COVID-19 pandemic and Brexit on the UK fashion and textile industry in the context of the ongoing climate crisis and emergent digital practices, as found in the literature, are outlined below.

Impact of Brexit on business

The UK's departure from the EU, often referred to as Brexit, has caused significant changes in the socio-political and economic landscapes within UK and EU member countries (Wood and Budhwar, 2016; Cumming and Zahra, 2016). Brexit has impacted the UK's Fashion and textile sector due to changes to tariffs, VAT, import duty, prices, regulations, and ease of buying as well as delivery of UK fashion goods to European consumers. There are issues relating to the skilled workforce in the sector as well as complications and uncertainties borne by Brexit, which have contributed to large scale change to the structure and operations of fashion and textile UK-based businesses (Moradlou et al., 2021b). Research suggests that Brexit is going to have substantial negative consequences on the UK economy (Brown et al, 2019).

There have been significant financial impacts related to new regulations since Brexit. For instance, EU consumers of UK products will now pay their own local VAT – a change which has led to an increase in orders from EU customers being cancelled. A survey conducted by the UK Fashion and Textile Association (UKFT) in 2021, found that 44% of the 138 participating UK fashion companies saw more returned or rejected items because of customs costs and VAT issues. 98% of participants stated they had been subject to increased costs after Brexit due to bureaucracy and paperwork, while 92% of the participants said they have had to pay increased freight cost. The rise in costs for paperwork for exports as well as longer lead times reduces the appeal of buying fashion and textile products from UK companies, financially impacting the UK fashion industry (RetailX, 2022). Additionally, quality and environmental standards set by UK are different and more stringent than those set by EU, making import into the UK more difficult (Xiong, 2022).

Trade policy uncertainty can be linked to Brexit and has altered companies' behaviours along the global value chain of the UK textile and apparel industry with impacts being traced back to 2016 when the referendum was announced (Casadei & Iammarino, 2021). Trade between EU member states and the UK is expected to decline in most manufacturing sectors, including textiles and fashion (Yi, 2022). The depreciation of the pound, however, has meant that countries such as China can buy from the UK at a lower cost, encouraging sales and alleviating some of the negative impact on the UK economy, which has arisen from the UK's reduction in the EU market after Brexit (Xiong, 2022). Trade with Japan is also predicted to increase; the export of textiles from the UK to Japan is predicted to rise by 55.40% (Yi, 2022).

According to a study by Moradlou et al. (2021a), business managers are re-evaluating where they choose to locate their supply chain so that disruptions due to Brexit as a geopolitical risk are reduced. Back-shoring, or reshoring, is also highlighted by Wan et al (2019) as a strategy for UK industries, enabling companies to meet UK demand, especially after Brexit. Cost efficiency was cited as a key driver for reshoring (Moradlou et al, 2022; Wan et al, 2019) with sustainability also highlighted as an important motivation (Moradlou et al 2021b; Orzes and Sarkis, 2019). Technological advances and Industry 4.0, such as digitalization and automation, are seen to be vital tools for the adaptation and implementation of reshoring decisions (Dey, et al., 2022). However, due to extensive offshoring of manufacturing over the past 20 years, there is a hugely reduced availability of skills from the UK (Bailey and De Propris, 2014; Benstead, Stevenson, and Hendry, 2017). As such, accessing a trained workforce remains the biggest barrier for the implementation of back-shoring strategies in the UK (Wan et al, 2019).

Impact of COVID-19 on business

The UK fashion and textile business is one of the most globalised trade groups and suffered significant negative impacts from the COVID-19 crisis (Leal Filho et al., 2023). During the first lockdown alone, UK clothing sales initially tumbled 34% in March 2020 due to the restrictions imposed as more than 80% of businesses relied on physical stores (McIntosh, 2020). For instance, Marks & Spencer experienced its first financial loss in 94 years, reporting a total loss of £87.6m before tax from March to September 2020, and consequently cutting over 7,000 jobs in August 2020 (Wood, 2020; Davey, 2021). According to the British Fashion Council (BFC), following the first lockdown in July 2020, the UK fashion industry was projected to endure a loss of over 240,000 jobs within the next 18 months, equating to approximately 27% of the fashion workforce (Indvik, 2020). Many tailoring businesses on Saville Row (Lawford, 2020) as well as high street shops, like Oasis and Warehouse, experienced damaging effects from the lockdown regulations, falling into administration and resulting in significant job losses (Butler, 2020).

However, the COVID-19 pandemic also brought about business opportunities, particularly for ecommerce. For instance, ASOS, an ultra-fast-fashion online retailer, had achieved rapid growth with documented average growth in sales of 20 to 25% per annum, affected by COVID-19 (Leal Filho et al., 2023). Sportswear brand Gymshark also benefited greatly from the shift in consumer behaviour toward online shopping, and an increased demand for wellness and casual attire, with yearly profits after tax increasing from £24.4 million in 2020 to £48.9 million in 2021 (Rydzek, 2022). During the pandemic, Gymshark gained significant attention on social media. Ben Francis, the founder and CEO of Gymshark, noted that their brand was one of the first to invest in social media and sponsor influencers to advertise their fitness clothing (Cook, 2020). Online fast-fashion retailer Boohoo and the online sales of Next also demonstrated resilience during the pandemic and reported soaring profits (Partridge, 2021; Wood and Jolly, 2021). As such, companies with a strong online presence demonstrated resilience during the COVID-19 pandemic.

Impact of climate change on business

The frequency of extreme weather events globally is on the rise, with significant implications also for the fashion industry's supply chain. The UN World Meteorological Organization's report on the state of extreme weather events indicates that climate change will only intensify if decisive action is not taken (United Nations, 2022). The cotton industry in Pakistan, for instance, has been severely impacted by floods caused by unusually strong monsoon rains, leading to a loss of approximately 40% of annual cotton yield (Ahmed & Farooq, 2022). The significant contributions of the global fashion and textile industry to climate change are well documented. Textile production alone generates around 1.2 billion tonnes of greenhouse emissions annually, exceeding those from all international flights and maritime shipping combined (Ellen MacArthur Foundation, 2017). With the world population forecast to grow, putting increased pressures on resources, the fashion and textiles industry is under increasing pressure to become more sustainable. In an industry survey conducted by the BFC with over 100 respondents, the top three main industry challenges were identified as: Policy and Regulation, Circular Design and The Post-Use Ecosystem (Bailey-Cooper et al., 2022), further highlighting the climate crisis as a key concern within the industry.

The UK government have set a goal to meet a net zero target by 2050 (GOV.UK, 2022). As part of the strategy to decarbonise, the UK government pledged £80m in June 2022 to support a 10-year sustainable change programme for the fashion industry, driving a circular fashion ecosystem in the UK (Douglass, 2022). The programme will be led by the BFC's Positive Fashion Institute, with a particular focus on recycling, innovation, and new circular business models. The fashion industry has taken a significant step towards tackling climate change by committing to the Fashion Industry Charter for Climate Action. The charter requires participating businesses to publicly report their sustainable progress, thereby increasing transparency and driving improvements in the industry. As of May 2023, the charter has garnered over 160 signatories worldwide, comprising major fashion brands, retailers, and industry associations. UK-based signatories include Burberry, Craghoppers, Mulberry Group plc, Primark, Regatta Group, Stella McCartney, and the British Fashion Council (United Nations Climate Change, n.d.).

Digital practices

The COVID-19 pandemic has accelerated the digital shift in the fashion industry, with an increased uptake of digital tools and platforms (Mesjar et al., 2023). The digital age has also been called the 'fourth industrial revolution', where we are seeing the adoption of smart technologies and networks as well a convergence of the physical and digital (Bertola & Teunissen, 2018). With earlier digital technologies, such as using CAD to produce technical flats, the technology would enable a streamlining of the design and production process. However, with the emergence of the metaverse, a parallel virtual reality, virtual fashion includes entirely new product categories, such as digital-only designs in the form of NFT (non-fungible tokens), enabled by blockchain technology. In general, innovations in digital fashion can be clustered into four themes: 1) Digital Design and E-Prototyping (incl. 3-D rendering, and AI for design) 2) Digital business and promotion (e-commerce and PLM system) 3) Digital human and the metaverse (digital 'skins' on avatars and VR/AR) and 4) Digital Apparel & smart e-technology (embedded electronics & smart clothing systems) (Sayem, 2022).

The creation of 3D modelling to prototype garments as part of physical production has the potential to reduce waste and cost as it enables fit testing and visualization of aesthetic features ahead of manufacture (Casciani et al., 2022). The 3D rendered design thus becomes part of the part of the manufacturing specification (Papachristou and Bilalis, 2015). This is particularly relevant considering the changed regulations post-Brexit whereby samples presented as a pristine product now require full duty and documentation at customs when crossing the continent (British Fashion Council, n.d.), therefore potentially avoiding additional costs and delays.

3D models may also be created for digital-only outputs, such as NFTs or skins for avatars in virtual worlds. These new digital types of fashion products and services can be produced at relatively low costs, which has enabled digital-only startups to emerge (Remane et al., 2017). Self-expression through digital-only designs also provides opportunities to dematerialize the fashion industry (Casciani et al., 2022), a key consideration during the ongoing climate crisis. A report by Roblox, an online gaming provider, in collaboration with Parsons New School (2022), found that Gen Z already values digital clothing as much as physical clothing, demonstrating the potential for digital-only designs.

While Augmented Reality (AR) is still an emergent technology (Liu et al, 2019), more than half of shoppers reported feeling more confident about a product after using AR (NielsenIQ, 2019). Virtual try-ons at home have the potential to reduce the return of online purchases, reducing textile waste (Hwangbo et al., 2020). The use of Artificial Intelligence (AI) to create designs is a quickly evolving innovation in digital fashion, which Särmäkari and Vänsa (2021) describe as a co-creation process, requiring the professional knowledge of the human designer when utilizing AI tools for design. Benefits of digital technologies, such as immersive technologies, include reduction in waste in design and manufacture by removing the need for physical products and reducing online returns, however, there are issues relating to high energy costs, consumer reluctance and skill shortages in technology and digital training (Mesjar et al., 2023; Harris et al., 2021).

Methodology

In order to collate quantitative data, to provide further insight into the pertinent impacts and changes which have emerged in the context of Brexit and Covid-19, a survey was designed and distributed to UK fashion and textile businesses. The survey also included questions related to changing digital practices, for example through operations and technology, as well as the importance of sustainability.

Survey Design

The survey was divided into 5 parts: demographics, climate change, Brexit, Covid-19 and digital practices. The survey began by gathering information on the general respondent profile covering several demographics such as trading name, SIC code, year of establishment (Casadei and Iammarino, 2021), annual turnover, number of employees (Casadei and Iammarino, 2021), female leadership (Brown et al., 2018) and attitude towards some general business characteristics (Black et al., 2019; Brennan and Vecchi, 2016; Colucci and Vecchi, 2021).

Thereafter, the respondents were directed to a set of questions covering the impact of climate change. They were asked about the importance of climate change on their strategic priorities, the likely impact on their business activities and future investment plans aimed towards reducing the carbon footprint.

Next, the respondents were presented with questions around Brexit which covered their operations in the EU and other countries, relocation plans, factors influencing offshoring decisions (Moradlou et al., 2021a) and reshoring decisions (Moradlou et al., 2021b), and challenges for reshoring (Moradlou et al., 2021b). We also captured the impact of Brexit on their business operations (Casadei and Iammarino, 2021) and future plans (Brown et al., 2019).

The Covid-19 section of the survey began by asking respondents how the pandemic influenced changes in their business strategy (Accenture, 2020; Pantano et al., 2020) prior to recording their relocation status owing to Covid-19 and the operations that would be impacted as a result. We also asked them to rate the importance of factors driving relocation due to the pandemic (Mazzarol and

Choo, 2003) and the type of innovations these companies had engaged with due to the pandemic (Tellis, 2013; Christensen et al., 2015). This section ended with some questions around the risks resulting from Covid-19 (Sheth, 2020).

The final part of the survey focused on digital practices and sought to understand how different technologies influence business operations. The respondents were also asked to identify the main challenge and main benefit stemming from each of the technologies that we surveyed.

Sampling

The survey was targeted at businesses using a purposive sampling approach. Micro, small, medium and large enterprises, from across the fashion and textile supply chain, were approached to take part in the survey through existing author networks, including the Future Fashion Factory (FFF) and the Business of Fashion, Textiles and Technology (BFTT) Creative Industries Clusters (funded by AHRC), and the UK Fashion and Textile Association (UKFT).

Findings

Sample Demographics

The survey recorded a total of 96 responses out of which 46 were complete, usable responses. Table 1 below summarises the demographics.

Accordingly, the majority of the companies that responded were established between 2015-2019 (39.1%), but there was representation across all periods which ensures the responses captured the attitudes of both recently established and legacy companies. In terms of annual turnover, the majority of the business that responded were turning over up to £50,000 per annum (41.3%), however, there was representation from businesses of all sizes with 26% of the businesses turning over £1m per annum. In terms of employment, using Brown et al.'s (2019) definition of firm sizes, the survey had captured responses from micro, small, medium, and large enterprises. However, 71.7% of the businesses that responded to the survey were micro businesses and the second largest representation was from small businesses (23.9%). We also sought to understand the distribution of female leadership across these businesses. Overall, 60.9% of the businesses in our sample had female ownership. A Fisher-Freeman-Halton Exact test was used to determine if there was a significant association between firm size and female ownership in the sample. We found evidence of a statistically significant association between firm size and female ownership *($p<0.01$)* with 75.8% of micro businesses being significantly associated with female ownership. These findings highlight the importance of female business leaders in micro-sized enterprises.

Table 1. Sample demographics

	Frequency	Percent
Period of Establishment		
1990s and earlier	10	21.7
2000-2009	4	8.7
2010-2014	10	21.7
2015-2019	18	39.1
2020 and after	4	8.7
Annual Turnover		
£0-£50,000	19	41.3
£50,000-£250,000	6	13.0
£250,000-£1m	5	10.9
£1m-£5m	7	15.2
£5m-£10m	2	4.3
£10m and above	3	6.5
Prefer not to say	4	8.7
Number of Employees		
1-9 (micro)	33	71.7
10-49 (small)	11	23.9
50-249 (medium)	1	2.2
250+ (large)	1	2.2
Majority Female Ownership		
Yes	28	60.9
No	16	34.8
I don't know	2	4.3

Next, we sought to understand the perception towards some business characteristics (Table 2). Most of the responding businesses were of the view that the market was growing rapidly (71.7%) with many segments (58.7%) with the product focus being mainly on physical attributes (69.6%) and online presence (71.7%). In terms of geographical focus, majority of the businesses were focused on the international market (76.1%) and they had many competitors (60.9%) leading to high intensity competition (69.6%) given that market entry was open to new players in most cases (89.1%). Interestingly, the responding businesses perceived that they had enhanced focus on innovation (84.8%) but limited supply chain automation (73.9%). There was more focus on sustainability (89.1%) and circular economy (89.1%) and high degree of business model innovation (78.3%) according to the respondents.

Table 2. *Most popular attitude towards selected business characteristics.*

Characteristic	Popular Attitude
Market Dynamics	Growing rapidly
Market Span	Many segments
Product Focus	Physical attributes
Online Focus	Online presence
Geographical Focus	International
Competition Intensity	High intensity
Market Concentration	Many competitors
Market Entry	Open to new players
Innovation	Enhanced focus
Supply Chain Automation	Limited automation
Sustainability Focus	More focus
Circular Economy	More focus
Business Model Innovation	High degree of innovation

Note: Popular attitude identifies the attitude that achieved a majority response (i.e., more than 50% of the businesses). N=46.

Impact of Brexit on business

A total of 72.7% of the active respondents had not planned to offshore from the UK to the EU due to Brexit. However, 9.1% of the businesses had planned to offshore to the EU due to Brexit whilst 4.5% had already implemented this move.

The top five challenges experienced by businesses due to Brexit were: increased market uncertainty, higher prices for products, increased delivery costs, customs delays and increased cost of imported raw materials. Likewise, the top five opportunities resulting from Brexit were identified as: UK government grants and funding, sustainability, support provided by Innovate UK, EDGE and the Department for International Trade, advancements in e-commerce and skills development schemes.

In terms of factors driving offshoring, the five most popular factors were delays in product delivery due to new border controls, higher costs through new tariffs for imports/exports to the EU, access to the EU market, access to international markets through EU trade deals and sustainability (e.g., climate change, carbon emissions, etc.). In contrast, the five most popular factors driving reshoring were identified as sustainability (e.g., climate change, carbon emissions, etc.), co-location of R&D/design/manufacturing, exchange rate fluctuations, communication/coordination and monitoring costs for foreign operations, and reducing lead times. The most important challenges for reshoring to the UK were high labour costs, high production costs, high raw material costs and access to raw materials.

In terms of remaining in the UK post-Brexit, the five most influential factors motivating businesses were proximity to UK customers, sustainability (e.g., climate change, carbon emissions, etc.), to avoid non-tariff costs (such as border delays and inventory levels) of importing to the UK, access to labour and high cost of transferring operations.

Finally in relation to Brexit we sought to identify how Brexit has impacted the future plans of businesses (Table 3). The findings further evidence that Brexit has considerably impacted the growth and future plans with all of the business decisions we queried on, with future plans either being largely scaled down or retained at constant levels due to Brexit.

Table 3. *The impact of Brexit on future plans.*

Business decision	Scaled down	Same level	Scaled up	I don't know
Capital investment (in premises, machinery etc.)	40.5%	32.4%	8.1%	18.9%
Developing and launching new products/services	35.1%	45.9%	10.8%	8.1%
Investments in R&D	35.1%	48.6%	5.4%	10.8%
Increase in export sales	48.6%	24.3%	8.1%	18.9%
Beginning selling to new markets	40.5%	29.7%	16.2%	13.5%

Note: 37 business responded to this question.

Impact of COVID-19 on business

We begin by analysing how the Covid-19 pandemic has impacted business strategy. **Figure 1** summarises the findings whereby 32 companies had selected one of the options shown below. Accordingly, the most popular change in strategy resulting from the pandemic was changes to, or introductions of, new product categories followed by the targeting of new customer segments. There were also a few companies that had entered new markets. In terms of relocating due to Covid-19, only 16.2% of the 37 businesses responding to this section of the survey had planned or implemented relocations.

We were able to identify seven factors that would potentially influence businesses to move location due to Covid-19. These were more emphasis on sustainability, changes in nearness to suppliers, changes in business climate, changes in access to transportation, changes in government incentives for Apparel and Technology Businesses, changes in access to labour and changes in access to training/education.

We followed this on by asking businesses to indicate the type of innovations they have engaged with because of Covid-19. These findings are reported in Table 4. Accordingly, the most important innovation businesses engaged in due to Covid-19 was further focus on online channels e.g., by promoting services online, website investment, new online technologies investments, etc.

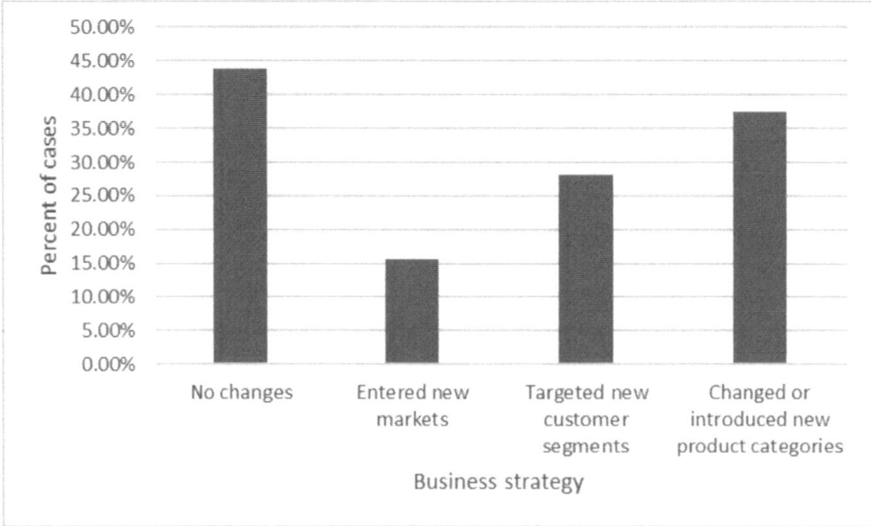

Figure 1. *Impact of Covid-19 on business strategy.*

Table. 4 *Innovations engaged with because of Covid-19.*

Innovation	N	Percent	Percent of cases
Further focus on online channel e.g., by promoting services online, website investment, new online technologies investments, etc.	21	30.9%	58.3%
Supply chain related innovation, e.g., automation, further outsourcing, new products, etc.	13	19.1%	36.1%
Sustainability and transparency related to product and distribution operations	13	19.1%	36.1%
Improved customer service provision (e.g., improved customer assistance online and offline, engagement with innovative technologies for product visualization and / or recommendations)	13	19.1%	36.1%
None of the above	8	11.8%	22.2%

Note: A total of 36 businesses responded to this question.

Finally, from a Covid-19 perspective, we sought to understand how businesses perceived the risks underlying certain factors. Interestingly, the only factors perceived as threats by most of the businesses were an uncertain future, new competitors, and trade policy. All the respondents agreed 100% that the increased demand for transparency and focus on sustainability were opportunities. In addition, the following were also seen as opportunities resulting from the pandemic: new products, consumers buying local, social media, changes in online/offline retail shopping, technology and innovation across shopping channels, new services, protection of brands, IP, design, and changing consumer behaviours e.g., spending/disposable income, changed personal goals e.g., fitness, health, experience, and work-life balance.

Impact of climate change on business

In terms of the importance of climate change in shaping strategic priorities, 86.9% of the businesses that responded to the survey agreed that it was either extremely important, very important or moderately important. The top three perceived impacts of climate change on business activities were: opportunities for satisfying environmentally conscious consumers, higher costs for raw materials, and higher prices for consumers. Next, we asked whether businesses were planning on implementing any specific investments to reduce their carbon footprint within selected areas. The most popular area of focus was R&D whilst the least popular area of investment to reduce carbon footprint was people.

Digital practices

In the final part of the survey, we asked businesses to let us know the extent to which operational activities were based on certain technologies. The findings showed that mobile commerce and digital marketing were technologies currently in high use. However, the following technologies have also become more important over the last year: body scanning, digital prototyping, 3D printing, machine learning and artificial intelligence (AI), virtual reality/augmented reality (AR), blockchain, mobile commerce, and digital marketing.

In terms of challenges and benefits of implementing technology, the respondents identified the cost of implementation, lack of know-how, lack of training, and integration costs as the main challenges whilst the main benefits of technology implementation were improving customer service and experience, reducing production costs, saving time, and improving sustainability.

Conclusion

The findings from this industry survey provides valuable insights into the UK fashion and textile industry's response to COVID-19, Brexit and climate change, whilst also considering digital practices. The findings highlight several important trends and challenges that the businesses are facing.

One significant finding is the priority placed on sustainability by industry respondents. This shift is driven by consumer awareness of environmental issues. Despite the uncertainties caused by Brexit and COVID-19, businesses are recognising the importance of addressing sustainability concerns and investing in research and development (R&D) to reduce their carbon footprint. The study also suggests that reducing carbon emissions is seen as a key factor for reshoring post-Brexit. Another noteworthy impact highlighted in the survey is the perceived increase in consumer demand for operational transparency due to the COVID-19 pandemic.

The survey also identifies an increased adoption of technology, particularly mobile commerce and digital marketing. However, the respondents identified several barriers to implementing technology, such as high implementation costs, lack of knowledge and training, and integration costs. Based on these findings, future research can explore ways to overcome the challenges associated with the implementation of new technologies. Additionally, further investigation could delve into the specific strategies and best practices that businesses can adopt to meet consumer demands for sustainability and transparency.

The authors recognise that the small sample size of participants in their study poses limitations to the generalisability of the results. However, the study still provides key insights into the industry's response to external factors, highlighting opportunities and challenges for businesses in the UK fashion and textile industry, and pointing to future avenues for research.

Acknowledgements

This project has been sponsored by the ESRC Impact Acceleration Account, the Future Fashion Factory (FFF) and the Business of Fashion, Textiles and Technology (BFTT) research clusters through the AHRC Creative Industries Clusters Programme, Leeds University Business School (LUBS) and supported by the UKFT.

References

Accenture (2020) *The Human Experience: How organizations should respond to the experience implications of COVID-19.* Available from: https://www.accenture.com/_acnmedia/Thought-Leadership-Assets/PDF-2/Accenture-COVID-19-New-Human-Truths-That-Experiences-Need-To-Address.pdf [Accessed 21 May 2020].

Ahmed, M. and Farooq, M. (2022) Pakistan Floods Raise Fears of Hunger After Crops Wrecked. Available from: https://thediplomat.com/2022/09/pakistan-floods-raise-fears-of-hunger-after-crops-wrecked/ [Accessed 10 May 2023].

Bailey, D. and De Propis, L. (2014) *Manufacturing reshoring and its limits: the UK automotive case.* Cambridge Journal of Regions, Economy and Society, 7 (3), November 2014.

Bailey-Cooper, N., Bartlett, S., Dubé, S. (2022) British Fashion Council (BFC), Positive Fashion Institute, The Circular Fashion Ecosystem Project: Phase 2 – Developing the Roadmap for Change. Available from: https://instituteofpositivefashion.com/uploads/files/1/BFC_CFE_PHASE2_V10[83229].pdf [Accessed: 12 May 2023].

Benstead, A.V., Stevenson, M. and Hendry, L.C. (2017) *Why and how do firm reshore? A contingency-based conceptual framework.* Operations Management Research 10, 85-103 (2017).

Bertola, P. and Teunissen, J. (2018) *Fashion 4.0. Innovating fashion industry through digital transformation.* Research Journal of Textile and Apparel, 22 (4), 352–369.

Black, S., Williams, D., Burcikova, M., Vecchi, A., Norton, Z., Brennan, G., Lyon, F., Colombi, C., D'Itria, E., Clutier-Leaute, D., Ruelle-Akl, N. (2019). Report - *Mapping Sustainable Fashion Opportunities for SMES.* European Commission Report. Directorate-General for Internal Market, Industry, Entrepreneurship and SMEs. ISBN 978-92-79-99648-1.

Brennan, L., and Vecchi, A. (2016). *Manufacturing Strategy in Time of Great Flux.* Operations Management, MEOP - Measuring Operations Performance, Springer, UK.

British Fashion Council (BFC) (2020) Annual Report & Accounts FY 2019/20. Available from: https://www.britishfashioncouncil.co.uk/bfcnews/4314/BFC-Annual-Report--Accounts-2019-2020 [Accessed 12 May 2023].

British Fashion Council (BFC) (n.d.) *Brexit.* Available at: https://www.britishfashioncouncil.co.uk/About/Brexit [Accessed: 12 May 2023].

Brown, R., Linares-Zegarra, J. M., and Wilson, J. O. S. (2018) *What Happens If the Rules Change? The* Impact of Brexit on the Future Strategic Intentions of UK SMEs. (1 March 2018) http://dx.doi.org/10.2139/ssrn.3066614.

Brown, R., Linares-Zegarra, J. M., and and Wilson, J. O. S. (2019). The (potential) impact of Brexit on UK SMEs: regional evidence and public policy implications. *Regional Studies,* 53(5), 761–770.

Butler, S. (2020) Oasis and Warehouse fall into administration with loss of 200 jobs. The Guardian, 15 April 2020.

Casadei, P., and Iammarino, S. (2021). Trade policy shocks in the UK textileand apparel value chain: Firm perceptionsof Brexit uncertainty. *Journal of Int*ernational Business Policy, 4, 262-285 (2021). https://doi.org/10.1057/s42214-020-00097-z.

Casciani, D., Chkanikova, O. and Rudrajeet, P. (2022), Exploring the nature of digital transformation in the fashion industry: opportunities for supply chains, business models, and sustainability-oriented innovations. Sustainability: Science, Practice and Policy, 18(1), 773-795.

Christensen, C. M., Raynor, M., and McDonald, R. (2015), *What is Disruptive Innovation.* Harvard Business Review, 93(12), 44–53.

Colucci, M., & Vecchi, A. (2021), Close the loop: Evidence on the implementation of the circular economy from the Italian fashion industry. Business Strategy and the Environment, 30(2): 876-873.

Cook, J. (2020) How Gymshark Became A $1.3 Billion Brand, And What We Can Learn. Forbes. 17 August 2020.

Crowe, D., Vecchi, A., Brennan, L., and Coughlan, P. (2006), *Manufacturing strategy and innovation in indigenous and foreign firms: an International Study.* International Journal of Manufacturing and Technology Management, Vol. 11, No. 1, pp. 3-27.

Cumming, D. J. and Zahra, S. A. (2016), *International business and entrepreneurship implications of Brexit.* British Journal of Management, 27, 687–692.

Davey, J. (2021) M&S clothing sales hammered by lockdowns in three months to Christmas. Reuters. 8 January 2021.

Dey, B.L., Alwi, S.F.S., Babu, M.M., Roy, S.K. and Muhammad, S.S. (2022), *Brexit or Brand it? The Effects of Attitude Towards* Brexit and Reshored Brands on Consumer Purchase Intention. British Journal of Management. https://doi.org/10.1111/1467-8551.12663

Douglass, R. (2022), *UK government pledges 80 million pounds to support circular industry model.* 10 June 2022. Available from: https://fashionunited.uk/news/business/uk-government-pledges-80-million-pounds-to-support-circular-industry-model/2022061063547 [Accessed: 10 May 2023].

Ellen MacArthur Foundation (2017), *Circular Fashion—A New Textiles Economy: Redesigning Fashion's Future.* Ellen MacArthur Foundation: Cowes, UK.

GOV.UK (2022) *Net Zero Strategy: Build Back Greener*. Department for Energy Security and Net Zero and Department for Business, Energy & Industrial Strategy. Available from: https://www.gov.uk/government/publications/net-zero-strategy [Accessed 10 May 2023].

Harris, J., Begum, L. and Vecchi, A. (2021), Business of Fashion, Textiles & Technology (BFTT): Mapping the UK Fashion, Textiles and Technology Ecosystem, University of the Arts London.

Hwangbo, H., Kim, E.H., Lee S., Jang. Y.J. (2020), *Effects of 3D Virtual 'Try-On' on Online Sales and Customers' Purchasing Experiences.* IEEE Access 8 (2020), 189479–189489.

Indvik, L. (2020), Up to 240,000 UK fashion jobs at risk, warns industry body. Financial Times 16 July 2020.

Janssen, M., and van der Voort, H. (2020), *Agile and adaptive governance in crisis response: Lessons from the COVID-19 pandemic.* International Journal of Information Management, 55, 102180.

Lawford, E. (2020), *How lockdown left Savile Row hanging by a thread.* Independent. 24 October 2020.

Leal Filho, W., Dinis, M.A.P., do Paço, A., Herédia-Colaço, V., Veiga Avila, L., Dennis, K., Tyler, D., Neiva, S. and Liakh, O. 2023. *COVID-19 and sustainability in textile, apparel and fashion use: An assessment of trends.* Textile research journal. 93(3-4), 674–690.

Liu, Y., Jiang, Z., Chan, H.C. (2019), Touching products virtually: facilitating consumer mental imagery with gesture control and visual presentation. Journal of Information Systems 36 (3), 823–854.

Mazzarol, T., and Choo, S. (2003), A study of the factors influencing the operating location decisions of small firms. Property Management, 21(2), 190–208.

Mesjar, L., Cross, K., Jiang, Y. and Steed, J. (2023) The Intersection of Fashion, Immersive Technology and Sustainability: A Literature Review. Sustainability, 15 (4), 3761.

McIntosh, S. (2020), Coronavirus: Why the fashion industry faces an 'existential crisis'. BBC. 30 April 2020.

McKinsey & Co. (2020), *The State of Fashion 2021 BOF McKinsey & Co.* Available from: https://www.mckinsey.com/industries/retail/our-insights/state-of-fashion. [Accessed 4 December 2020].

Moradlou, H., Reefke, H., Skipworth, H., and Roscoe, S. (2021a), Geopolitical disruptions and the manufacturing location decision in multinational company supply chains: a Delphi study on Brexit. International Journal of Operations & Production Management, 41(2), 10–130.

Moradlou, H., Fratocchi, L., Skipworth, H., and Ghadge, A. (2021b), *Post-Brexit back-shoring strategies: what UK manufacturing companies could learn from the past?* Production Planning & Control, 33(14). DOI: 10.1080/09537287.2020.1863500.

National Endowment for Science, Technology and the Arts (NESTA) (2017), *The State of Small Business – Putting UK entrepreneurs on the map.* Available from: https://media.nesta.org.uk/documents/the-state-of-small-business-uk.pdf [Accessed: 10 May 2023].

NielsenIQ (2019), *Augmented retail: The new consumer reality.* Available from: https://nielseniq.com/global/en/insights/analysis/2019/augmented-retail-the-new-consumer-reality-2/ [Accessed: 10 May 2023].

Orzes, G. and Sarkis, J. (2019) *Reshoring and environmental sustainability: An unexplored relationship?* Resources, Conservation and Recycling, 141, 481–482.

Papahristou E. and Bilalis, N. (2016) *Can 3D Virtual Prototype Conquer the Apparel Industry?* J Fashion Technol Textile Eng 4:2. doi:10.4172/2329-9568.1000134.

Partridge, J. (2021), Boohoo profits soar as Covid turns customer focus to loungewear. The Guardian. 5 May 2021.

Pantano, E., Gabriele, P., Scarpi, D., and Dennis, C. (2020), *Competing during a pandemic? Retailers' ups and downs during the COVID-19 outbreak.* Journal of Business Research, 116, 209-213.

Remane, G., Hanelt, A., Nickerson, R.C., Kolbe, L. (2017), *Discovering Digital Business Models in Traditional Industries.* Journal of Business Strategy 38(2).

RetailX (2022) *Fashion 2022: Ecommerce* Sector Report. Available from: https://retailx.net/product/fashion-2022-ecommerce-sector-report/ [Accessed 12 May 2023].

Roblox & Parsons (2022), *Metaverse Fas*hion Trends 2022. Available from: https://blog.roblox.com/wp-content/uploads/2022/10/FINAL_2022-Metaverse-Fashion-Trends-report_Roblox-x-Parsons.pdf [Accessed 12 May 2023].

Rydzek, C. (2022), Gymshark more than doubles revenues in 2021. *TheIndustry.Fashion*. 23 May 2022.

Sayem, A.S.S. (2022), *Digital fashion innovations for the real world and metaverse.* International Journal of Fashion Design, Technology and Education. 15 (2).

Särmäkari, N. and Vänsa, A. (2021) *,'Just hit a button!' – fashion 4.0 designers as cyborgs, experimenting and designing with generative algorithms.* International Journal of Fashion Design Technology and Education. 15 (4), 1–10.

Sheth, J. (2020). Business of business is more than business: Managing during the Covid crisis. *Industrial Marketing Management*, 88, 261-264.

Tellis, G. J. (2013). Unrelenting Innovation: How to Build a Culture for Market Dominance. San Francisco: John Wiley & Sons.

UK Fashion & Textile Association (UKFT) (2020), *Fashion & Textiles post Brexit.* Available from: http://textilehouse.co.uk/wp-content/uploads/UKFT-Brexit-Position-Paper.pdf [Accessed 12 May 2023].

UK Fashion & Textile Association (UKFT) (2021) Brexit bureaucracy hits UK fashion and textile companies hard. Available from: https://www.ukft.org/brexit-bureaucracy-survey-july-2021/ [Accessed 12 May 2023].

United Nations World Meteorological Organization (2023) WMO annual report highlights continuous advance of climate change. Available from: https://public.wmo.int/en/media/press-release/wmo-annual-report-highlights-continuous-advance-of-climate-change [Accessed: 15 May 2023]

United Nations Climate Change (n.d.) *Participants in the* Fashion Industry Charter for Climate Action. Available at: https://unfccc.int/climate-action/sectoral-engagement/global-climate-action-in-fashion/fashion-industry-charter-for-climate-action/participants-in-the-fashion-industry-charter-for-climate-action#Supporting-Organizations- [Accessed: 10 May 2023].

Vecchi, A., Evans, G. (2018), *The Changing Nature of the Fashion Industry and its Impact on Place-Making.* Proceedings of 3rd International Conference on Clusters and Industrial Districts, Valencia, Spain, May 24-27, 2018.

Vecchi, A. (2017), *Reshoring of Manufacturing - Drivers, Opportunities, and Challenges.* Operations Management, MEOP - Measuring Operations Performance, Springer, UK.

Wan, L., Orzes, G., Sartor, M. and Nassimbeni, G. (2019), *Reshoring: Does home country matter?* Journal of Purchasing and Supply Management, 25 (4).

Wood, G. and P. Budhwar (2016), *Brexit and beyond: the BJM and unforeseen events*, British Journal of Management, 27, 680–681

Wood, Z. (2020) Marks & Spencer falls to first loss as Covid crisis hits clothing sales. The Guardian. 4 November 2020.

Wood, Z. and Jolly, J. (2021), Next shares hit all-time high as Covid crisis fuels online sales surge. The Guardian. 1 April 2021.

Xiong, W. (2022), *The Impact of Brexit on UK-China International Trade.* Proceedings of the 27th International Conference on Social Sciences and Economic Development (ICSSED 2022), Atlantis Press. 29 April 2022.

Yi, C. (2022), Economic impacts of UK's free trade agreements with Korea, Japan, and EU as a breakthrough of Brexit. Manchester School, University of Manchester, September 2022, 90(5), 541–564.

TRANSITIONING TO ZERO CARBON AND ZERO WASTE: LESSONS FROM BIOLOGY

Veronika Kapsali[1], *Cathryn A Hall*[1] *and Steven E Naleway*[2]

[1]*London College of Fashion, University of the Arts London, London*
[2]*Department of Mechanical Engineering, University of Utah, Utah, USA*

Abstract

The fashion and textile sector look to biology, whether it is directly, via the development of new biobased, biofabricated or biosynthetic materials, or indirectly, via the introduction of system-based frameworks such as the circular economy model, for solutions that will support the delivery of zero waste and net zero targets. Within this milieu, the agency of the textile designer is limited to selection and implementation of said bio-related materials into their practice.

This paper discusses the development of the Bio-Inspired Textile (BIT) Design framework, intended to extend the textile designer's agency towards these targets, by drawing on broader design lessons in resource efficiency, longevity, and recovery (Kapsali, Hall, 2022) encoded within studies of the properties of biological materials by material scientists (such as, Naleway et al. 2015).

Efforts to identify an approach to develop accessible content for a textile design audience, result in the creation of a new perspective for practice-based modelling of findings from biology and materials science. We demonstrate how textile-based tools and methods can facilitate knowledge exchange between disciplines. Furthermore, we identify the potential for new forms of knowledge, emerging from this perspective, to enhance a design-led contribution to the sector's zero-waste and net-zero targets.

Keywords

Circular economy, textile design, sustainable design, material science, biological materials

Introduction

The urgent need to reduce greenhouse gas emissions was one of the key messages resulting from COP26. Despite this, conclusions from COP27 highlight commitments such as achieving global Net Zero by 2050, fall short of meeting the medium and long term goals set out by the Paris Agreement (2022). Although there are many factors that contribute to the challenges facing a net zero trajectory, the sheer complexity of globalised manufacturing value and supply chains constitute key challenges to implementing and evaluating new approaches. The fashion and textile (FT) sectors constitute a prime example: combined they form complicated supply chains, beginning with agriculture and/or petrochemical production (raw materials) and subsequently incorporate convoluted manufacturing steps, logistics and retail. The FT industry contributes about 8-10% of annual global CO2 emissions (Niinimäki et al., 2020) ranking it among the most harmful.

Fashion designer Suzanne Lee pioneered new thinking around the nature of raw materials for the textile industry and potential, alternative sustainable sources of material. Her work Biocouture (Lee et al., 2005) inspired a new class of textile designer who swap conventional studios for a studio/lab hybrids. These new spaces enable experimentation with tools, methods, and materials from microbiology to create alternative sources of materials for fashion and textile applications.

The FT sector already looks to biology, whether it is directly via the development of new biobased materials or indirectly via the introduction of system-based frameworks such as the circular economy and regenerative design models, to support the delivery of zero waste and net zero targets. However, within this milieu, the agency of the textile designer is limited to selection and implementation of said sustainable and/or bio-related materials into their practice. The role of design is limited to efforts in reducing the weight of products and identifying alternative materials within the context of the NetZero agenda, a narrative solidly embedded within contemporary design culture (Kapsali and Hall, 2022).

The work discussed in this paper constitutes key elements of interdisciplinary discourse that underpin a novel Bioinspired Textile (BIT) design framework[1] developed to contribute towards a net zero future for the textile and fashion sector. BIT presents an alternative and complementary view that takes a design rather than materials led approach to learning from biology. The key difference between a design and materials lens is the scale and nature of one's basic building blocks. A microbiology approach draws on microorganisms and/or their derivatives as basic components typically spanning nano to micro scale. BIT focuses on what we can learn from biology that can help us improve the environmental impacts of our designs from the fibre and beyond (micro to macro). It also challenges the established perception that mechanical properties are primarily *intensive*, in other words implemented at chemistry level. Biology teaches us that an *extensive* approach is also possible, but barriers to knowledge from biology and materials science typically exclude the creative textile design community from these ideas.

The STEM sector currently holds two approaches to biomimetic design (applying ideas from biology into engineering), bottom-up and top-down. The bottom-up approach is solution orientated (Cohen and Reich, 2016) and requires multiple, integrated competencies (Santulli and Langella, 2011). The work requires the identification of a design problem, a relevant paradigm from biology and interpretation of said biological example to address the design problem. This approach is not currently feasible at scale, primarily due to the interdisciplinary nature of the practice that requires individuals to

---

[1] http://www.bioinspiredtextiles.com

navigate complex language, knowledge and cultures from different disciplinary domains (Jacobs et al., 2022).

The top-down approach focuses on generating understanding of a particular biological mechanism, without necessarily having an application in mind. This approach can involve creating models of how the mechanism in question works using the tools of physics and engineering. However, dissemination of both approaches targets discipline specific audiences i.e. material scientists which excludes textile designers by default.

This paper focuses on lessons from biological materials made from relatively weak substances that demonstrate extraordinary mechanical properties. Access to relevant information requires specialist terminology, and concepts from materials science which constitutes a barrier for designers. Interestingly, complex terminology used by scholars is also problematic within scientific communities.

During a comprehensive review of hundreds of journal articles dedicated to the topic of structure/ function relationship of strong biological materials, scientist Steven Naleway and his team observed a range of inconsistent terms used to describe a set of eight reoccurring *biological structural design elements* (BSDEs): fibrous, helical, gradient, layered, tubular, cellular, suture and overlapping. Naleway et al. developed a graphic tool (fig.1) to express individual BSDE structures and mitigate terminological disharmony among scholars. This inadvertently, enabled access to concepts embedded in biology and material science for a creative design community. This paper foregrounds the interdisciplinary discourse that underpinned the development of the 8 original BSDEs into a practical design framework that could facilitate the FT industry's transition to NetZero.

Biological structural design elements (BSDE) for textiles

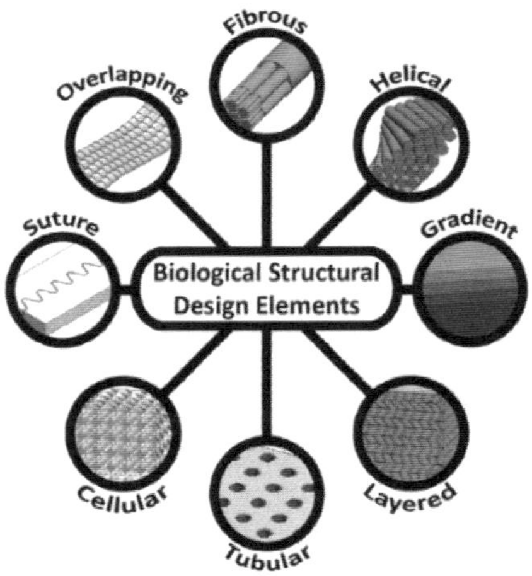

Figure 1. Naleway et al. (2015) graphical representation of the eight, most common biological structural design elements (BSDE).

Naleway's model of BSDE is composed of eight reoccurring archetypes that emerged from a comprehensive review of studies on the relationship between the structure and mechanical properties (i.e., stiffness, toughness and tensile strength) of various biological materials such as mother of pearl and the giant squid beak. Figure 1 illustrates the graphical representation of each structure created by Naleway and his team for a material science audience.

We observed that these archetypes are not exclusive to biology; the same structures are evident across several design disciplines including textiles. In effect, this visual (fig.1) represents a portfolio of shared structural archetypes that form an abutment connecting the material science of biological materials with textile design. Although textile designers are not commonly concerned with advanced material properties, the lessons emerging from the structure/ property relationships expressed by the BSDE model can help designers manage the drape (stiffness), stretch (elastic recovery), etc of their practical outcomes. It is without question that the gap between the BSDE framework in its current form and its application to textile design, is wide. The work described in this paper constitutes the first steps in attempting this, as such is to be regarded as indicative and not an exhaustive study of the topic.

BSDE Fibrous

'Fibrous structures offer high tensile strength when aligned in a single direction, with limited to nil compressive strength'(Naleway et al., 2015)

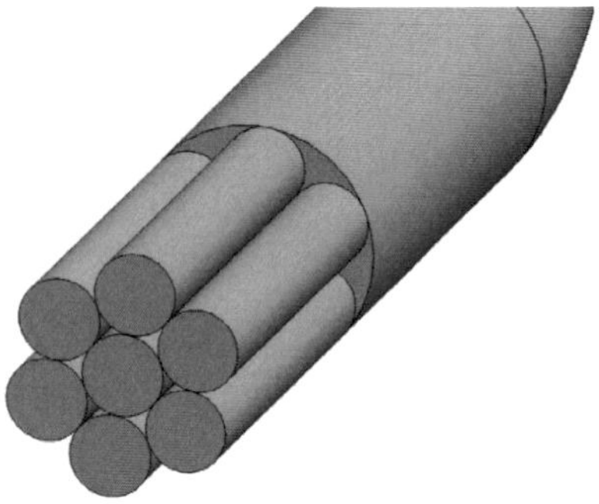

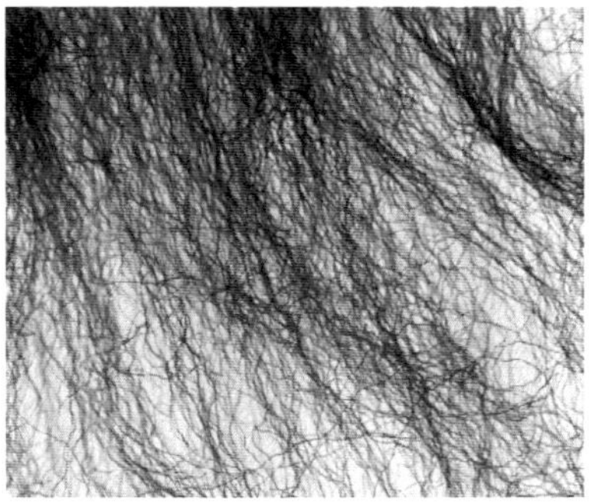

Figure 2. Left: BSDE Fibrous. Right: arial image of semi carded wool fibres

BSDE Fibrous describes structures that are composed of fibres (fibrils or filaments at smaller spatial scales) that are aligned in a single direction and often as a hierarchy across multiple length scales. Muscle, skin, tendons and silk are all examples nano- to micro scale fibrous structures. Physical properties such as shape (wave/kink of the individual fibres), length, type of alignment and the movement of the fibres (i.e. sliding between fibrils or friction between fibres) are factors that contribute to the mechanical behaviour of fibrous structures. Given the wide range of fibre shapes and sizes present in biological materials (from nano to micro), it is not always possible to distinguish individual fibres (i.e., collagen fibres in skin) as it is possible with textile fibres (i.e. silk, cotton, wool).

In the context of textiles, a (textile) fibre is the single unit that acts as the basic element (building block) to create textile structures (Hatch, 1993). Textile designers convert loose fibres into fibrous *structures* via a process called carding which aligns the textiles fibres into parallel form called a web (non-woven context) or sliver (yarn spinning). This is the first step for introducing *structural hierarchy*[2] into a textile paving the way for further orders of hierarchy, for example, a sliver of aligned fibres (n=1) is further processed into a yarn by adding twist into the z-axis (n=2) forming a helical structure (see section 2.2). The same principle applies to the formation of non-woven materials. Aligned sheets of fibre (n=1) are layered positioned on top of each other but laid in different directions (n=2).

Textile fibres are selected by a textile designer indirectly via yarn or fabric sourcing and are determined by the design brief which outlines the properties the fabric needs to deliver, i.e., aesthetic, functional, economic etc. Mechanical properties such as stretch, strength etc. are intensive i.e., attributed to the fibre's chemistry. It is not common practice for textile designers to work directly with raw fibres, this practice is primarily observed at the two ends of the sector's spectrum; among craft-based felt makers

[2] Hierarchical structures contain structural elements which themselves have structure. The hierarchical order of a structure or a material may be defined as the number n of levels of scale with recognized structure.

or technical non-woven textile producers. However, designers could find new opportunities to introducing properties via an extensive approach and reduce the range of fibre types included within their designs.

BSDE Helical

> *'The helical configuration provides increased strength and toughness in multiple directions to create puncture resistance and disrupt crack propagation' (Naleway et al., 2015)*

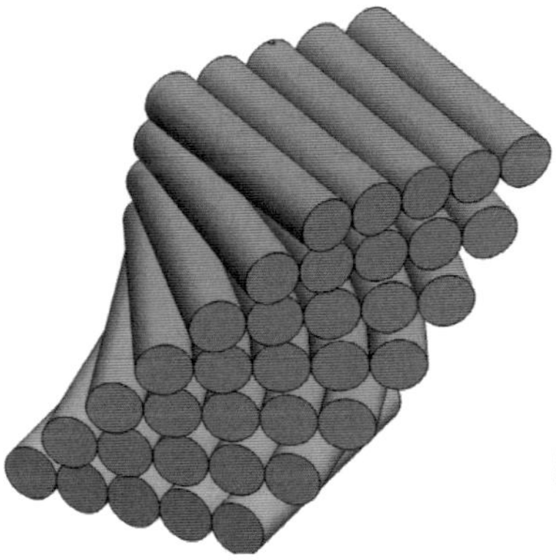

Figure 3. Left: BSDE Helical. Right: side image of ply yarn featuring helical twist.

BSDE Helical (fig. 3, left), refers to 'twisted-ply' structures present in nature at nano scale (a-helix keratin, DNA) where they often act as springs or coils and at micro-meso-macro (spicules of glass sea sponge, narwhal tusk) primarily used for reinforcement.

In textiles, 'twisted-ply' describes the twisting or spinning fibres into yarn which can then be plied together (fig. 3, right). Following the alignment process of carding and sliver production, twist is added to create a single yarn (n=2). When two or more single yarns are combined this is described as plying. Furthermore, plied yarns can also be twisted together to form 'cabled' yarns (n=3). Thus, a hierarchy of second and third order structures is created. Twist, in its different forms generally impacts the strength of the yarn; plying has the same effect whilst also adding volume.

Yarn twisting is implemented in the z-plane and constitutes an obvious link to BSDE helical. Helical configurations also occur in the x, y planes as in the layers of orientated fibre webs laid in different directions in non-woven textile technology to implement strength (under tensile stress) in multiple directions.

Designers specialising in constructed textiles (knit and weave) work with yarns to build their fabrics. However, designers do not have any agency in the way yarns are constructed, unless they have access to hand spinning techniques which tend to occur in craft-based practice. New understanding on the importance and role of helical structures in textiles can present opportunities for extensive approaches to the introduction of properties into fabrics.

BSDE Gradient

'...materials and interfaces that accommodate property mismatch, this is created through a gradual transition with the benefit of increased toughness, wear resistance and arrest crack growth'. (Naleway et al., 2015)

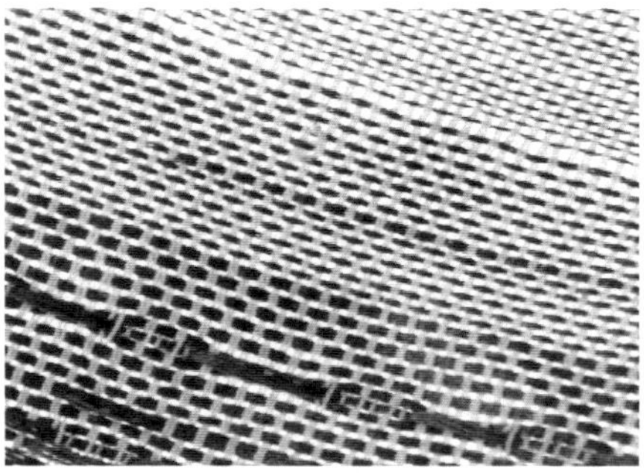

Figure 4. Left: BSDE Gradient. Right: arial image of semi carded wool fibres Palavi Verna (2021)

The mechanical properties most associated with gradient structures in biology is crack resistance (ibid) measured in terms of toughness. However, terminology typically pertaining to the mechanical properties of a material (tough, strong, ductile, brittle, hard) do not feature in the creative textile lexicon simply because the job of the designer is not to create textiles with advanced mechanical properties, this is the scope of a chemist or materials scientist. If *toughness* describes the ability of a material to absorb energy and deform without fracturing, then textiles are *tough* because they are flexible and can withstand quite a bit of force without cracking or ripping.

Gradient in textiles usually refers to a gradual transition of colour. *Ombre* is a name given to textiles that demonstrate a gradual transition from one colour to another, this can be achieved via a technique known as *ombre dying*. A similar effect can also be achieved structurally via the design of the textile construction. An *ombre effect* can be engineered by gradually transitioning from one coloured yarn to another during the knitting or weaving process (visual property). Another variation is to alter the density of the knit or woven structure across the length of the textile (fig. 4, right) moving from a more densely woven structure to a looser one or more heavily manipulated (i.e., pleated, smocked, embroidery, laser cutting) to a lightly manipulated surface creating more or less flexibility (mechanical property). Any number of textile techniques can be applied in this way.

The lesson for a designer, here, is not how to create a super tough textile but understanding how to apply one's own skills and knowledge to manage the level of stiffness via design. In turn, this opens up a completely new conceptual space for textile practice; the precise management of textile stiffness on a local/ topical scale during textile manufacture. For example, working within this framework, knitwear specialist Nicci James (NJ) drew on gradient structures such as those observed in the beak of the giant Humboldt squid.

The beak is composed of 4 substances, water, chitin, protein and pigment. These are distributed across the beak in different ratios, resulting in a range of stiffnesses from super tough at the tip of the beak used to hook and slice the flesh of its prey, to super compliant at the points where the beak interfaces with the squid's soft body (Miserez et al., 2008). This inspired NJ to restrict the material base of her BIT practice to wool and develop via knit stich and felting, an approach to managing the stiffness of her designs.

Figure 5. Left: Detail of Nicci James (2021) knitted biker jacket. Right: a gradient structure is applied via yarn twist and knit stich to implement a flexible line along the length of the collar to enable folding (right). The composition is 100% wool

NJ initially produced an experimental range of swatches focused on understanding the relationship between level of twist in wool yarn, knit stich and felting finish on stiffness. NJ applied this new tacit knowledge to recreate a 'denim' jacket archetype that foregrounds the implementation of a level of stiffness in specific regions within the garment, typically attributed to the type of weave (twill) and fibre (cotton) using just knit, wool and felting finishing (fig. 5). Interestingly, the outcomes of NJ's textile design practice embody innovations that have the potential to reduce the number of materials required to manufacture of complex garments such as a denim jacket, and a shorter value chain. Both elements (reduced range of substance and number of manufacturing steps) contribute to the potential for reducing the energy invested in the formation of the product.

BSDE Tubular

'...structures that consist of arrays of long aligned pores (tubules) within a bulk material' (Naleway et al., 2015)

Tubular structures (fig.10, left) are typically observed in hard biological materials such as animal hooves and teeth to prevent the propagation of cracks caused by impact. The presence of tubules within a solid material is proven to stops crack growth when the tip of the fracture reaches the tubule e.g. in alligator gar (Atractosteus spatula) and armoured fish scales (Yang et al., 2013). The presence of tubules also improves the energy absorption of the material such as in the horn of a male sheep during seasonal flighting.

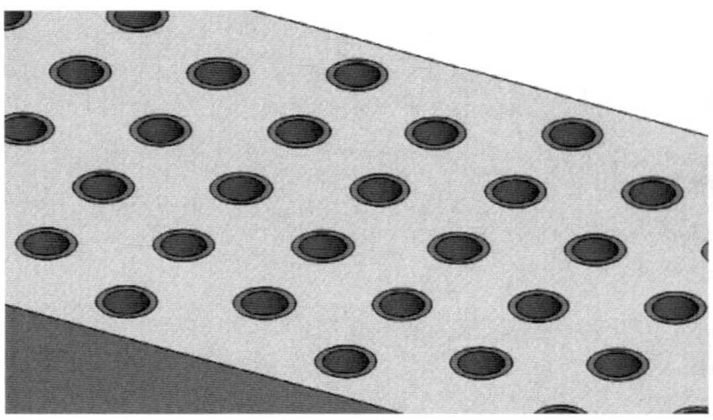

Figure 6. Left: BSDE Tubular. Right: arial image Anna Bruun Kristiansen (2019) tubular weave.

Textile fibres used in the garment and apparel sectors are typically soft and flexible. Solid textile structures, such as those observed in the high-performance composite sectors include strong filaments such as carbon fibre embedded in specialist resins used in transport vehicles. This structural design element was excluded from the original bioinspired textile design framework because the project's primary audience is designers working with conformable, fibre-based materials. Figure 6 demonstrates that although tubular structures are possible, they remain conformable even in bulk. Further research is required to understand the potential for this structure in other textile sectors such as automotive, construction etc that utilise stiff textile composites.

BSDE Texture

Naleway' s review focuses on the mechanical properties of biological materials, in other words how materials behave when different forces are applied to them. As such, the resulting range of BSDE do not feature surface morphology or texture. Examples such the lotus leaf, gecko foot, sharkskin and numerous other examples were beyond the scope of Naleway' s original study. However, novel technologies informed by properties linked to surface morphology of biological materials have permeated the textile technology vernacular. For example, The role of surface contact area in the management of adhesion as observed in the lotus leaf is a narrative that is relatively known to textile designers as the 'lotus effect' (Ensikat et al., 2011). Also, past developments in swimwear technology namely Speedo's Sharkskin (Gamble et al., 2013; Lee, 2014) introduced the concept of a ribbed surface for managing drag forces generated as a swimmer moves through water. The accuracy of these technologies is inconsequential to the discussion at hand, what is important is the opportunity to extend our understanding of texture as a function of surface contact area and how this can be applied to textile design.

Examples such as the lotus leaf and shark dentils describe bumps (lotus leaf) and ridges (shark dentils) at micro scale, as discussed earlier, this is not within a textile designers' sphere of influence. However, what is increasingly interesting about the BSDE archetypes is that they can often transcend scale; lotus papillae are observed at micro scale, so is the distance between ridges of shark dentils. However, pangolin scales feature texture similar to the shark's dentils but can be measured in mm, and protects the scale from abrasion from rocks and soil (Tong et al., 2007).

Abrasion of textiles through wear is a challenge that occurs in many applications from clothing to technical textiles. This new structural design element could support textile designers to consider the numerous methods they use to create texture towards reduced abrasion or even how to include the effects of abrasion within their design. For example, Alice Timmis (AT) an experienced weaver, engaged with the BIT framework motivated by the waste produced by the fashion industry through poorly sized garments. Prior to engagement with the BIT project, AT had developed a method for reviving pre-consumer textile deadstock using industrial brushing techniques, she had also experimented with shaping woven textiles from 2D to 3D by brushing.

AT identified a conceptual link between the BSDE texture (Fig 7) and the techniques she had developed previously and was able to develop this practice further by thinking about the relationship between surface contact area and weave design. AT created woven textiles that feature a ribbed surface of loose floats composed of a low friction weft yarn designed to pull during brushing causing the fabric to contract, giving it 3D qualities. By experimenting with the scale and positioning of the ribs, AT identified an opportunity to design woven garment pieces that could be shaped to fit any body size simply via brushing (fig.8). This new approach has the potential to extend the use of the product through personalised fit to mitigate poor fit in mass produced garments. The concept also demonstrates additional potential for reducing the requirement for seams and subsequent waste during garment production and reducing the number of steps required for the garment formation phase.

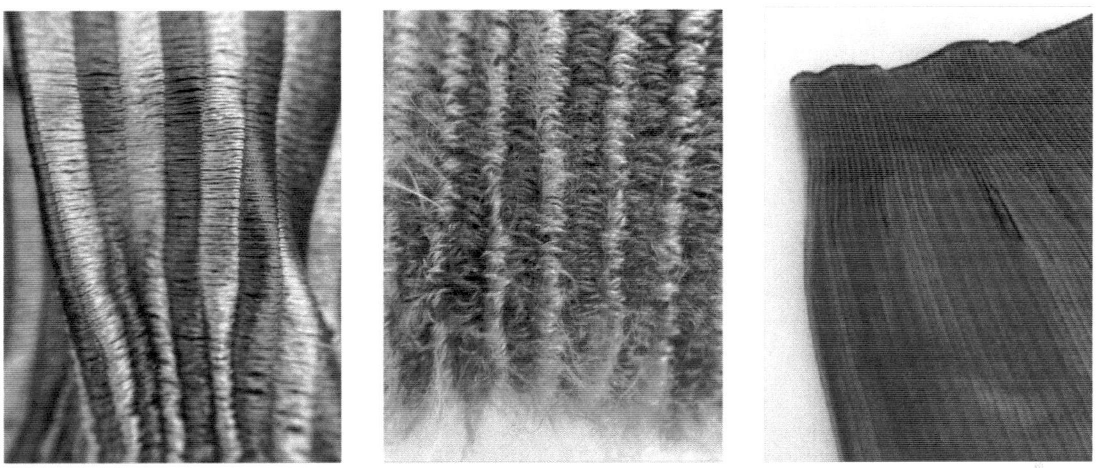

Figure 8: Detail of Alice Timmis 2021 woven textile that can be brushed to fit the wearer. Left: detail of woven rib structure before brushing, middle: after brushing, right: macro view of half skirt with shaped waist induced by localised brushing

BSDE Cellular

"...include open and closed cell foams, scaffolds, or other highly porous materials (e.g., honeycombs) This results in a high-strength but conversely low-weight structures which increases toughness and can resist buckling or bending. In biology these structures exist in a composite material (nano to micro level) or in bulk material (macro level)". (Naleway et al., 2015)

BSDE cellular structures are often observed in sandwich configurations i.e., between two dense walls. For example, the shell of the turtle contains a cellular middle and dense walls either side (fig 9). Cellular structures can contain cells that are either open; cell walls are created by interconnect pathways across the structure or closed; self-contained cell units (ibid).

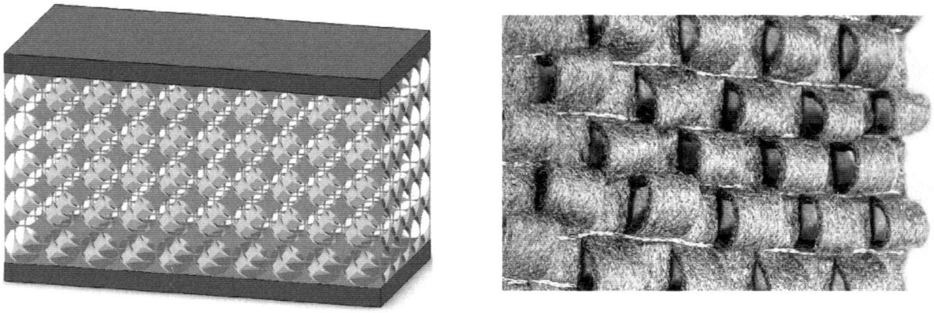

Figure 9: Left: BSDE Cellular . Right: top view of cellular textile by Lisiqi Xie (2021)

Cellular textiles are designed to feature close and orderly distribution of hollows or holes, these can be created using numerous textile techniques in constructed textiles (honeycomb knit or weave), finishing (printing, chemical treatment) and fabric manipulation (smocking and pleating). Conventionally, textile designers have little need to resist buckling or bending. However, there is potential to reimaging cellular textiles as structures that can reduce the amount of material used in a particular textile product, manage volume and control stiffness.

Textile designers can introduce volume in the x, y axis that impacts length and width of the fabric using conventional techniques such as dropping stitches in knitting or lace work. This allows the creation of a textile that uses less yarn or raw material when compared to the same size solid piece of fabric (fig.10). Woven cellular structures with volume designed into the z axis are normally used to introduce thermal resistance to a garment system such as cellular pockets of padding used to line a jacket, however this remains the provenance of technical textiles. More recently, textile designers such as Arantza Villas (Fig.10) have developed novel techniques for creating textiles able to stand unaided, a process draws on the traditional technique of pleating can result in textile systems that can resist compressive forces in excess of their own weight.

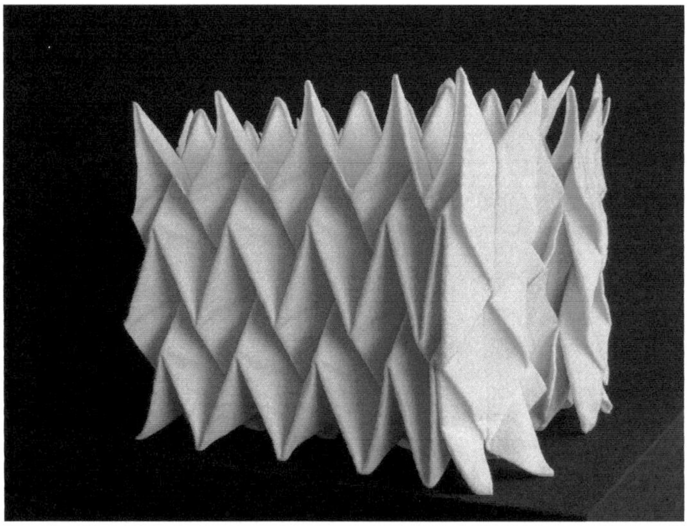

Figure 10. Cellular textile cube made from 100% wool felt and pleating by Arantza Villas (2021)

BSDE Layer

"complex composites that consist of multiple layers or interfaces". (Naleway et al., 2015)

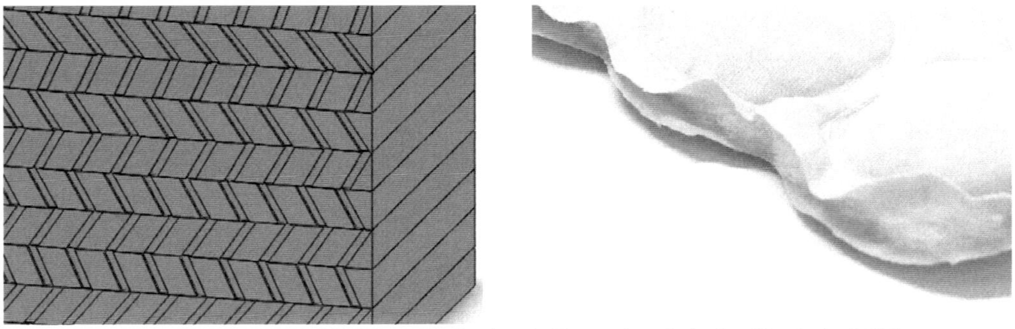

Figure 11. Left: BSDE Layer. Right: detail of layered textile by Sue Hee Parks (2021)

Typically, layered structures (fig. 11, left) are observed in biological materials when toughness is required in systems primarily composed of brittle materials, such as the iridescent lining of the Abalone shell (Gastropoda), commonly known as *mother-of-pearl*. The shell is made up of 95% calcium carbonate (commonly known as chalk) known to be a brittle and weak material. Tiny plates of calcium carbonate are layered on top of each other and held in place by tiny amounts of chitin (Currey, 1977; Yang et al., 2011). Upon impact, the tiny plates absorb the energy by sliding across each other rather than cracking. The resulting structure is seven times tougher than the calcium carbonate alone (Naleway et al., 2015).

The use of brittle materials is uncommon in textiles, except for embellishments such as beads and mirrors. However, composite textiles are common and defined as 'a product formed by intimately combining two or more discrete physical phases, usually a solid matrix and a fibrous material' (Textile Institute, 2021). This configuration provides ways of creating fabrics that combine the properties of each constituent as in composite textiles such as a quilt (fig. 11 right).

Layered textile structures can be created using numerous techniques such as laminating, stitch bonding etc. Localised layers can be introduced via screen printing; a process in which print paste is forced through the unblocked areas of a woven mesh onto a textile substrate. This process is primarily used for aesthetic purposes by designers specialised in the textile print discipline, to introduce decorative features onto a plain substrate. Although not conventionally regarded as a composite structure, this perspective opens opportunities for creative textile designers to think about combining materials of different phases in a way that utilises the mechanical and physical properties afforded by the material combination, the impacts of strategic positioning and amount of material used.

BSDE Overlap

'composed of a number of individual plates or scales that can slide or shift past each other, forming a flexible protective surface.'(Naleway et al., 2015)

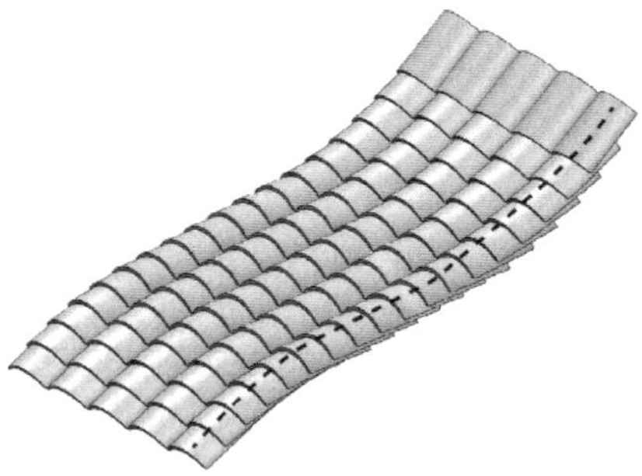

Figure 12. Left: BSDE Overlap. Right: detail of overlapping textile by Katy De Beer (2021)

Overlapping structures (fig. 12, left) are often observed in scaled or armoured skins of animals to offer protection against predator attack. The scales of a pangolin constitute a prime example, they are composed of stiff/tough material which enable the animal to carry a protective shield against tigers, but also allow for sufficient levels of flexibility that does not interfere with the range of motion the animal needs to grow and carry out other life processes.

During the Middle Ages, the development of full plate steel armour technology accommodated the combat culture of the period, especially during the 'Hundred-year war' (Rogers, 1993; Curry, 2014). Figure 13 presents a gauntlet that embodies the state of the art in armour construction circa 1585. Gauntlets were a type of glove designed to protect the hands but at the same time allowing grip and flex motions necessary to wield a sword and hold onto the reigns of a horse. There is no evidence to suggest that the techniques used to produce the flexible elements of gauntlets were informed by the pangolin or other scaled creatures, but there is an obvious similarity.

Overlapping structures in modern textiles are common in fabrics embellished with sequins type components (fig.12, right), here the purpose is purely decorative. However, overlapping allows stiff and hard materials such as steel, bone, and plastic to be included in materials structure that bend. The opportunity for textile designers is to understand how controlled bending can be implemented into the design of textile products e.g., for adaptability or multifunctionality.

Figure 13. Detail of metal gauntlet ca. 1585 Source: V&A archive. https://collections.vam.ac.uk/item/O97506/pair-of-gauntlets-piccinino-lucio/

BSDE Suture

"wavy or interdigitating interfaces that are found within a variety of plates, scutes, and bones, and generally consist of two phases: rigid suture teeth and a compliant interface layer". (ibid)

Figure 14. Left: Suture structure graphic. Right: detail of suture textile by Dr Laetitia Forst (2021)

BSDE suture (fig.14, Left) refers to examples in biology where stiff plates create joins that are both strong and flexible to provide a small amount of movement between the plates as well as strategic points of failure to avoiding serious cracks along then stiff plates (Li, Ortiz and Boyce, 2011) as observed in box fish scutes and sculls of several mammals including humans.

In a surgical context, surgeons use sutures to hold together (stich) a cut or wound. Within textiles, we have a variety of methods to join materials together to create garments and other products. The most common is a simple *seam* in which thread is used to stitch pieces of fabric together, seams can be implemented by hand or machine and embody different forms depending on their role, i.e. if they are load bearing or ornamental. The difference between *suture* in biology and our textile joining techniques is that textile seams are designed to be strong and not fail, in essence the opposite to their role in nature.

Flexibility is a natural property of textiles and failure in the seam is considered a fault. However, the concept of a *fallible* seam can contribute new thinking in design strategies for the disassembly of garment and other textile products. For example, Dr Laetitia Forst (LF), a multimedia textile design applied the BIT framework to review the outcomes of her doctoral study on textile design for disassembly (Forst, 2020b). LT observed the similarities between the laser cutting techniques she developed as part of her doctoral practice to identify new pathways for creating the properties and behaviours of composite textile systems using reversible, non-permanent seams/ joins (fig14, right).

Discussion

'There are not many professions more harmful to the environment than design.' Papanek and Fuller (1972)

We argue that lessons from biology can facilitate the FT industry's transition towards this goal by elevating the importance of textile design and practice. The revised BSDEs constitute a key link that enables access to knowledge from biology and materials science pertaining to the way structure and design enable materials made from simple and abundant substances to exhibit extraordinary, emergent behaviours.

The objective of this approach is not to impart new textile techniques to designers, instead the BSDE invites designers to reimagine their existing tool kits, via a new lens, informed by knowledge from biology interpreted by material scientist. In turn, this stimulates new design thinking and practice that embraces mechanical (and other) properties within the creative design process that could address issues of material and energy waste within the FT sector.

We acknowledge that this approach can appear simplistic. However, we found that designers who engaged with the BIT design framework developed new and innovative textile-based concepts in design for disassembly, mono-material strategies, and consolidation of manufacturing processes, that have the potential to advocate lower-carbon solutions. The practical implementation of concepts emerging from NJ, AT and LF's practice, informed new combinations of established textile and apparel manufacture processes (weaving, knitting sewing, felting, pleating, milling, die-cutting etc.) albeit on a small, craft-based scale. However, translating these novel concepts to an industrial context would disrupt the FT industry's established value chains and economic models.

For approaches such as the BSDE to have impact on FT industry practice, and contribute to a Zero Waste/ Net Zero transition, they must be applicable at scale to ensure industry buy in. Two key obstacles prevent the implementation of a bioinspired textile design framework, emerging from the BSDE, and render it high risk. First, the novel combination of established industry methods required to implement said new concepts disrupts the established value chain currently used in textile and apparel manufacture processes. The untested nature of the new value chains would require new manufacture skills and associated costing models to scale. Secondly, the absence of a credible evaluation system capable of measuring the potential environmental impacts of the BIT design framework in aspects such as the use of narrower range of substances and shorter value chains. Both require further study.

Conclusion

Biology can contribute to innovation and support the transition to Zero Carbon and Zero Waste for the FT industry. Much work is currently underway developing bio-based, -fabricated, -synthesised materials from waste streams, that are biodegradable and/or can replace synthetic counterparts within the value chain (Goldsworthy, 2014; Pervez, 2017; Forst, 2020a; Ribul, Goldsworthy and Collet, 2021). This paper explores how design principles from strong materials in biology, show us an extensive (rather than intensive) approach to introducing properties and behaviours to textile systems, and can be implemented by drawing on the tacit skills and knowledge of creative textile designer/makers.

The foundation informing this approach comes from a comprehensive review of studies into emergent properties of strong/ tough biological materials made from weak, simple substances in ambient conditions by material science scholars. The findings revealed a pattern of eight reoccurring structural design elements that underpin the emergent properties of strong biological material systems. The author (Naleway et al) devised a system of visual graphics to represent each structure, which enabled textile-based researchers to associate with textile structures and techniques and develop a framework to enable knowledge exchange between STEM and textile design.

Biology, within the context of this work, does not teach designers new techniques or methods for making alternative materials. It simply provides a new lens from which to regard one's individual skills and knowledge and ultimately deepen understanding of one's own practice. This has the potential to empower creative design practitioners, by enabling them to challenge the status-quo which tells us that we need advanced chemistry to introduce key properties and functions to our products. Of course, we are not suggesting that this particular framework can eradicate the FT's negative impacts on the environment or mitigate the need for advanced chemistry. But it does give back some agency to the craft of textile design and making in a way that complements efforts in other research and industry sectors to achieve Zero Waste/ Net Zero targets.

Acknowledgements

This paper and the research behind it would not have been possible without the exceptional support of my colleagues Dr Kate Goldsworthy (co-director, Centre for Sustainable Design, University of the Arts London) and Professor Carole Collet (Director, CSM LVMH Sustainable Innovation, University of the Arts London) for challenging and insightful discussions on circular and sustainable design within the creative sector. I would also like to thank the biologically inspired textiles advisory group for their contributions to discussions on this topic. Finally, I would like to acknowledge the AHRC for funding this work (grant AH/T006412/1) and the University of the Arts London for hosting the research.

References

Cohen, Y.H. and Reich, Y. (2016) *Biomimetic design method for innovation and sustainability*. Springer.

Curry, A. (2014) *The Hundred Years' War: 1337–1453*. Bloomsbury Publishing.

'Emissions Gap Report 2022: The Closing Window — Climate crisis calls for rapid transformation of societies.' (2022). United Nations Environment Programme. Available at: https://www.unep.org/emissions-gap-report-2022.

Ensikat, H.J., Ditsche-Kuru, P., Neinhuis, C. and Barthlott, W. (2011) 'Superhydrophobicity in perfection: the outstanding properties of the lotus leaf', *Beilstein journal of nanotechnology*, 2(1), pp. 152–161.

Forst, L. (2020a) 'Disassembly discussed: creative textile sampling as a driver for innovation in the circular economy', *Journal of Textile Design Research and Practice*, 8(2), pp. 172–192.

Forst, L. (2020b) Textile design for disassembly: a creative textile design methodology for designing detachable connections for material combinations. PhD. University of the Arts London. Available at: https://ualresearchonline.arts.ac.uk/id/eprint/16332/.

Gamble, L., Lang, A., Bradshaw, M. and McVay, E. (2013) 'Surface Patterning: Controlling Fluid Flow Through Dolphin and Shark Skin Biomimicry', in. *APS Division of Fluid Dynamics Meeting Abstracts*, pp. 1A – 034.

Goldsworthy, K. (2014) 'Design for Cyclability: pro-active approaches for maximising material recovery', *Making Futures*, 3.

Hatch, K.L. (1993) *Textile science*. Minneapolis/Saint Paul: West Publishing Company.

Jacobs, S., Eggermont, M., Helms, M. and Wanieck, K. (2022) 'The Education Pipeline of Biomimetics and Its Challenges', *Biomimetics*, 7(3), p. 93.

Kapsali and Hall, C. (2022) 'Sustainable approaches to textile design: Lessons from biology', in. *DRS2022*, Bilbao: Design Research Society. Available at: https://doi.org/10.21606/drs.2022.199.

Lee, M. (2014) 'Shark skin: taking a bite out of bacteria', *Remarkable natural material surfaces and their engineering potential*, pp. 15–27.

Lee, S., Du Preez, W., Jones, N.T. and Preez, W. du (eds) (2005) *Fashioning the future: tomorrow's wardrobe*. Repr. London: Thames & Hudson.

Li, Y., Ortiz, C. and Boyce, M.C. (2011) 'Stiffness and strength of suture joints in nature', *Physical Review E*, 84(6), p. 062904. Available at: https://doi.org/10.1103/PhysRevE.84.062904.

Miserez, A., Schneberk, T., Sun, C., Zok, F.W. and Waite, J.H. (2008) 'The transition from stiff to compliant materials in squid beaks', *Science*, 319(5871), pp. 1816–1819.

Naleway, S.E., Porter, M.M., McKittrick, J. and Meyers, M.A. (2015) 'Structural Design Elements in Biological Materials: Application to Bioinspiration', *Advanced Materials*, 27(37), pp. 5455–5476. Available at: https://doi.org/10.1002/adma.201502403.

Niinimäki, K., Peters, G., Dahlbo, H., Perry, P., Rissanen, T. and Gwilt, A. (2020) 'The environmental price of fast fashion', *Nature Reviews Earth & Environment*, 1(4), pp. 189–200.

Pervez, W. (2017) *DESIGN FOR DISASSEMBLY-A CIRCULAR APPROACH*. Virginia Commonwealth University in Qatar.

Ribul, M., Goldsworthy, K. and Collet, C. (2021) 'Material-Driven Textile Design (MDTD): A methodology for designing circular material-driven fabrication and finishing processes in the materials science laboratory', *Sustainability*, 13(3), p. 1268.

Rogers, C.J. (1993) 'The military revolutions of the Hundred Years' War', *The Journal of Military History*, 57(2), p. 241.

Santulli, C. and Langella, C. (2011) 'Introducing students to bio-inspiration and biomimetic design: A workshop experience', *International Journal of Technology and Design Education*, 21, pp. 471–485.

Tong, J., Lü, T., Ma, Y., Wang, H., Ren, L. and Arnell, R.D. (2007) 'Two-Body Abrasive Wear of the Surfaces of Pangolin Scales', *Journal of Bionic Engineering*, 4(2), pp. 77–84. Available at: https://doi.org/10.1016/S1672-6529(07)60017-1.

Yang, W., Gludovatz, B., Zimmermann, E.A., Bale, H.A., Ritchie, R.O. and Meyers, M.A. (2013) 'Structure and fracture resistance of alligator gar (Atractosteus spatula) armored fish scales', *Acta biomaterialia*, 9(4), pp. 5876–5889.

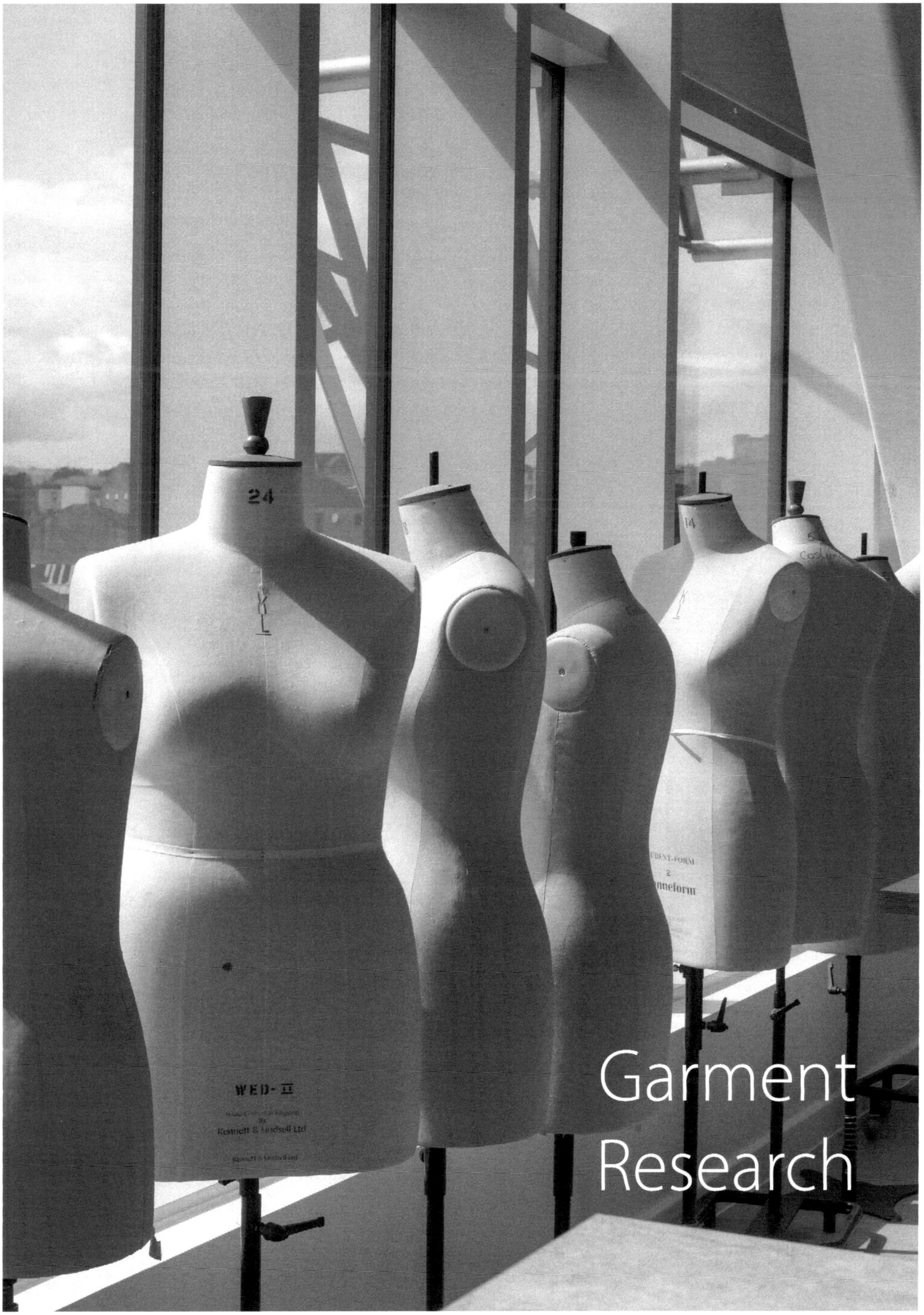

Garment Research

ANALYSIS OF ALGORITHMIC CLUSTERING OF 3D BODY SCAN DATASET TO DEVELOP GARMENT SIZE

Ruqey Alhassawi, *Simeon Gill, Steven Hayes and Kristina Brubacher*

Department of Materials, University of Manchester, Oxford Road, Manchester M13 9PL

Abstract

As there is no perfect standard clothing size system that provides fit satisfaction, or indeed accurate and precise fit, for all customers across numerous geographical regions, this research seeks to analyse the exact body dimensions of individuals, extracted via 3D body scanning, to identify similarities and disparities in the body sizes and shapes of the research sample. This research evaluates a body dimensions dataset of different ethnicities by applying dimensionality-reduction and clustering methods to create body-shape clusters and identify potential sizing systems that are applicable to them. The PCA results indicate that four identified components can explain over 84% of the variability of female body shape characteristics. The K-means clustering algorithm and the analytic results highlight that shoulder circumference and mid-neck-to-waist distance measurements should be taken into account in addition to the bust, waist, and hip measurements to classify the female participants' bodies into three body types. A new body sizing system can help enhance the design and sizing of clothing products to ensure a good fit of the garment to the body. Likewise, the proposed methodology in this study can be applied to any type of worn product where an understanding of body shape and acknowledgement of those variations in body proportions both play a primary role in comfort, fit, and performance.

Keywords

3D body scan, anthropometric measurements, data mining, clustering, sizing system, body shape.

Introduction

Customers tend to possess garments that adapted their body form and meet their aesthetic desires (Cui, Chattaraman and Sun, 2022). Purchasing clothes online, on the other hand, has highlighted certain challenges, such as the inability to try clothes before purchasing, misleading representations with regard to quality, inaccurate style, wrong colour expression, and sometimes bad after-sales service (Silva *et al.*, 2021; Fu *et al.*, 2022). Fit and comfort, on the other hand, are two essential and closely linked variables in garment consumption (Hugo & Van Aardt, 2012). The issue of fit, in particular, became increasingly apparent when consumers bought clothes without first trying them on (Howarton & Lee, 2010; Fu *et al.*, 2022). For instance, discontent with the fit of the apparel accounts for over 30% of all clothing online shopping rates of return, making it the most common cause for returns (Gallino & Moreno, 2018; Silva *et al.*, 2021).

Traditional anthropological measurements are used to construct uniform size systems that can be employed to design apparel that fits a variety of body types (Kouchi, 2019). Designers can make clothes that are customised to fit the proportions of the body by taking measurements such the chest, waist, and hip circumferences. In some cases, traditional measurements may result in a poor fit because take into account a few key dimensions of the body and may not capture the full complexity of an individual's body shape (Ma *et al.*, 2020). In contrast, the generality of body form studies is based on data captured from 3D human models. The 3D body scanning technology supports classic anthropometric measurements extraction of body structure by taking out information on cross-sectional areas, segmental volumes, and some measurements that are manually difficult to take (Chi *et al.*, 2022). This results in a more accurate, well-fit and acceptable garment size. Consequently, it is crucial to be able to analyse and have a deep understanding of body shapes to compare their accuracy relations with traditional anthropometric measurements, which are still widely used in the apparel industry (Paqueta, Peña & Viktor, 2011).

The study of anthropometry and its related data analysis provides garment designers with information about body measurements and shapes for their target population. Once the body scale of specific large samples of population data has been accurately defined, it becomes possible for apparel manufacturers and designers to obtain the appropriate structure of clothes (Hua-e., 2005; Zou *et al.*, 2006).

It has been found from the previous surveys (Kuo, Wang & Lu, 2020) that there were significant differences between different populations on body shapes. This is a view acknowledged by Xiaohui et al. (2018) and Brownbridge, Sanderson and Gill (2014), who mention that choosing only four essential measurements, height, breast, waist, and hip circumferences and permit regular fit, could support size practice. Women's dimensions and body proportions differ from one to the other even though they share the same measurement scale. Therefore, focusing on and identifying the population's body dimensions is crucial to help designers and manufacturers achieve better apparel production performance and meet the requirements of customers. Accurately measuring a person's body size is necessary for appropriate clothing fitting, which makes it essential for the manufacture of clothing. This is anticipated to increase consumer satisfaction and save the business the costs associated with returns as customers find that the clothes are uncomfortable.

Earlier research used multivariate approaches such as principal component analysis, cluster analysis, and discriminate analysis in the related study of breast, face, foot, and body shape categorisation with 3D anthropometric measurements (Lee & Wang, 2015; Pei, Park & Ashdown, 2019; Chi *et al.*, 2022; Parker *et al.*, 2022). However, no previous studies of 3D body scan datasets used AI techniques to cluster participants to determine size variation and body shape categorisation.

This study aimed to analyse a body anthropometric database for a specific population to enhance sizing systems. Moreover, the study intended to determine how anthropometric characteristics cluster in the given anthropometric dataset and describe each cluster's characteristics. Thus, the research uses analytic techniques to study female body anthropometries. It also presents results from the analysis and clarifies their concepts and objectives based on statistical concepts. It then utilises clustering analysis to define human body features in relation to the anthropometric measurements of the study sample. Body type analysis is very significant in made-to-measure (MTM), a phrase used to describe bespoke apparel cut and made from a standard-sized basic design (Simmons, Istook, & Devarajan, 2004). Body type analysis is also a critical method for combining 3D body scanners with digital product manufacturing. This is discussed in greater detail in Section 4.

Research Method

The present study employs a categorisation method for recognising the body size using various multivariate statistical methods and data-mining techniques.

Data Collection

Typically, this research **is classified as** quantitative research type which relies on several data collection methods such as **surveys**, data generated by lab equipment, **analytics** software or existing datasets (Seeram, 2019). In this research, the database is extracted by 3D body scanning technology and Size Stream (Cary, NC, USA) software. The data was collected by the Apparel Design Engineering group at the University of Manchester and includes data from a variety of ethnic groups (Figure 1 A), involving 33 body measurements of 677 female participants.

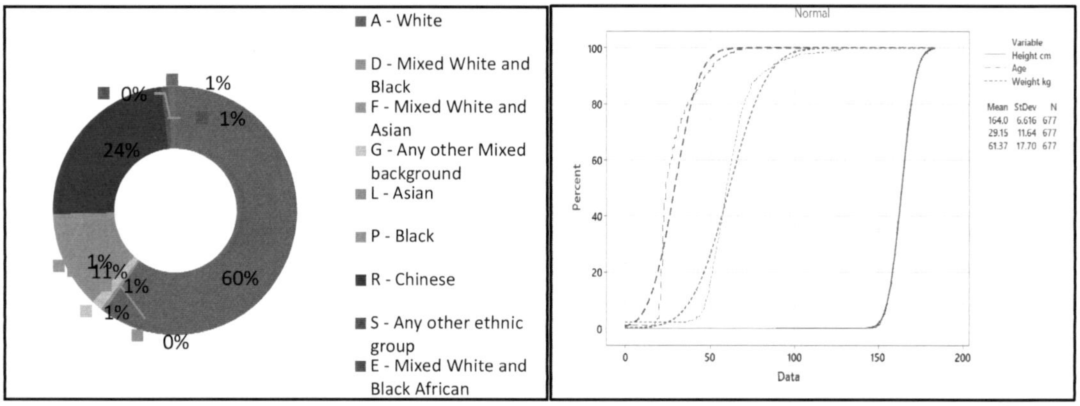

(A) (B)

Figure 1: (A) The chart shows the percentage of each ethnicity of the women data sample, (B) Empirical Cumulative Distribution Function of Height (cm), Age, Weight (kg)

The details of the participants are as follows (Figure 1 B):

- The participants ranged from 18 to 77 years old, with an average of 29 years old, and the median age was 24.00± 11.64 years old.
- The stature of the participants ranged from 124.50 cm to 183.60 cm with an average of 163.98 cm, and the median height was 164.00± 6.62 cm.
- The participants' weight ranged from 37.10 to 163.65 kg, averaging 61.37 kg, and the median weight was 59.50± 17.70 kg.

Statistical Analysis Methods

The research used numerical methods to find relationships and correlation coefficients between variables in the data. The Kaiser-Meyer-Olkin (KMO) and Bartlett's test (or p-value) were applied for statistical analysis. The statistical analysis was performed using MINITAB Version 21.2.0.0 (Minitab Inc, State College, Pennsylvania, USA) and Orange Data mining software (Orange3-3.32.0-Miniconda, Developer: University of Ljubljana). For data visualisation, and data analysis, MATLAB (The MathWorks Inc, Massachusetts, USA) and SPSS for Windows Version 28.0.1.0 (SPSS Inc., Chicago, IL, USA) was used.

Many previous studies have proposed exploring the significant body dimensions and aimed at classifying populations into sub-groups of similar body dimensions to develop the sizing systems. The present study initially reviewed many works of literature involved in the study and analysis of 3D anthropometric data for much-related body parts by utilising AI and data mining techniques (Hsu, Lin & Wang, 2007; Choi & Ashdown, 2011; Gill, 2015; Dianat, Molenbroek & Castellucci, 2018; Xiaohui *et al.*, 2018; Kuo, Wang & Lu, 2020; Chi *et al.*, 2022). Using insights from these studies, the present study's methodology flowchart was constructed and modified to fit this study's data and aim. Figure 2 displays the statistical analysis flowchart for the present study.

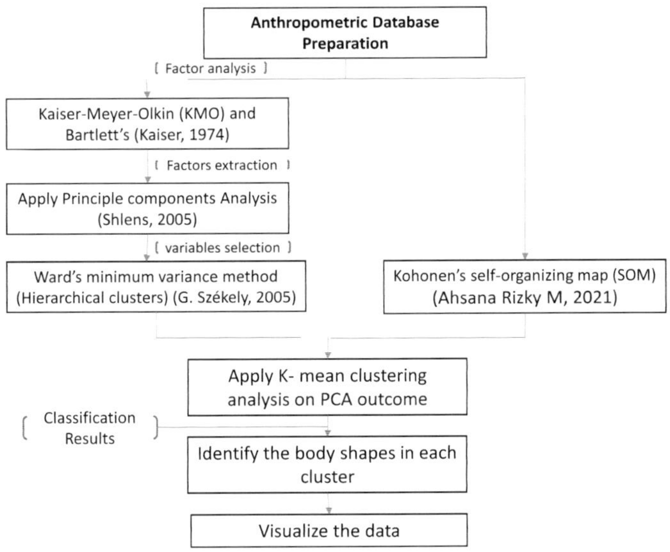

Figure 2: Methodology flow chart

Principal component analysis

Principle component analysis (PCA) is a multivariate statistical analysis method for identifying the key factors in the anthropometric data. This method converts a data set into a new set called principal components, which are orthogonal (linearly uncorrelated) variables (Chung, Lin, & Wang, 2007). PCA was applied to identify important factors for body shape representation to develop the size system.

Ward's minimum variance method

Ward's minimum variance hierarchical cluster analysis method was used to identify the number of clusters for the subsequent non-hierarchical analysis.

Self-Organizing Map (SOM) algorithm

The most popular AI clustering method is Kohonen's self-organising map (SOM) neural network, a useful tool for visualising the characteristics during the discovery stage of data mining. One component of SOM is a competitive layer that divides a dataset of vectors with any set of dimensions into as many classes as the layer's neurons. Since the neurons are placed in a two-dimensional (2D) topology, the layer would represent the distribution and approximate the topology of the dataset in two dimensions (Ahsana, 2021).

K-means clustering analysis

K-means cluster analysis splits the participants into groups for the non-hierarchical process, a relatively functional technique. However, it requires a pre-specified value for the number of clusters to preface the algorithm. The outcomes are sensitive to the specific number of clusters, and the algorithm often ends at a local optimum (Kuo, Wang, & Lu, 2020). Diverse ways are available for defining the number of clusters, such as principal component analysis, Ward's minimum variance method, and hierarchical clustering. The K-means analysis is performed multiple times for each k, and the best ending was decided regarding the best sum of squared errors.

Findings

Statistical analysis results

Dimension reduction PCA

Initially, the KMO of the dataset was evaluated. This is a statistical test to evaluate the suitability of data for factor analysis (Kaiser, 1974). The range of the KMO measure is between 0-1; it is important to ensure its value is 0.6 or greater (Kaiser, 1974). The step that followed involved examining Bartlett's test (or p-value), ensuring that it was less than 0.05 to confirm the validity of using factor analysis for feature detection. In this case, Table 1 shows a reasonably high KMO measure of the sampling adequacy (KMO = .944) and Bartlett's test = 0.000 – less than 0.05 – as the Sig value.

Table 1 KMO and Bartlett's Test value

KMO and Bartlett's Test		
Kaiser-Meyer-Olkin Measure of Sampling Adequacy.		.931
Bartlett's Test of Sphericity	Approx. Chi-Square	53128.907
	Df	496
	Sig.	.000

Figure 3 below illustrates the scree plot, revealing that the slope is descending with the eigenvalues, and it is clear that selecting the first four principal factors is convenient.

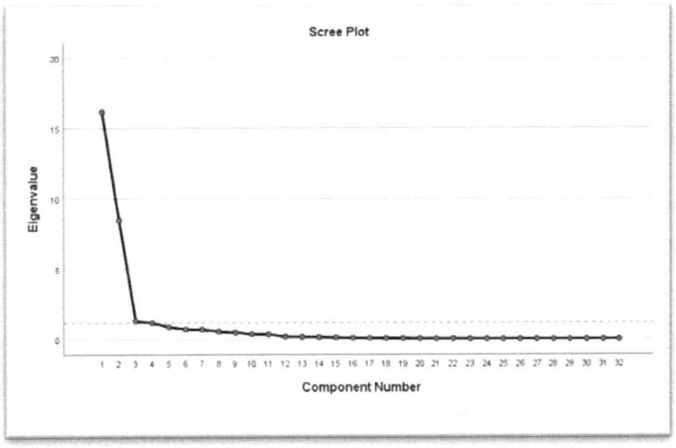

Figure 3: Scree plot of principal components analysis for body shapes.

The four principal components (eigenvalues: 16.150, 8.475, 1.285 and 1.168) cumulatively account for 84.616 % of the total variance, as illustrated in Table 2 below:

Table 6 Total variance explained.

Component	Initial Eigenvalues			Extraction Sums of Squared Loadings			Rotation Sums of Squared Loadings		
	Total	% of variance	Cumulative %	Total	% of variance	Cumulative %	Total	% of variance	Cumulative %
1	16.150	50.468	50.468	16.150	50.468	50.468	14.777	46.177	46.177
2	8.475	26.483	76.950	8.475	26.483	76.950	9.413	29.416	75.593
3	1.285	4.016	80.967	1.285	4.016	80.967	1.486	4.644	80.238
4	1.168	3.650	84.616	1.168	3.650	84.616	1.401	4.379	84.616

The results of PCA in this study indicated that the four components could explain over 84.616 % of the variability of female body shape characteristics (see Table 3). These principal components (PCs) were named Main measurements, Body length–related measurements, Shoulder length measurements, and Bust-related measurements.

Table 7 Results of principal component analysis for body dimensions.

Name	Anthropometric dimensions	PC1	PC2	PC3	PC4
Main measurements	[84] Abdomen Circ [Tape]	0.946			
	[76] Low Waist Circ Pitched [Tape]	0.943			
	[74] Bust Circ with Drop [Tape]	0.940			
	[72] Chest/Bust Circ [Tape]	0.934			
	[105] Opt SoB Waist [Tape]	0.932			
	[86] Seat Circ [Tape]	0.922			
	[153] Hips 8 inches Down from SOB OPT Waist	0.920			
	[75] Underbust Circ [Tape]	0.913			
	[133] Axilla Chest Circ [Tape]	0.912			
	[85] Hip Circ [Tape]	0.905			
	[154] Hips 2 inches Above the Crotch	0.877			
	[178] Across Chest Arm to Arm Length	0.873			
	[13] SideNeck to Bust R [Tape]	0.833			
	[37] Crotch Length Full	0.820			
	[0] Collar/Mid Neck Circ	0.818			
	[12] SideNeck to Bust L [Tape]	0.813			
	[11] Bust Width 1	0.702			
	[1] Neck Base Circ	0.675			
Body length–related measurements	[40] Inseam Length L		0.961		
	[41] Inseam Length R		0.960		
	[36] Crotch Height		0.958		
	[105h] Opt SoB Waist [Tape] CB Height		0.955		
	[43] Outside Leg Length R		0.948		
	[42] Outside Leg Length L		0.940		
	[99] Subject Height		0.901		
	[16] Arm Length L		0.859		
	[17] Arm Length R		0.854		
	[21] Sleeve Length R		0.817		
	[20] Sleeve Length L		0.811		
Shoulder length measurement	[3] Shoulder Length L			0.667	
	[4] Shoulder Length R			0.630	
Bust related measurements	[95] CB Length [CB Nk - Waist] [Tape]				0.746
	[88] X Bk Pit-to-Pit [Tape]				0.497
Eigenvalue		16.150	8.475	1.285	1.168
Variance explained (%)		50.468	26.483	4.016	3.650
Cumulative explained (%)		50.468	76.950	80.967	84.616

The Ward's minimum variance (Hierarchical clustering) method

Ward's minimum variance method and classification (Hierarchical clustering) techniques were performed in this step to obtain the optimal clustering number.

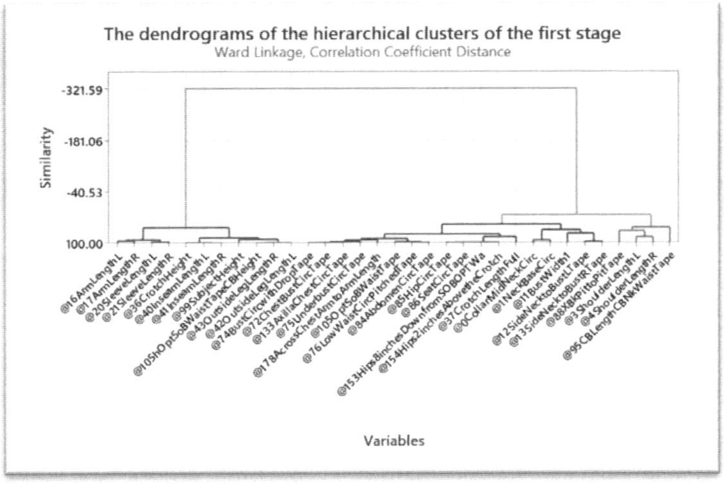

Figure 4: Dendrograms of Hierarchical clustering

The variable with the highest factor loading of each principal component from the previous step was considered the independent variable for this cluster analysis. This means the 33 body measurement variables of 677 females were grouped into three clusters (Figure 4) and used in the following handling step.

Self-organizing Map (SOM) analysing techniques result.

The self-organising feature map (SOM) is the most used network for clustering applications. Self-organising maps acquire the ability to group related data collectively. To create the network, the map size needs to be identified; this corresponds to the number of rows and columns in the grid. For this example, the optional map size value was set to a 12-by-12 grid, so there are 144 neurons. The default topology of the SOM is hexagonal. Figure 5 shows the neuron locations in the topology and indicates the quantity of training data associated with each of the neurons (cluster centres).

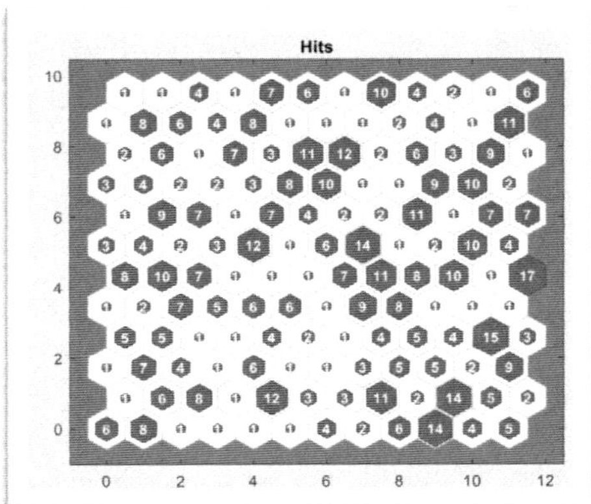

Figure 5: The SOM topology of this project data

It is possible to visualise the SOM by displaying weight planes, also referred to as *component planes* (four inputs in this case), as illustrated in Figure 6).

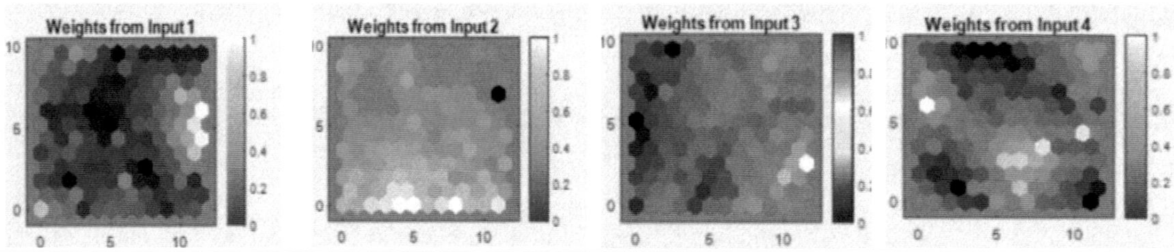

Figure 6: SOM component planes

Figure 6 shows a weight plane for each input vector element connected to each neuron (Darker colours represent larger weights). If the connection patterns of the four inputs were very similar, it is possible to assume that the inputs are highly correlated. In this case, Input 2 is mismatched. The SOM network clustered the anthropometric data into three distinct groups. Figure 7 shows a clear representation of the clusters.

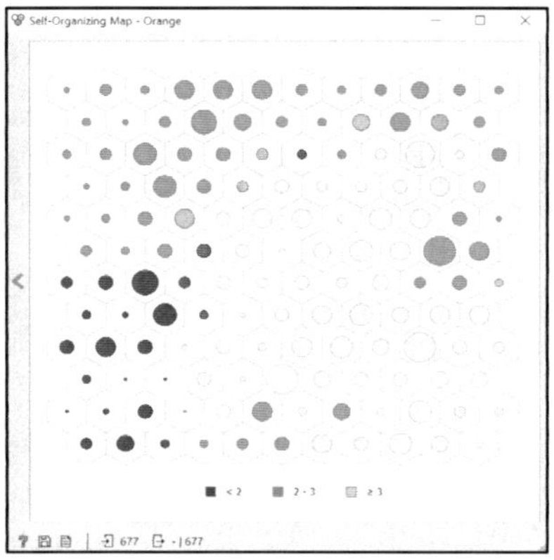

Figure 7: the SOM clustering for given data

The three clusters resulting from the Hierarchical and SOM Clustering mean that there are three different body shapes with three different sizing dimensions for the studied data for sizing system developments.

K-means cluster analysis

Using the K-Means method (Lee & Wang, 2015) for cluster analysis, the body shape data can be clustered and the observed variables can be categorised into distinct groups depending on the outcomes of factor extraction, as shown in Table 3. In the second stage of non-hierarchical clustering, the K-means method was applied to 3 clusters. Each body type was described by its cluster centroid data in total main measurements, body length–related measurements, shoulder length measurements, and bust-related measurements. Figure 8 shows the contribution of the elements to the distribution of participants in each cluster.

The distribution of the data cases in each cluster was 85, 261, and 331 participants in clusters 1, 2, and 3, respectively. The lowest number of participants is in the first cluster, whereas the most significant number is in the third cluster. To assess the significance of each feature of the entered data – since the experiment is carried out with the outcomes of the principal components step – the four principal components were utilised as inputs of the two-step algorithm through the SPSS program. This made it possible to test the importance level of each of the PCs to visualise the data for sizing system development.

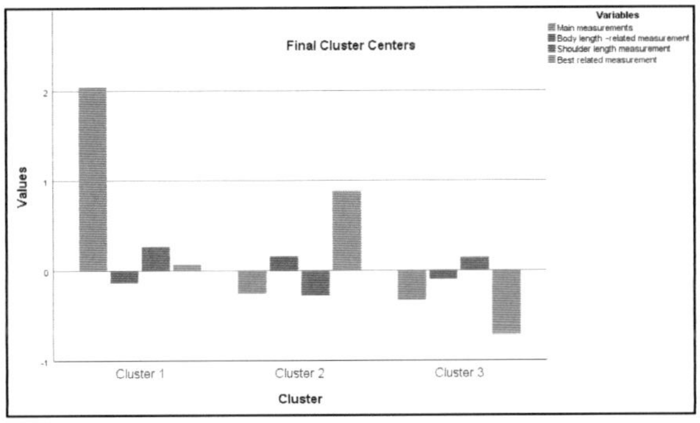

Figure 8: Contribution of variables in each cluster

The first (main measurements) and fourth components (Bust related measurements) were given a high amount of relevance in determining garment sizes, as illustrated in Figures 9 A and B. The third factor (Shoulder length measurement) is next in significance. Finally, the second component (Body length–related measurement) has the lowest impact on the effectiveness of the distribution of the data.

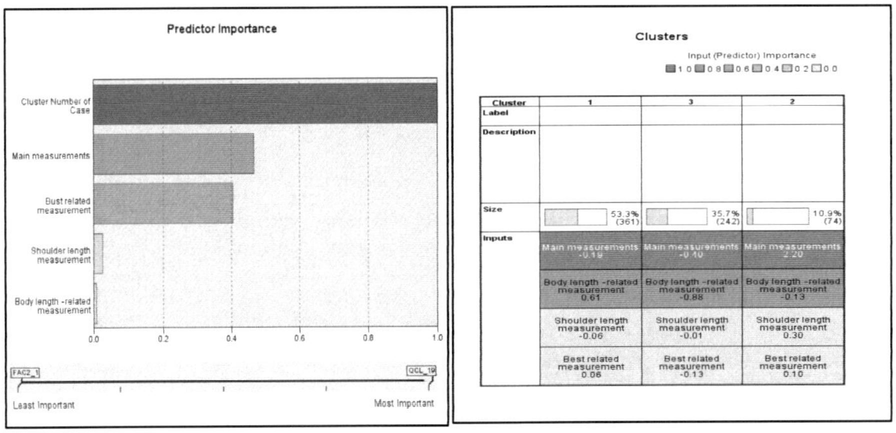

(A) (B)

Figure 9: (A) shows the level of importance of characteristics in the data distribution, and (B) shows the significance of each component's involvement in each cluster.

The analytic result visualisation

The distribution regarding the most critical principal components results, which are main measurements (x-axis) vs. bust-related measurements (y-axis) vs. shoulder length measurements (z-axis), is illustrated in Figure 10 A for all three clusters. In comparison, the 2D plot (Figure 10 B) shows the data distribution of the first crucial measurements regarding the PC analysis outcomes. The clusters are distributed based on the basic variables that describe body sizes according to correlated variables.

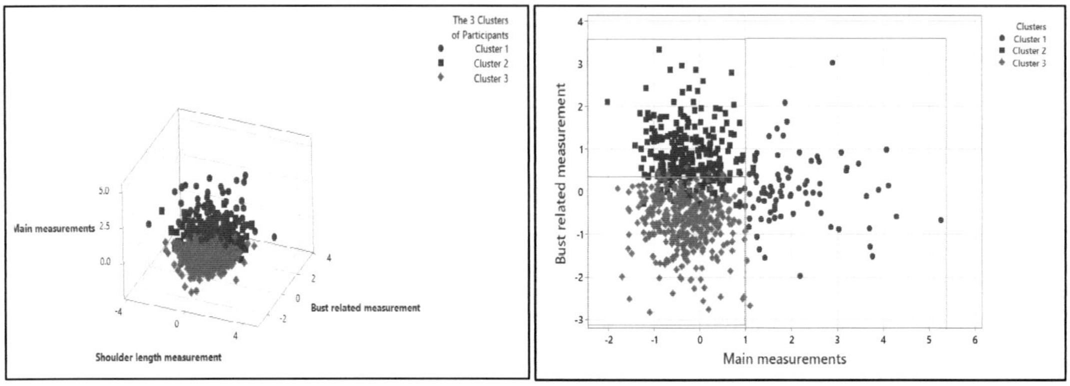

(A) (B)

Figure 10 (A): 3D scatter plot data distribution of Main body measurements vs. Bust-related measurements vs. Shoulder length measurements for the three clusters, (B) 2D scatter plot of Main measurements vs. Bust-related measurements.

Table 4. Body dimensions of the four body shape types in cm

	Cluster 1	**Cluster 2**	**Cluster 3**
Number of observations	85	261	331
Average distance from the centroid	1.85198	1.482283	1.48595
Average Hip measurements	60.80	50.13	49.26
Average pit-to-pit measurements	34.78	30.71	27.97
Average shoulder length R and L	26.02	22.70	22.90
Average subject height	164.92	74.72	73.57

The variables from each principal component [85] Hip Circ [Tape] 0.905 (value of factor loading), [88] X Bk Pit-to-Pit [Tape] 0.497, the average of Shoulder Length L 0.667, Shoulder Length R 0.630, and [99] Subject Height 0.901 were chosen. It has been observed that these variables are visually related to each other. Therefore, each body type will be described by its cluster centroid data (Table 4) in Hip measurements, Pit-to-Pit measurements, average shoulder length R and L, and subject height.

It is clear that the extracted variables from the PCA and the importance level for each variable's components aid in the conceptualisation of the variety of female bodies. For instance, the PCA results allow focusing on some extracted body dimensions, such as the height from the centre neck point to the centre waist circumference, besides the primary measurements (bust, waist, and hip) that could produce well-fitting clothes. Figure 11 illustrates the data sample distribution on a 2D scatter plot across one of the direct measurements from the PC1 (Hip measurements) vs. one of the variables from PC4 (Pit-to-pit measurements).

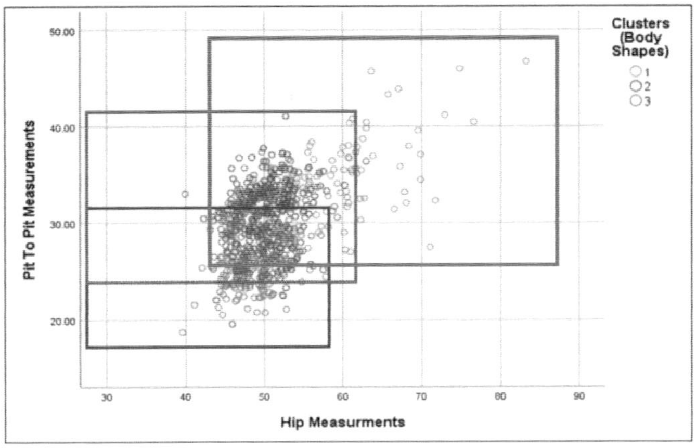

Figure 11: 2D scatter plot Hip measurements vs. Pit-to-pit measurement orange frame is for LAW body shape, the blue frame is for AAA body shape, and the purple frame is for SSN

The first body shape type (Cluster 1) was characterised by its large hip size, average Pit-to-pit measurements, and wide shoulder length (LAW type). The second body shape type (Cluster 2) was characterised by its average Hip measurements, average Pit-to-pit measurements, and Average shoulder length (AAA type). The third body shape type (Cluster 3) was characterised by its small Hip measurements, small Pit-to-pit measurements, and narrow shoulder length (SSN type). The highest proportion was found with the SSN body shape type (48.89 %), followed by AAA (38.55 %) and LAW (12.55%).

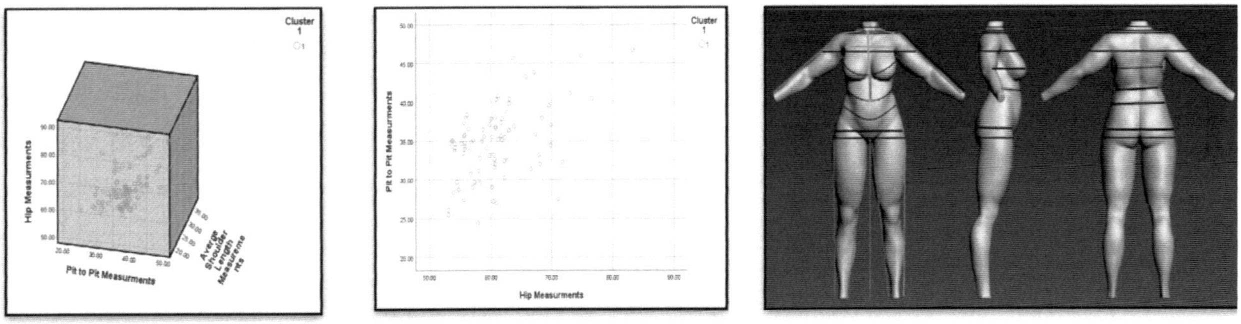

Figure 12: Cluster 1 LAW body shape information with the identification of the body dimensions that have a high loading factor.

The scatter plot (Figure 12) of cluster 1 through Hip measurements vs. Pit-to-pit measurements describes LAW body type due to the Cluster 1 data distribution on the 3D plot graph through the three variables. The interval of distribution for the Hip measurements is between 53.04 and 83.27, and the average is 60.80. For Pit-to-pit measurements, it is between 24.45 and 46.67, and the average is 34.78. For the Shoulder length, it is between 19.03 and 36.21, and the average is 26.02, and for the subject, height is between 125.88 and 182.76, and the average is 164.92.

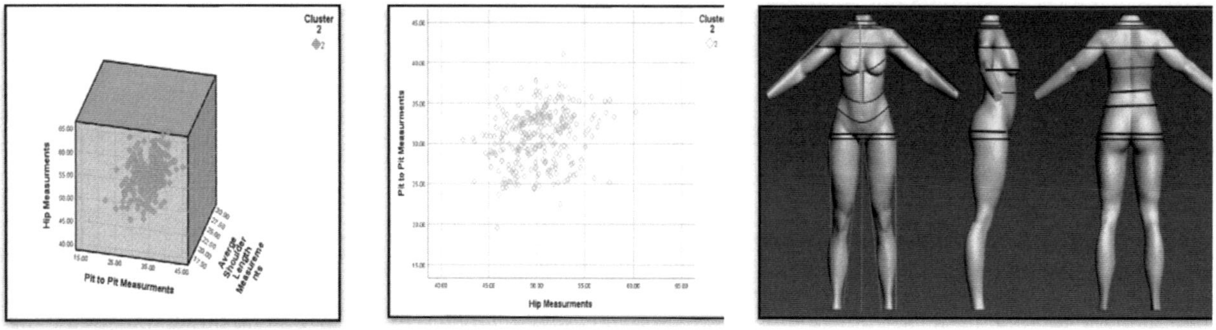

Figure 13: Cluster 2 AAA body shape information with the identification of the body dimensions that have a high loading factor.

The scatter plot (Figure 13) of Cluster 2 via Hip measurements vs. Pit-to-pit measurements describes (AAA body type), due to the Cluster 2 data distribution on the 3D plot graph through the three variables. The interval of distribution for the Hip measurements is 42.26 to 60.24, and the average is 50.13; for Pit-to-pit measurements, it is 19.55 to 41.07, and the average is 30.71. For the Shoulder length, it is between 17.06 and 28.46, and the average is 22.70, and for the subject, height is between 149.50 and 185.19, and the average is 166.85.

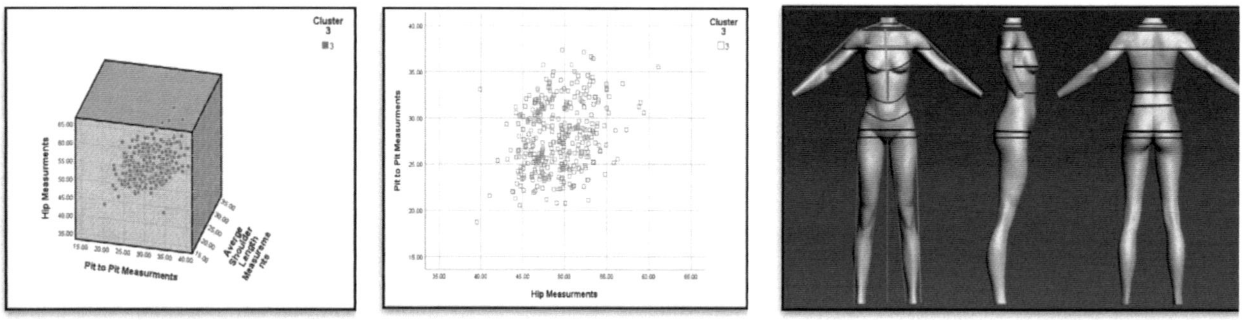

Figure 14: Cluster 3 SSN body shape information with the identification of the body dimensions that have a high loading factor

The scatter plot (Figure 14) of Cluster 3 via Hip measurements vs. Pit-to-pit measurements describes SSN body type due to the Cluster 3 data distribution on the 3D plot graph through the three variables. The interval of distribution for the Hip measurements is between 39.54 and 61.10, and the average is 49.26. For Pit-to-pit measurements, the range is between 18.72 and 37.35, and the average is 27.97. For the Shoulder length, the interval is between 18.22 and 32.41, and the average is 22.90, and for the subject, height is between 146.95 and 181.96, and the average is 163.77.

Each body shape type can be classified into several size subgroups based on the control dimensions and size interval. As intense, based on clustering analysis techniques results, the size interval was considered as the real distribution in each cluster. It has been found that there are three sizes in each body type, meaning there are nine sizes. (Table 5) can be used to determine the fit size when considering the garment product size.

Table 5 The proposed size chart from clustering analysis results (unit: cm).

Size	SSN Body Type			AAA Body Type			LAW Body Type		
	Hip one side measurement	Bust one side (pit to pit) measurement	Shoulder length	Hip measurement	Bust (pit-to-pit) Measurement	Shoulder length	Hip measurement	Bust one side (pit to pit) measurement	Shoulder length
S	39.5-46.72	18.63-25	17.6-20.5	42-48	19.5-26.5	18.22-23	53-63.1	24-31.8	19-24.5
M	47-54	25.5-31.5	21.24.5	48.5-54.5	26.6-33.6	23.5-28	63.2-73.3	32-39.4	24.7-30.5
L	55-61	32.00-37.5	25-28.5	54.6-60.24	34.00-41.1	28.5-32.5	73.4-83.27	39.5-46.67	31-36.21

An efficient sizing system should provide fewer sizes, serve a large proportion of society, and provide a better fit. Creating a standard-size system is challenging since these three requirements contradict one another (Kuo, Wang & Lu, 2020). This study used Hierarchical, k-mean and SOM clustering methods to develop sizing systems for classifying body shapes. These algorithms have high efficiency for clustering massive data into groups within similar information through many dimensions.

Discussion

Analytical Results

PCA was used to analyse 33 anthropometric measurements of 677 female subjects, which produced new variables that were constructed as linear combinations or mixtures of the original variables called principal components (PCs). The contribution rate of four PCs was approximately 84%, which seems to be approximately a high value that describes the original variables. This means that the four components used were found to explain more than 84% of the variability within dataset body shapes.

In the PCA stage, the variables with the largest loadings considered suitable measurements for garment size were used in the Hierarchical cluster analysis. The Hierarchical method was used to identify the optimal number of clusters for the female body. Furthermore, the SOM approach was employed to project 33 anthropometric variables onto a 2D map without reducing dimensions to extract the optimal clustering number. The SOM method was an effective analysis tool, as it can cluster data characteristics within large datasets, such as a sizeable database of anthropometric dimensions. The performance of the k-means cluster analysis method was applied next. Figure 15 displays the scatter plot of nearly non-overlapping PC1 via PC2 data. This means that the characteristics and body dimensions of the PC1 and PC2 components give distinct classifying results for the participants' body shapes.

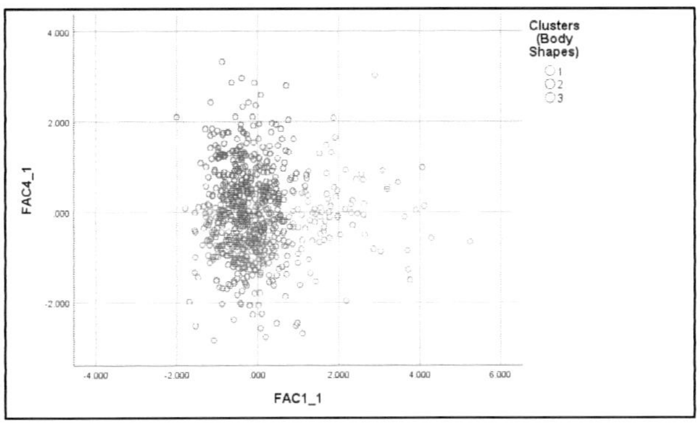

Figure 15: the distribution of the data across PC1 via PC2

There are multiple main body measurements that have a high loading factor. This study examined the key body measurements such as ([84] Abdomen Circ [Tape], Chest/Bust Circ [Tape], [85] Hip Circ [Tape]) and their factor loading (0.946, 0.934, and 0.905), respectively. On the other hand, there are only two variables for PC4 ([95] CB Length [CB Nk - Waist] [Tape] and [88] X Bk Pit-to-pit [Tape]) and their factor loading (0.746 and 0.497), respectively. The PC2 variables are related to the bust to waist, and arm hole right to arm hole left region that could be considered for the level of clothes tightness or smoothness on the body. This suggests that these variables could be important for clothes sizing.

Statistical analysis results in visualisation software.

To visualise the data analysis results, Zbrush software was used to illustrate the average measurements for each of the three cluster results (see Figure 16). Furthermore, the Marvelous Designer v2.9.3 (Tokyo, Japan) software was used for virtual fitting on the avatar, describing the three body shape clusters with effective body dimensions (Figures 17 A and B).

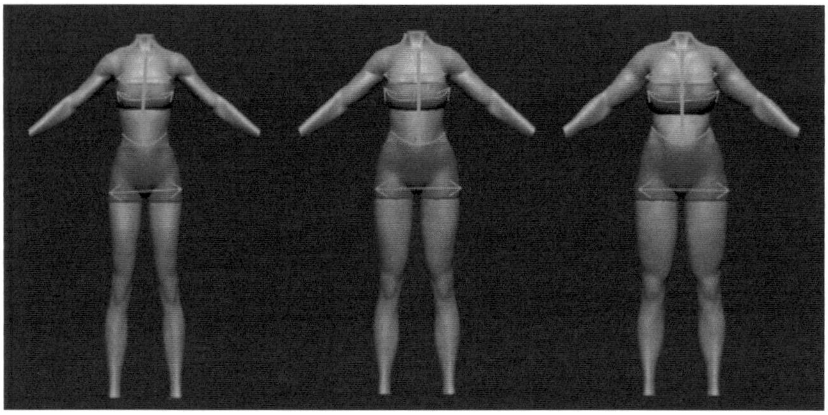

Figure 16: 3D human models for proposed body shapes with the examined body measurements

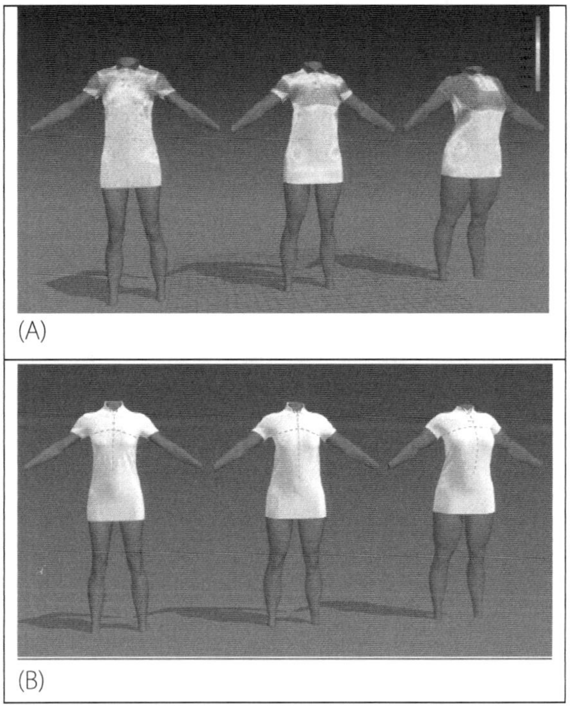

Figure 17: (A) design for virtual fitting on three body shapes left (SSN), middle (AAA), and right (LAW), (B) 3D body model with the support measurements on the garment.

Figure 17 B demonstrates the virtual clothes fitting outcome that used the proposed three body shapes with the extracted measurements. This clearly explains how tightness/smoothness on bust area distribution on the blocks ensures the garment fits. In recapitulation, the higher number of clusters for participants could affect their respective group sizes. The highest proportion was found in this study with the proposed SSN body shape type (48.89 %), with 331 out of 677 participants.

Conclusion

This analytical research study employed the extracted 3D body measurements to find correlated body attributes. Three cluster algorithms (Hierarchical, SOM, and K-mean) were used to classify population characteristics of body shapes within comparable sizing. The methods (Hierarchical and SOM clustering methods) have been used to compare their results in exploring the optimal number for data clustering. Both methods provide an optimal number of 3, enabling the K-mean algorithm to cluster the data set into three distinct body shape types regarding body anthropometry data. Each body type is described according to hip, pit-to-pit, and shoulder length measurements. Each body type has been divided into three sizes depending on the value of the relevant cluster intervals, and it extracted nine different sizes. Those outcomes can provide beneficial information for designing and processing body-related products. Considering the exact and comfortable fit would lead to customer satisfaction and trust in clothes makers, which is a crucial aim for the sustainability of the apparel production line and reduced purchasing returns.

References

Ahsana Rizky M (2021) *Self Organizing Maps (SOM) Using R & Kohonen (case study: Wines dataset).* Available at: https://ahsanarizkym.medium.com/self-organizing-maps-som-using-r-kohonen-case-study-wines-dataset-4bdbad42e41 (Accessed: 8 October 2022).

Chi, C. *et al.* (2022) 'A study on segmentation and refinement of key human body parts by integrating manual measurements', *Ergonomics*, 65(1), pp. 60–77. doi:10.1080/00140139.2021.1963489.

Choi, S. and Ashdown, S. p. (2011) '3D body scan analysis of dimensional change in lower body measurements for active body positions', *Textile Research Journal*, 81(1), pp. 81–93. doi:10.1177/0040517510377822.

Chung, M.J., Lin, H.F. and Wang, M.J.J. (2007) 'The development of sizing systems for Taiwanese elementary- and high-school students', *International Journal of Industrial Ergonomics*, 37(8), pp. 707–716. doi:10.1016/j.ergon.2007.05.004.

Cui, T., Chattaraman, V. and Sun, L. (2022) 'Examining consumers' perceptions of a 3D printing integrated apparel: a functional, expressive and aesthetic (FEA) perspective', *Journal of Fashion Marketing and Management*, 26(2), pp. 266–288. doi:10.1108/JFMM-02-2021-0036.

Dianat, I., Molenbroek, J. and Castellucci, H.I. (2018) 'A review of the methodology and applications of anthropometry in ergonomics and product design', *Ergonomics*, 61(12), pp. 1696–1720. doi:10.1080/00140139.2018.1502817.

Fu, B. *et al.* (2022) 'An improved clothing size recommendation approach based on subdivision of female body types', *Ergonomics*, 0(0), pp. 1–15. doi:10.1080/00140139.2022.2069867.

Gallino, S. and Moreno, A. (2018) 'The value of fit information in online retail: Evidence from a randomized field experiment', *Manufacturing and Service Operations Management*, 20(4), pp. 767–787. doi:10.1287/msom.2017.0686.

Gill, S. (2015) 'A review of research and innovation in garment sizing, prototyping and fitting', *Textile Progress*, 47(1), pp. 1–85. doi:10.1080/00405167.2015.1023512.

Howarton, R.D. and Lee, B. (2010) 'Market analysis of fit preferences of female boomers', *Journal of Fashion Marketing and Management*, 14(2), pp. 219–229. doi:10.1108/13612021011046075.

Hsu, C.H., Lin, H.F. and Wang, M.J. (2007) 'Developing female size charts for facilitating garment production by using data mining', *Journal of the Chinese Institute of Industrial Engineers*, 24(3), pp. 245–251. doi:10.1080/10170660709509039.

Hua-e., W. (2005) 'Research on body form parameters of women's trunk.', *Beijing Textile Journal.*, 26(5), pp. 59–61.

Hugo, S.H. and Van Aardt, A.M. (2012) 'Evaluative criteria applied by South African female fashion consumers when purchasing casual daywear', *International Journal of Consumer Studies*, 36(4), pp. 460–471. doi:10.1111/j.1470-6431.2011.01073.x.

Kaiser, H.F. (1974) 'An index of factorial simplicity', *psychometrika*, 39, pp. 31--36.

Kouchi, M. (2019) Anthropometric methods for apparel design: Body measurement devices and techniques. 2nd edn, Anthropometry, Apparel Sizing and Design. 2nd edn. Elsevier Ltd. doi:10.1016/B978-0-08-102604-5.00002-0.

Kuo, C.C., Wang, M.J. and Lu, J.M. (2020) 'Developing sizing systems using 3D scanning head anthropometric data', *Measurement: Journal of the International Measurement Confederation*, 152, p. 107264. doi:10.1016/j.measurement.2019.107264.

Lee, Y.C. and Wang, M.J. (2015) 'Taiwanese adult foot shape classification using 3D scanning data', *Ergonomics*, 58(3), pp. 513–523. doi:10.1080/00140139.2014.974683.

Ma, Q. *et al.* (2020) 'Learning to dress 3D people in generative clothing', *Proceedings of the IEEE Computer Society Conference on Computer Vision and Pattern Recognition*, pp. 6468–6477. doi:10.1109/CVPR42600.2020.00650.

Paqueta, E., Peña, I. and Viktor, H.L. (2011) 'From anthropometric measurements to three-dimensional shape', *Indian Journal of Fibre and Textile Research*, 36(4), pp. 336–343.

Parker, C.J. *et al.* (2022) 'A method for increasing 3D body scanning's precision: Gryphon and consecutive scanning', *Ergonomics*, 65(1), pp. 39–59. doi:10.1080/00140139.2021.1931473.

Pei, J., Park, H. and Ashdown, S.P. (2019) 'Female breast shape categorization based on analysis of CAESAR 3D body scan data', *Textile Research Journal*, 89(4), pp. 590–611. doi:10.1177/0040517517753633.

Petrova, A. and Ashdown, S.P. (2008) 'Three-Dimensional Body Scan Data Analysis', *Clothing and Textiles Research Journal*, 26(3), pp. 227–252. doi:10.1177/0887302x07309479.

Seeram, E. (2019) 'An overview of correlational research', *Radiologic technology*, 91(2), pp. 176–179.

Silva, S.C. *et al.* (2021) 'Need for touch and haptic imagery: An investigation in online fashion shopping', *Journal of Retailing and Consumer Services*, 59(November 2020), p. 102378. doi:10.1016/j.jretconser.2020.102378.

Simmons, K., Istook, C.L. and Devarajan, P. (2004) 'female Figure Identification Technique (FFIT) for Apparel. Part 1: Describing Female Shapes', *Journal of Textile and Apparel, Technology and management*, 4(1), pp. 1–16.

Xiaohui, T. *et al.* (2018) 'Automatic human body feature extraction and personal size measurement', *Journal of Visual Languages and Computing*, 47(February), pp. 9–18. doi:10.1016/j.jvlc.2018.05.002.

Zou, F.Y. *et al.* (2006) 'Application of neural network to identification of young females' body type', *Conference Proceedings - IEEE International Conference on Systems, Man and Cybernetics*, 3, pp. 2257–2261. doi:10.1109/ICSMC.2006.385197.

SUSTAINABLE SPORTS COMPRESSION GARMENT PRODUCTION THROUGH SEAMLESS KNITTING

Zeeshan Azam[1], *Kristina Brubacher*[1], *Simeon Gill*[1] *and Anura Fernando*[1]

[1] *Department of Materials, The University of Manchester, Oxford Rd, Manchester, M13 9PL*

Abstract

Compression garments are designed to exert an engineered amount of external compression on a particular body part or whole body by releasing kinetic energy stored in elastomeric yarn, supporting underlying tissues, positively affecting performance and improving muscle recovery. Seams exert undesirable compression, and stitching produces waste in the cutting operation. Body sized weft knitting machines create garments with minimum or no stitching and with fewer operations. This provides more possibilities for aesthetic designs and more control over the knitting parameters to have the required compression, resulting in the freedom of body movement and a good fit. This study investigated the variation in garment dimensions and compression (mmHg) applied by three knitted structures manufactured with two feeding tensions to two sizes of wooden legs. Results found compression gradually decreases towards the thigh from the ankle, measured using a Picopress device, and is significantly affected by the elastomeric materials, feeding and body dimensions. The authors suggest including material properties and body dimensions for compression garments to result in better fit, reduced product failures, improved garment functionality, and cost savings.

Keywords

Seamless knitting, Compression garments, Sizing, Sustainability

Introduction

Sports compression garments (SCGs) are tight-fitting clothes designed with negative ease to apply pressure to the body parts they cover, worn by athletes particularly for post-activity benefits (Atkins et al. 2020). Stretchable knitted structures provide the required extension, and elastomeric fibres offer recovery (Brubacher et al. 2021). Key reported advantages of SCGs when correctly worn are ease of muscular pain and fatigue, optimized blood flow, improved muscle oxygenation decreasing blood lactate, and injury reduction (Atkins et al. 2020; Duffield and Portus 2007; Born et al. 2013; Fu et al. 2013; MacRae BA, Cotter JD 2011; Ashdown 2011; Sang et al. 2015). These factors can increase overall endurance and performance during activity or exercise.

SCG fit is a correlation between a garment's size and body dimensions depending upon required compression. When one measurement fluctuates dramatically from the assumed standard, the correlation is weak. As a result, the resulting clothing won't always offer the desired level of compression and won't always fit the body well (Gupta et al. 2006; Robinette 1986; Ashdown 1998).

The sustainability core principle is to meet the current needs without harming future generations' needs, and the first concept in sustainability is to develop sustainability defined by the world commission on Environment and Development. Sustainable production is defined in many ways, including products with increased lifetime and utilization, using methods and procedures that are less harmful to the environment (Fung and Liu 2019). On the other hand, parallel to sustainable development, sustainable consumption should be adopted. Products should meet basic needs and improve life quality while reducing natural resource consumption (Lavorata 2014; Norwegian Ministry of Environment 1994).

Waste reduction is a core element of sustainability, and studies reported 31.2% (Power 2012) of a waste factor in the knitted product market, which highlights the need for new technologies like 3D and seamless knitting, which has already been adopted in different product categories especially small products like gloves, socks, stockings. These are produced with minimum or no waste, unlike cut and sewn products, where each product produces waste without any post-manufacturing operation (Choi and Powell 2005). This study focusses on improving the quality of sustainable compression garments addressing the required fit and compression, which will improve the product's functionality, use, and comfort to the wearer.

Circular knitting creates tubular fabric and can be utilized to develop different hosiery products, semi-seamless and cut and sewn garments. The diameter of the machine has a significant impact on the circumference/width of the knitted tube. Still, other elements such as fewer knitting needles, different stitch patterns, experimenting with stitch length, and tension of the knitting yarns also impact the garment size (Maity et al. 2022; Santoni 2022; Merz 2022). As the diameter remains the same, it is challenging to make any significant changes in the dimension of products and requires different machines to produce different sizes. If different sizes are produced by the same machine changing other factors, only small dimensional changes can be achieved, which is a major limitation. However, it may also lead to changes in the compression level and comfort properties of the products. When an athlete has varying body dimensions at one or more points, it presents the same problem of varying compression, which may lead to unhealthy compression exertion on the wearer. This study aims to examine how the feeding and tension of an elastomeric knitting pattern affect compression and dimension variation. The study also discusses the difficulties with seamless compression hosiery stockings in terms of their uneven compression caused by varying body measurements.

Materials and methods

Materials

The main yarn was Nylon 6 (NH) (CH2) 5 CO) n, double ply, 70 denier, 24 filaments. It belongs to the family of synthetic textiles comprising polyamides with repeating amide link units. The inlay yarn was made of 180 deniers Lycra® with two coverings of 40 denier nylon (Double covered (DCV)). In a double covering, an elastic yarn (Lycra®) is stretched before being wrapped twice with nylon yarns facing opposite directions.

Experimental factors

This study evaluates compression variation using the experimental factors of a knit structure, feeding tension of elastomeric yarn and the wooden leg sizes with three, two and two levels of each factor. The experimental factors and their levels are shown in Table .

Table 1 Experimental factors and their levels

Experimental Factors	**Levels**		
	1	**2**	**3**
Knitting structure 3 *3 (Rib)	Plain (P)	Cross (C)	Random (R)
Knitting tension of elastomeric yarn	Medium (M)	Tight (T)	
Size of measuring mannequin	24	27	

Weft knitting

Weft knitting is the interloping of knitting loops in a horizontal direction (K.F. 2011; Chandrasekhar, Iyer;Bernd, Mammel ;Wolfgang 2004). In circular weft knitting, the fabric production is done in tubular form. These knitting machines are categorized into three: small, medium, and large diameter knitting machines. Small or medium-diameter knitting machines are used to produce hosiery and body-size garments.

In circular weft knitting, the size dimension of a garment is limited to changes made through knitting parameters and feeding tensions in circular knitting, as the number of knitting needles/diameters of the knitting machine remains the same. Knitted compression garments have a minimum of two or more yarns knitted together through a different stitch. Knitting loops using a different combination of knitted stitches creates a different aesthetical look and is called a different knitted structure. Depending upon the knitting machine, type of yarn, and position, feeding yarns are categorized as main, plaiting and inlay yarn. The main yarn gives the structure required strength, the inlay yarn gives the required compression and plaiting is used to improve the recovery properties. The aesthetic look is mainly changed through the variation of the main yarn. For compression garments, aesthetic look and compression can be altered through the knitting pattern of the inlay yarn. The compression product's focus featured is compression, and this different combination of stitches to inlay yarn and its knitting tension/feeding was selected as a variable for this study. Three different knitting structures with varying patterns of inlay yarns were used in this study as shown in Table 2.

The three knitted structures produced a 3*3 rib, also known as mock rib on a single cylinder knitting machine. Some wales are not knitted in the rib structure as those needles miss throughout the structure. In mock rib, the elastomeric yarn is knitted on wales through tuck stitches, and these wales are pushed inside, looking as if the wale is missed. In the 3*3 rib structure, the inlay yarn produces three tucks and three missed stitches, as shown in Figure 1. In plain rib (P), the inlay yarn creates the same stitch every time on a specific needle, i.e., tuck stitch on three needles and then miss stitch on three needles. In the Cross (C) rib structure, the inlay yarn tuck and miss stitch position is increased by one needle in every course, which produces the look of cross/diagonal inlay yarn running in the fabric and can be seen in the fabric view created through APEX-IV, a programming software of the Shima Seiki. In the random (R) rib structure, three tuck and a miss stitch pattern is random in a repeat of six courses which does not allow the formation of diagonal or straight lines in the produced specimens.

Samples were produced on a small diameter knitting machine from Merz with 4.75 inches diameter, 351 needles and four feeders from which main, plaiting and inlay yarn can be fed. The machine is equipped to control the stitch size of each knitted loop through an electronic cam system where stitch cam height is controlled through a motor, and one millimetre of stitch height can be controlled by up to 100 units. It helps to control the stitch length in the required length and can be uniformly increased or decreased. The elastomeric yarn was fed through Merz Furnisher; tension can be increased or reduced uniformly through its rotation.

Table 2 Knitting notations and actual knitted specimens

Knitting structure 3 *3 (Rib)		Front	Back
Plain (P)	Diagrammatic notation	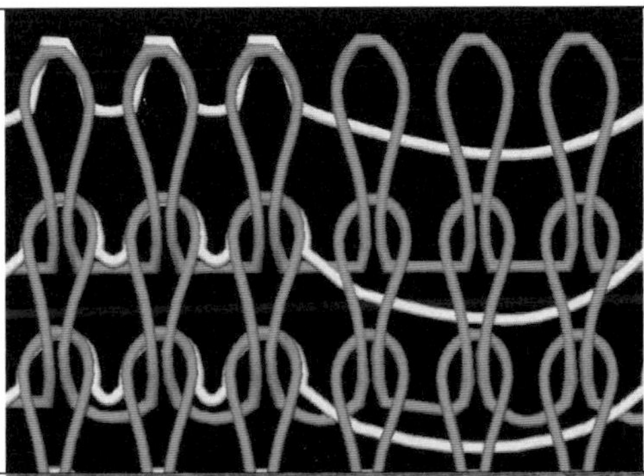	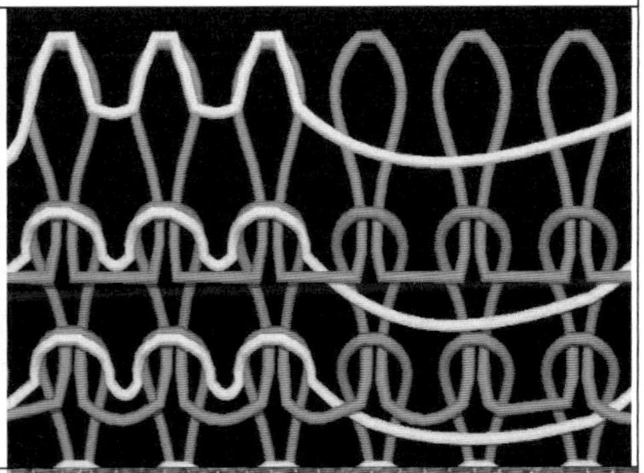
	Fabric view	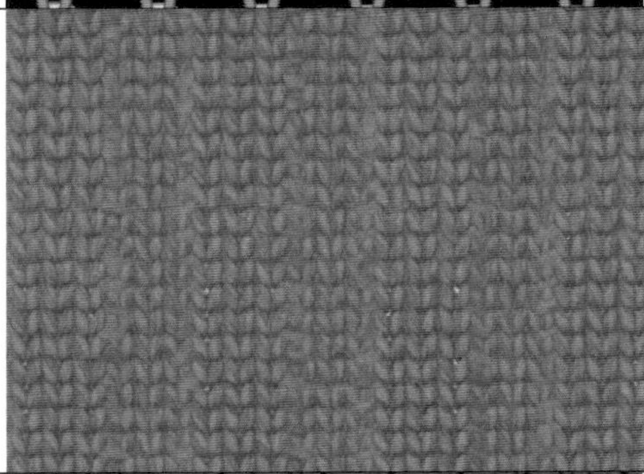	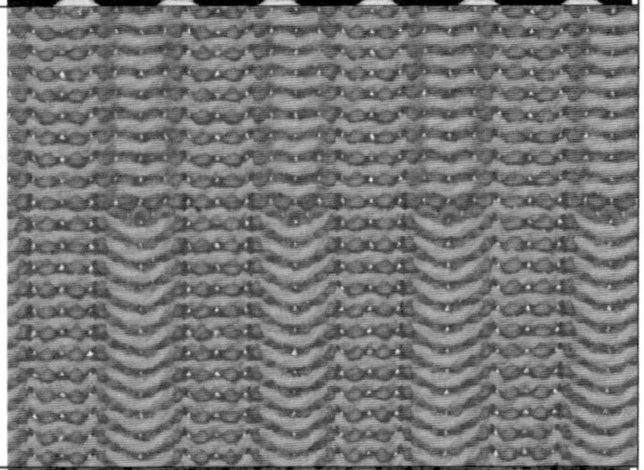
	Actual specimens	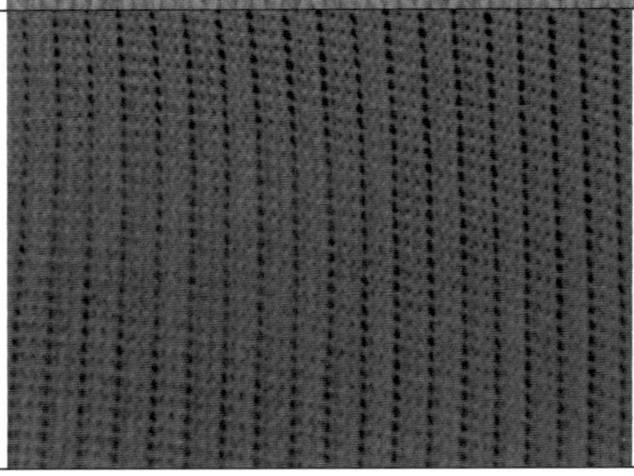	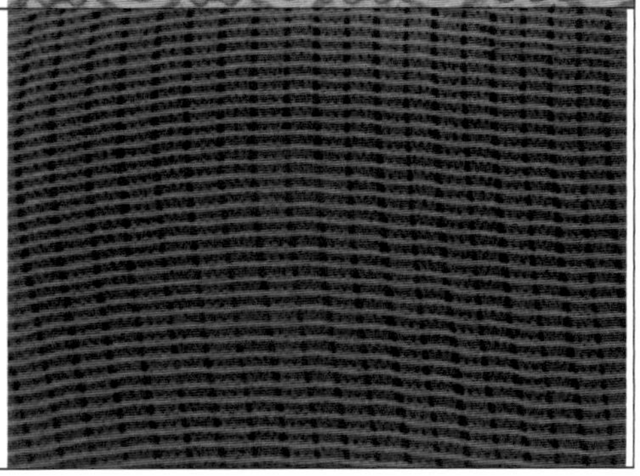

Cross (C)		
Actual specimens	Fabric view	Diagrammatic notation
	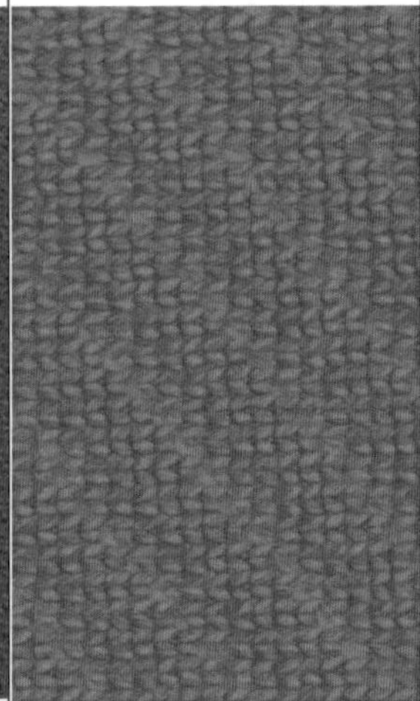	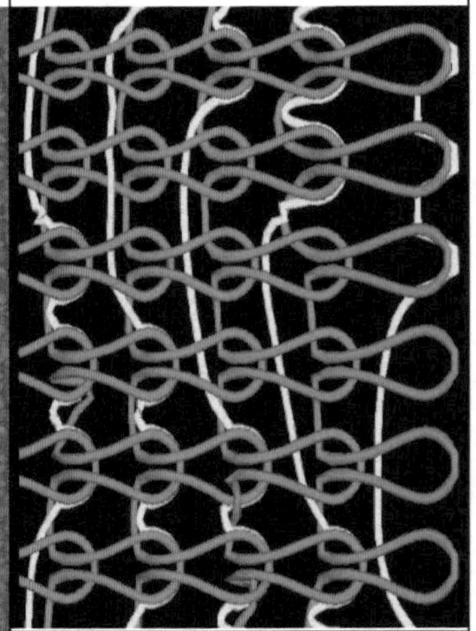
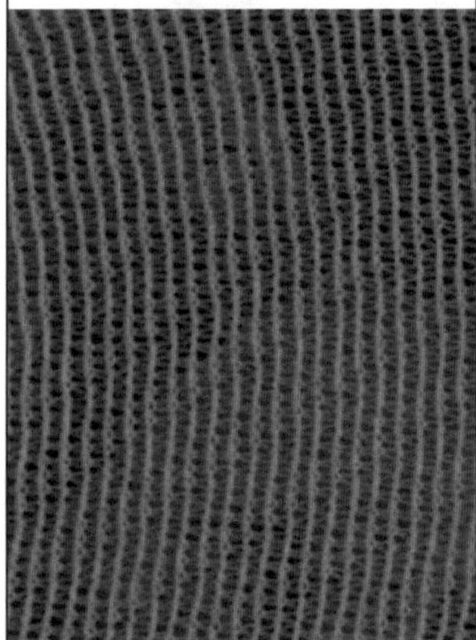	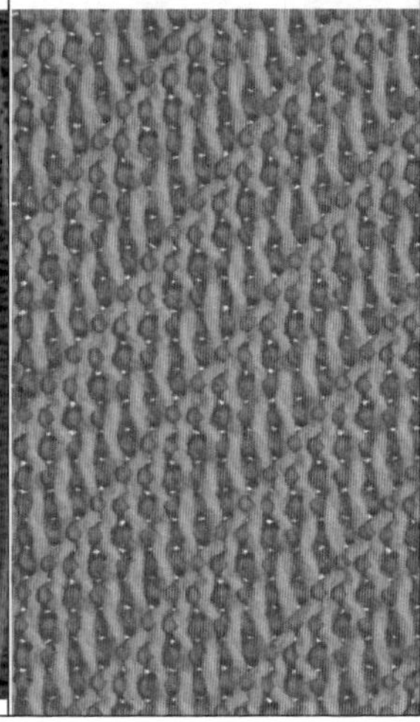	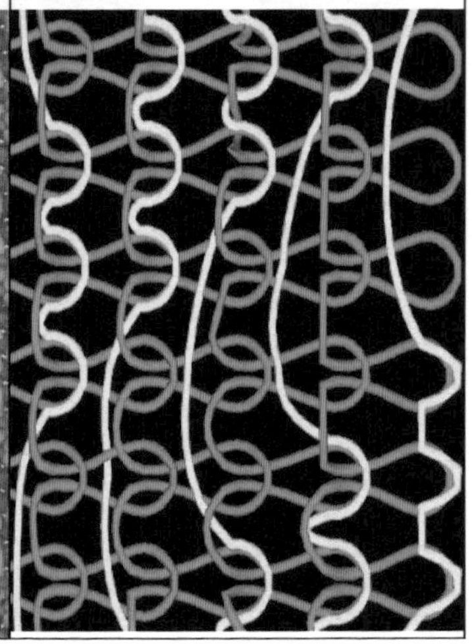

Random (R)			
	Diagrammatic notation	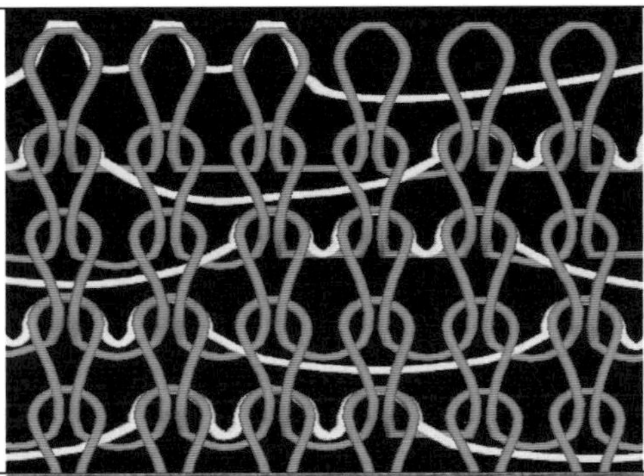	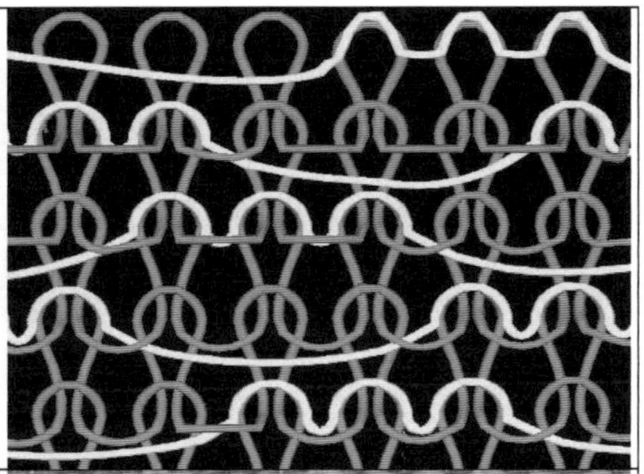
	Fabric view	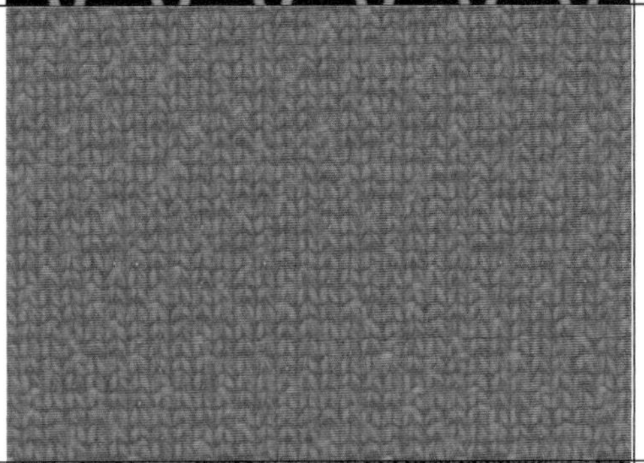	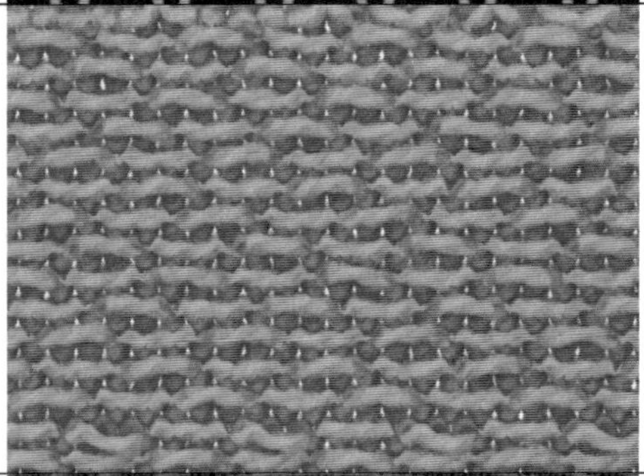
	Actual specimens		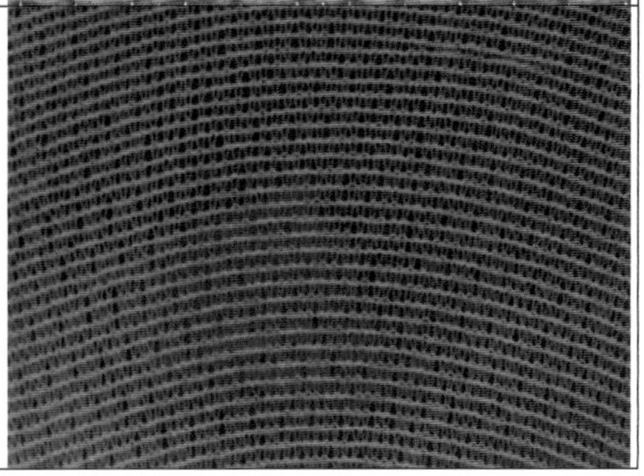

Testing

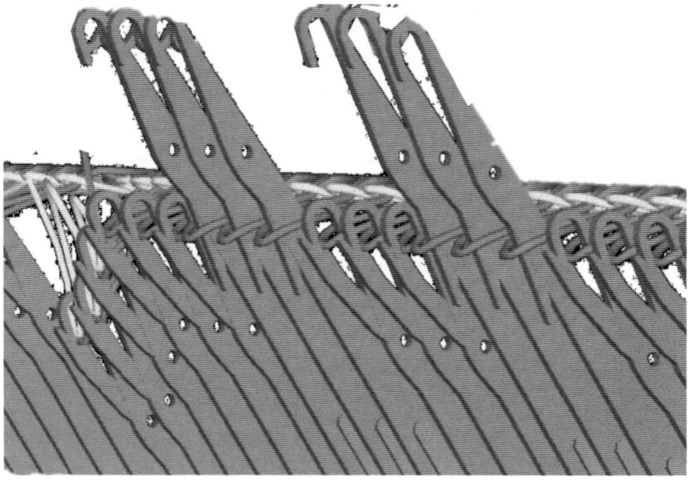

*Figure 1 Knitting needles for 3*3 rib structure*

All the samples were washed and relaxed in standard environmental conditions before analysis 22°C & relative humidity 65% for 24 hours as found in (Basra et al. 2020; Qadir et al. 2014; Ogulata and Mavruz 2010; Siddique 2018). Firstly, the physical properties, i.e., circumference, wales, and height between different measuring points, were measured physically, while the courses between other measuring points were the same in all specimens and found through machine programming software. Compression testing was done on a solid wooden leg of two sizes, 24 and 27; the details of circumferences and height for these sizes are shown in Table 3 and given in RAL 387/1 German standard for compression stocking (Ral-gz 2008). Picopress ® compression testing equipment was used to measure the compression in mmHg at all seven points. The human body leg circumference increase is not uniform, requiring measuring the compression at several points. These measuring points are at different height positions depending upon circumference variation. The measurement was taken three times at every point, and the average was used for analysis.

Table 3 Circumference and heights of size 24 and size 27

	Circumference (cm)		**Increasing height from the ground for both sizes (cm)**
	Size 24	**Size 27**	
B	24	27	12
B1	30	33	22
C	37	40	32
D	36	39	40
E	41	44	47
F	52	56	62
G	60	64	74

Results and Discussion

Physical properties

Table 4 describes the physical properties of all six produced specimens. The circumference was measured by doubling the width of the specimen at the measuring point. As knitting parameters of the main and inlay yarn changed from point B towards point G, the main yarn stitch length increased, and the inlay yarn tension decreased, or its feeding increased. This affects the dimensions, which increase following the leg measurements. The inlay yarn tightness affected the circumference, and inlay yarns with tight feeding were found to reduce the circumference. Inlay yarn knitting patterns were also found to affect the dimension/circumference. Specimens with inlay knitting patterns in cross or random directions in neighbouring courses had the same and higher circumference compared with the specimen in which the inlay knitting pattern remained the same in all courses.

Moving up from point B to G, the number of wales per centimetre decreases as the circumference increases. The number of knitting needles remains the same in circular knitting, and more dimensions/circumferences will have the same number of wales, leading to a decrease in the wales per centimetre. Specimens with higher knitting inlay yarn tightness or low dimension had a higher wale number per centimetre.

The difference in height between the measuring point of the legs, as shown in Table 3, leads to engineering the number of courses to achieve that length. The courses offered in Table 4 are the courses of the main yarn only, and the same number of inlay yarns were also used as it was used after every course. This course * number represents the number of main yarn courses knitted from one measuring point to the next measuring point. The number of knitting courses was maintained the same in all six specimens. Still, due to different knitting patterns, it leads to a slight difference in the height between the measuring points which was recorded and mentioned. Specimens with higher tightness decrease the circumference, as discussed earlier but were found to increase their vertical dimensions between two points.

Table 4 Physical properties of produced specimens

Sr. #	Measuring Point	B	B1	C	D	E	F	G
PM	Circumference (cm)	14.4	17	19.4	20	22.4	25	25.4
	Wales (cm)	25	21.18	18.56	18	16.07	14.4	14.17
	Courses *	156	136	106	88	174	140	
	Height (cm)	10.7	10.4	8.7	7.8	16.7	13.7	
PT	Circumference (cm)	12.6	15.2	17.6	18	20.6	22.4	23
	Wales (cm)	28.57	23.68	20.45	20	17.48	16.07	15.65
	Courses *	156	136	106	88	174	140	
	Height (cm)	11.2	11	9	8.4	17.8	14.4	
CM	Circumference (cm)	15	17.2	19.4	20.2	22.8	25.4	25.6
	Wales (cm)	24	20.93	18.55	17.82	15.79	14.17	14.06
	Courses *	156	136	106	88	174	140	
	Height (cm)	9.9	9.8	7.8	7.4	16.1	13.1	
CT	Circumference (cm)	13	15.4	18	18.2	20.8	23.2	23.4
	Wales (cm)	27.69	23.38	20	19.78	17.31	15.52	15.38
	Courses *	156	136	106	88	174	140	
	Height (cm)	11.5	11.2	9.2	8.3	18.5	13.4	
RM	Circumference (cm)	15	17.2	19.4	20.4	22.8	25.4	25.6
	Wales (cm)	24	20.93	18.56	17.65	15.79	14.17	14.06
	Courses (cm)	156	136	106	88	174	140	
	Height (cm)	10.1	10	7.9	7.2	16.2	12.3	
RT	Circumference (cm)	13	15.4	17.8	18.2	20.8	23	23.4
	Wales (cm)	27.69	23.38	20.22	19.78	17.31	15.65	15.38
	Courses *	156	136	106	88	174	140	
	Height (cm)	10.3	10.1	8.5	7.8	17.6	14.5	

The dimensions of knitted products are essential for effective compression and fit. Modifying the examined parameters of knitted structures are mainly used to give a garment a different aesthetic appearance or effect. Still, they can affect the garment's dimensions, changing the amount of compression needed, as discussed in the next section. To have varied compression, the inlay yarn tightness is adjusted. However, this alters the dimensions of the products, which could lead to an uncomfortable fit if the necessary adjustments are not made to ensure that they have the exact measurements. The ability to alter the garment's dimension is likewise constrained when making seamless garments, unlike cut and sewn garments, where the necessary pattern may be cut from the fabric. As there are no seams or stitches needed, seamless clothing offers sustainable production. Still, there are also more significant challenges regarding the garments' dimensions, higher skill requirements, depth of material knowledge, challenging production processes, and complex knitting machines. On the plus side, however, it offers advanced control over the knitting parameters, stitch selection, and the possibility of producing complex structures, enabling the creation of complex garments.

Compression results

Table 5 represents the compression results of all specimens measured on wooden legs size 24 and size 27. Gradual compression was found in all specimens in which higher compression was found at the lowest ankle point B, and compression was lowest at the upper thigh point G. All three knitting patterns have the same number of tuck and miss stitches to inlay yarn, but their knitting pattern affects compression. Compression was higher with the random knitting pattern of inlay yarn followed by the cross and plain pattern. According to what was previously stated regarding the distribution of the knitted stitch pattern in adjacent courses across all courses, it arbitrarily distributes the miss and tuck stitch throughout the randomly knitted pattern, increasing compression compared to the plain and cross. In the plain knit pattern, tuck and miss are always on the same needles, which resulted in wales reducing the circumference due to the tension of the inlay yarn but increased for plain specimens.

In contrast, in the randomly knitted tuck pattern, the tension of the inlay yarn is distributed through all the knitted wales resulting from having a higher circumference and less height between measuring points but higher compression. In the cross pattern, the tuck and miss stitch runs in a cross pattern, distributing the tension across all wales as compared with plain, and this was found to elicit higher compression but lower than the cross pattern. A general understanding exists that smaller dimensions increase compression, which is found to be incorrect as plain specimens had the lowest circumference, but their compression was the lowest. In this case, the number of knitting needles were the same, and the dimensional decrease was due to the building of more tension on inlay tucked stitched wales as the tuck was always on the same needle/wale.

When knitted specimens are produced at high tension, more energy is required stretching them based on a higher stress-strain peak as a result of a higher tensile modulus of fabric (Teyeme et al. 2021; Jariyapunya and Musilová 2019; Jariyapunya and Musilová 2020), and this energy is stored in the garment, which results from applying high compression on the body when worn. High stretching of knitted specimens requires more power and vice versa. In Table 4 the specimen's dimension was lower at high tension than those produced with medium tension. Stretching the garment requires more force and more energy is stored due to lower dimension and to wear on the same size. High stretch also requires more force thus all specimens produced with high inlay yarn tightness results in high compression. For size 27, higher compression results were found in all specimens as the knitted specimens were stretched higher to be worn, requiring higher pulling force and higher potential energy stored in the garment, which is then exerted on the wooden leg. An important finding is that specimens' compression remains gradual when worn on a larger size. It shows that if specimens are worn on a larger body, the garments will exert higher gradual compression on all measuring points compared to the original size dimension. More importantly, if body dimension between different measuring points varies in different ratio to the sizes mentioned in Table 2, then the garment may exert high or low compression but not gradually, which may be a disadvantage. Gradual compression is essential to achieve the health benefits of SCGs for the wearer.

Different anthropometric studies (Brubacher et al. 2021; Brubacher 2020; Brubacher et al. 2017; Bougourd 2007; MacRae et al. 2016; Gorea and Baytar 2016) show that human body dimension variation is irregular, not linear, and one body measurement can vary despite other measurements remaining the same. In these cases, the effectiveness of the compression garment will be challenging. In high-compression medical and sports compression garments, the structural knitting variation, the yarn tension, and garment dimensions can affect compression. This study did not consider the radius of curvature from the Laplace law relationship as all the testing was done on circular wood legs where pressure applied at every point is considered uniform.

Table 5 Compression results measured in mmHg for all specimens

Sr. #	Size 24							Size 27						
	B	**B1**	**C**	**D**	**E**	**F**	**G**	**B**	**B1**	**C**	**D**	**E**	**F**	**G**
PM	10	9	7	6	5	5	4	12	8.5	7.5	6.5	5.5	5	5
PT	11	9.5	7	6	6	5	5	14	9	7.5	7	6.5	6	5.5
CM	10.25	9	7	5	5	4	4	12.5	9	7.5	7.5	6.5	5.5	5
CT	11	7	7	6	6	5	5	13	9	8	7	7	6	6
RM	12	8	7	6	6	6	5	13.5	9.5	8	7	7	6	6
RT	11	7	7	6	6	5	5	14.5	10	8	7.5	7	6.5	6

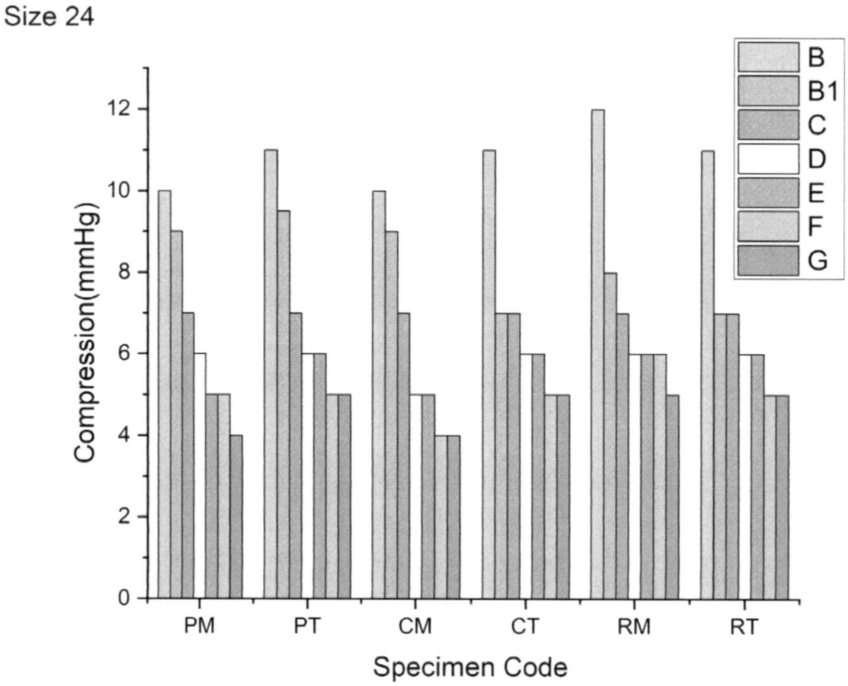

Figure 2 Compression results of size 24

Size 27

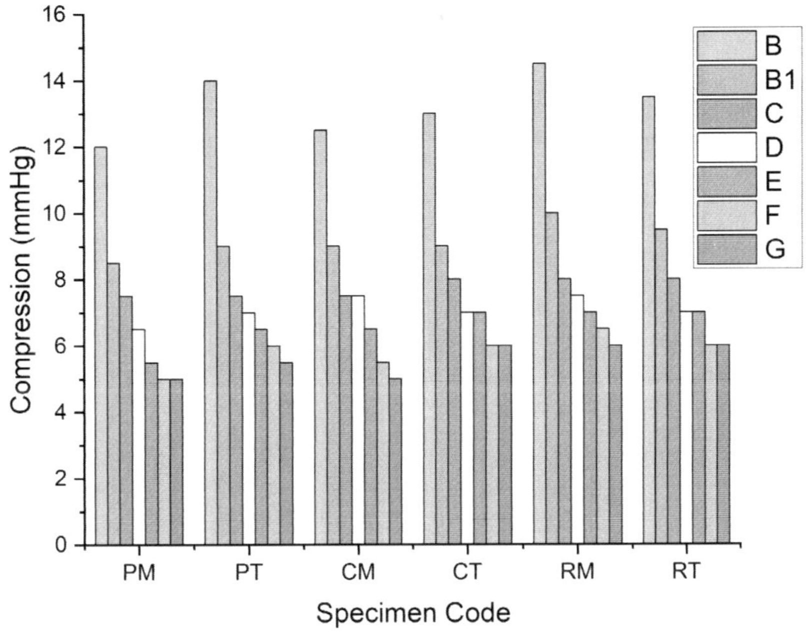

Figure 3 Compression results of size 27

Three-dimensional (3D) body scanners are helpful for anthropometric research since they can extract data in a variety of forms and sizes, as well as measure subjects (Parker et al. 2022; Simmons and Istook 2003; Troynikov and Ashayeri 2011; Parker et al. 2017). The approach is contactless and uses 3D body scanning, which can function utilizing optical laser, infrared light, white light, and photogrammetry to determine the subject's position and location (Fan et al. 2004). According to the study's findings, it can be recommended that 3D body scanning can be utilized to extract athletes' measurements to create sports compression apparel if their body dimensions vary from the standard sizing.

Conclusion

An enhancement in product quality will result in a longer lifespan, better use, and improved garment functionality, making the product more sustainable. The demand for seamless hosiery is driven by environmental concerns and higher knitting quality with some products challenges which this study focused on. Exerting the correct amount of compression on the body is a challenge for compression garment design and development, as variations in knitting, materials, and body dimensions lead to different compression. It was found that if leg dimensions follow RAL-387/1 (Ral-gz 2008), the exerted compression is gradual. However, compression will be higher or lower depending on the body dimension variation. Still, if at some point the body dimension variation is different to the standard, then compression may not remain gradual, and garment effectiveness will be reduced. The tension/feeding rate of the inlay yarn results in changes in the compression and knitting pattern of the inlay yarn. Retaining the same number of tuck and miss stitches but changing their knitting order with the arrangement of random tuck and miss stitches in which tuck repeats in every wale found to exert higher compression than the plain and cross pattern. The effectiveness of the compression garment can be improved if produced considering the material's properties, knitting parameters and body dimensions. However, this study has limitations because it only used a small number of input variables, and its findings may not be generalizable to all levels of the input component. Because body dimensions might vary, the author suggests the needs to investigate employing digital 3D scan data to alter garments to provide the necessary progressive compression.

References

Ashdown, S.P. (1998). An investigation of the structure of sizing systems. A comparison of three multidimensional optimized sizing systems generated from anthropometric data with the ASTM standard D5585-94. *International Journal of Clothing Science and Technology*, 10(5), pp.324–341.

Ashdown, S.P. (2011). Improving body movement comfort in apparel. *Improving Comfort in Clothing*, pp.278–302.

Atkins, R. et al. (2020). Lower-body compression garments worn following exercise improves perceived recovery but not subsequent performance in basketball athletes. *Journal of Sports Sciences*, 38(9), pp.961–969. [online]. Available from: https://doi.org/10.1080/02640414.2020.1737387.

Basra, S.A. et al. (2020). Development of interlock knitted seersucker fabric for better comfort properties. *Journal of Engineered Fibers and Fabrics*, 15.

Born, D.P., Sperlich, B. and Holmberg, H.C. (2013). Bringing Light into the Dark: Effects of Compression Clothing on Performance and Recovery. *International Journal of Sports Physiology and Performance*, 8(1), pp.4–18. [online]. Available from: https://journals.humankinetics.com/view/journals/ijspp/8/1/article-p4.xml [Accessed May 10, 2022].

Bougourd, J. (2007). Sizing systems, fit models and target markets. In *Sizing in Clothing*. pp. 108–151.

Brubacher, K. (2020). Comparison and Evaluation of Sizing Systems Used in Commercial Women's Compression Sportswear. *Proceedings*, 49(1), p.140.

Brubacher, K. et al. (2017). Design of sports compression garments: Exploring the relationship between pressure distribution and body dimensions. *8th Asia-Pacific Congress on Sports Technology*.

Brubacher, K. et al. (2021). Evaluation of the Accuracy and Practicability of Predicting Compression Garment Pressure Using Virtual Fit Technology. *Clothing and Textiles Research Journal*, pp.1–18.

Chandrasekhar, Iyer;Bernd, Mammel ;Wolfgang, S. (2004). *Circular Knitting*.

Choi, W. and Powell, N.B. (2005). Three dimensional seamless garment knitting on V-bed flat knitting machines. *Journal of Textile and Apparel, Technology and Management*, 4(3), pp.1–33.

Duffield, R. and Portus, M. (2007). Comparison of three types of full-body compression garments on throwing and repeat-sprint performance in cricket players. *British Journal of Sports Medicine*, 41(7), pp.409–414. [online]. Available from: https://bjsm.bmj.com/content/41/7/409 [Accessed May 10, 2022].

Fan, J., Yu, W. and Hunter, L. (2004). Clothing appearance and fit: Science and technology. *Clothing Appearance and Fit: Science and Technology*, pp.1–239.

Fu, W., Liu, Y. and Fang, Y. (2013). Research Advancements in Humanoid Compression Garments in Sports: *International Journal of Advanced Robotic Systems ARTICLE*, 10. [online]. Available from: https://journals.sagepub.com/doi/full/10.5772/54560 [Accessed May 10, 2022].

Fung, E. and Liu, R. (2019). Sustainable Sports Fashion and Consumption.

Gorea, A.C. and Baytar, F. (2016). Towards Developing a Method for Identifying Static Compression Levels of Seamless Sports Bras using 3D Body Scanning. , pp.22–28.

Gupta, D. et al. (2006). Developing body measurement charts for garment manufacture based on a linear programming approach. *Journal of Textile and Apparel, Technology and Management*, 5(1).

Jariyapunya, N. and Musilová, B. (2020). Analysis of Stress and Strain to Determine the Pressure Changes in Tight-Fitting Garment. *Autex Research Journal*, 20(1), pp.49–55.

Jariyapunya, N. and Musilová, B. (2019). Predictive modelling of compression garments for elastic fabric and the effects of pressure sensor thickness. *Journal of the Textile Institute*, 110(8), pp.1132–1140.

K.F., A. (2011). *Advances in Knitting Technology*. [online]. Available from: http://www.sciencedirect.com/science/article/pii/B9781845693725500127.

Lavorata, L. (2014). Influence of retailers' commitment to sustainable development on store image, consumer loyalty and consumer boycotts: Proposal for a model using the theory of planned behavior. *Journal of Retailing and Consumer Services*, 21(6), pp.1021–1027. [online]. Available from: http://dx.doi.org/10.1016/j.jretconser.2014.01.003.

MacRae, B.A., Laing, R.M. and Partsch, H. (2016). General considerations for compression garments in sports: Applied pressures and body coverage. In *Compression Garments in Sports: Athletic Performance and Recovery*. Springer International Publishing, pp. 1–32.

MacRae BA, Cotter JD, L.R. (2011). Compression garments and exercise: garment considerations, physiology and performance. *Sports Med*, pp.815–845.

Maity, S., Singha, K. and Pandit, P. (2022). Production of seamless knitted apparels. *Advanced Knitting Technology*, pp.203–218.

Merz. (2022). Merz Maschinenfabrik - Knitting Machines. [online]. Available from: https://www.merz-mf.de/en/products.php [Accessed May 31, 2022].

Norwegian Ministry of Environment. (1994). Symposium: Sustainable consumption. Oslo, Norway: Ministry of the Environment. In *Norwegian Ministry of Environment. Oslo, Norway: Ministry of the Environment*. Oslo, Norway, pp. 1–180. [online]. Available from: https://www.worldcat.org/title/symposium-sustainable-consumption-19-20-january-1994-oslo-norway/oclc/79024864 [Accessed May 16, 2023].

Ogulata, R.T. and Mavruz, S. (2010). Investigation of porosity and air permeability values of plain knitted fabrics. *Fibres and Textiles in Eastern Europe*, 82(5), pp.71–75.

Parker, C.J. et al. (2022). A method for increasing 3D body scanning's precision: Gryphon and consecutive scanning. *Ergonomics*, 65(1), pp.39–59. [online]. Available from: https://doi.org/10.1080/00140139.2021.1931473.

Parker, C.J., GillL, S. and Hayes, S.G. (2017). 3D Body Scanning has Suitable Reliability: An Anthropometric Investigation for Garment Construction. In *International Conference and Exhibition on 3D Body Scanning and Processing Technologies*. pp. 298–305.

Power, E.J. (2012). Sustainable developments in knitting. *International Journal of Business and Globalisation*, 9(1), pp.1–11.

Qadir, B., Hussain, T. and Malik, M. (2014). Effect of Elastane Denier and Draft Ratio of Core-Spun Cotton Weft Yarns on the Mechanical Properties of Woven Fabrics. *Journal of Engineered Fibers and Fabrics*, 9(1), pp.23–31.

Ral-gz, Q.A. (2008). Medical Compression Hosiery Quality Assurance.

Robinette, K.M. (1986). Anthropometric methods for improving protection. *ASTM Special Technical Publication*, pp.569–580.

Sang, J.S., Lee, M.S. and Park, M.J. (2015). Structural effect of polyester SCY knitted fabric on fabric size, stretch properties, and clothing pressure. *Fashion and Textiles*, 2(1), pp.1–9. [online]. Available from: https://link.springer.com/articles/10.1186/s40691-015-0047-4 [Accessed May 7, 2022].

Santoni. (2022). SM8-TR1 | Santoni. [online]. Available from: https://www.santoni.cn/en/sm8-tr1?Sportswear [Accessed July 13, 2022].

Siddique, H.F. (2018). Effect of Elastic Elongation on Air-Permeation at Ankle and Calf Portion of Compression Socks. , (1), pp.0–14.

Simmons, K.P. and Istook, C.L. (2003). Body measurement techniques: Comparing 3D body-scanning and anthropometric methods for apparel applications. *Journal of Fashion Marketing and Management*, 7(3), pp.306–332.

Teyeme, Y. et al. (2021). Predicting compression pressure of knitted fabric using a modified laplace's law. *Materials*, 14(16), pp.1–13.

Troynikov, O. and Ashayeri, E. (2011). 3D Body Scanning Method for Close- Fitting Garments in Sport and Medical Applications. In *47th Annual Conference 2011. Ergonomics Australia - Special Edition Methodological*. pp. 239–248.

EXPLORING MOTION CAPTURE TECHNOLOGY TO TRANSFORM BOTH THE METHOD OF CREATING DESIGN AND PATTERN CUTTING FOR SUSTAINABLE FASHION DESIGN

Devabrata Paramanik[1] *and Debbie Moorhouse*[1]

Department of Music & Design Arts, School of Arts and Humanities, University of Huddersfield

Abstract

This paper examines Motion Capture Technology's (MOCAP) potential to transform both the methods of creating design and the pattern cutting to explore sustainable fashion design. This research uses an experimental design approach of two-dimensional and/or three-dimensional (2D/3D) practices and reflects on both the authors personal design processes, their experience within the fashion design industry and academia.

MOCAP is used to generate body data during a performance or a performative act. This data is then visualised, manipulated, and translated to create new patterns for fashion design and finally made into a garment.

The paper is organised in three sections which are: 1) How MOCAP is used to generate body data? 2) How the MOCAP generated body data is then explored to transform both the design and the pattern cutting method. 3) Creating sustainable fashion products

The outcomes of these experiments are a capsule collection of three garments aimed to inform the advancement of digital practices in fashion design.

Keywords

Fashion design, motion capture, pattern cutting, sustainability, diversity, zero-waste

Introduction

This practice-based research is informed by Christopher Frayling's model of "research through art and design" (Frayling 1993, p.5). It is concerned with a new developmental and innovative Creative Pattern Cutting that engages with customizing motion capture technology (MOCAP) to transform both the methods of creating design and the pattern cutting to explore sustainable fashion design. Also, our research is aimed to encourage both our Postgraduate and Undergraduate students to challenge the conception of pattern cutting through practice and research.

The garment chosen for this experiment is a formal shirt, an item of clothing which has traditionally been worn by all genders, in various styles and various degrees of fit. For this experiment a medium sized semi fitted formal shirt was chosen in a classic style cut to a basic shirt pattern. As the outcome of this experiment is a study of how pattern design can be explored through motion capture technology there was no need to classify the garment [shirt] as women's wear or menswear but will be referred to in this study as unisex.

To keep the process within the realms of sustainable good practice the aim from the outset was to produce outcomes with zero-waste or minimal waste at the production stage. Within the time frame of this study further consideration of zero-waste beyond the design and pattern cutting stage were not taken into consideration. Fabric with sustainable credentials was purchased from within the UK and manufacture of the final shirts was achieved locally.

Pattern cutting

Traditionally pattern cutting methods have primarily used a two-dimensional (2D) approach using flat horizontal and vertical measurements of the body in an upright static position (Luible and Charbonnier, 2016), in effect working from the outside towards the body (Aldrich,2013, p.5; Aldrich, 2015, p.5; Lindqvist, 2015, p.7; Almond and Power, 2018, p.34). A block is then created from these measurements and manipulated through measurement adjustments, slash and spread manipulation and dart displacement to simulate the 2D design initially presented to the pattern cutter by a designer. Traditionally this process of creating 2D pattern is done by hand, and still is in many of the smaller independent ateliers, but more increasingly in larger companies done digitally for speed and efficiency with the accessibility of software such as Lectra and Gerber (lectra.com), and Clo3d (clo3d.com). This pattern is then used to manufacture a three-dimensional (3D) garment (toile) for testing of fit and proportion, which can be a lengthy process before the correct fit is achieved.

Another more design led approach to pattern cutting is the creative process of 3D draping which involves the creative pattern cutter working with a design idea or a loose sketch to drape fabric directly onto a mannequin creating a garment in the first instance in 3D rather like sketching the design but using fabrics (Amaden-Crawford, 1989; Duburg and van der Tol, 2009; Kiisel, 2020). In some of the couture ateliers this approach of working directly with the fabric with all its draping properties and use of different grainlines, is most often done directly onto the human form (live model) to allow for consideration of working from the body outwards to create a final piece which considers the whole kinetic structure and proportion of the body (Lindqvist, 2015, p.5) a process which designers such as Lacroix and Dior favoured.

Garment Fit

Producing clothing for an unknown customer, as in ready to wear rather than made to measure or bespoke clothing, designers and pattern cutters have had to draw on their understanding and experience of the human form in all its variations and estimate the dimensions of the garments they are hoping to sell successfully (i.e., a good fit), either by body measurements (with no ease) or by garment measurements (with ease). The difficulties in achieving a generalised sizing system become apparent when analysing this anthropometric data (measurement of humans) with differences in body shape for each size and across different cultures (Cassidy, 2017, p.1). Poor fit in a mass market results in heavy returns and over stocks through lost sales due to poor fit and discrepancies in sizing. Added to this problem is the resurgence of the resale and vintage market with all its size and fit inconsistencies. By contrast made to measure garments (made from blocks adjusted to individual client measurements) and bespoke garments (made from patterns drafted to a client's measurements, fit and body shape) provide a final garment with a more exacting fit. This more personalised approach is also now becoming once again more popular because of its slow approach to manufacture, less waste with sample testing, more client involvement in the design process creating emotional attachment and more appeal as a designer/manufacturer to create something of quality and long lasting (Brewer, 2019, p.1; Ozdamar Ertekin and Atik, 2015, p.54). What this research proposes rather than creating patterns for a body which remains constant (Aldrich, 2004) is to create new patterns for a body which is a moving variable (Burnham, 1973; Lindqvist, 2014, p.5) to learn more about the possibilities of pattern design and to open the conversation to more innovative pattern cutting for sustainable best practice. Lindqvist's research (2014 and 2015) on using the break lines of a body, considering all its expression and movement to create pattern, builds on the transformative cutting of a single piece of fabric by Sevin-Doering (2007), and highlights the need to understand the body in garment making.

Motion Capture Technology (MOCAP)

Motion capture technology is used widely in a variety of industries including the military, film making and video game development (e.g., VFX), medical applications (e.g., neuroscientists), and the sports industry (e.g., sports therapists). It is primarily a recording of the actions of human models or performers from different perspectives through physical sensor points (marker based) placed on certain points throughout the body (Zong and Lee, 2011, p.92) or in some cases with other software sensor points can be marker less (magnetic sensors). The technology was originally explored for the life science market for gait analysis. This information (data) is then used to animate digital models in 2D or 3D avatars (Figure 1). In the healthcare industry motion capture offers significant benefits in disease prevention particularly as the technology becomes more accessible. motion capture assists in the diagnosis and treatment of physical conditions and can track and analyse a patient's motor activity and deterioration/improvement overtime. In the sports and sports therapy industry motion capture is used to evaluate physical limits and optimisation of movements. By collection data through motion capture sports men and women, trainers and clinicians can learn more about how to prevent injuries. Studies in biomechanics use motion capture technology to gain kinematic data and body skeletons that can help to analyse movement, measuring the bony segments during various activities (Chakraborty,2021). Filmmaking and video gaming is an industry that has become heavily dependent on motion capture as a tool to improve animation through the recording of human activity through various body movements and facial expressions. Advances in this area clearly apparent and have advanced rapidly in recent years. While Fashion & Textile design usually tends to seek opportunities in new technologies it has been found that MOCAP is minimally explored and even less so where the outcome aims to provide a physical garment. (Paramanik, 2013). This research study aims to look at the possibilities of using motion capture to generate data from a clothed body to transform the design and pattern cutting method of a specified garment. The path of the motion capture in a rectilinear motion (straight line) and curvilinear motion (curved line) will be translated to create new patterns for fashion design with sustainable responsible practice being a major consideration in the outcomes. The research study as Lindqvist (2014) suggests is a pattern cutting method, contrary to more widely used 2D and 3D methods of pattern cutting, which starts with the body and ends with the body as one would expect. The limitations within this process of motion capture for pattern cutting and design might be in that motion capture collects data only about joints and skeletons and not about skin and muscle (Zong and Lee, 2011, p.92) which would be a major consideration of pattern for garments as change in body measurements through dynamic posture would affect pattern design. Apart from in the sportwear industry (Cheung and Ng, 2007) motion capture is not known to be widely used in the apparel industry.

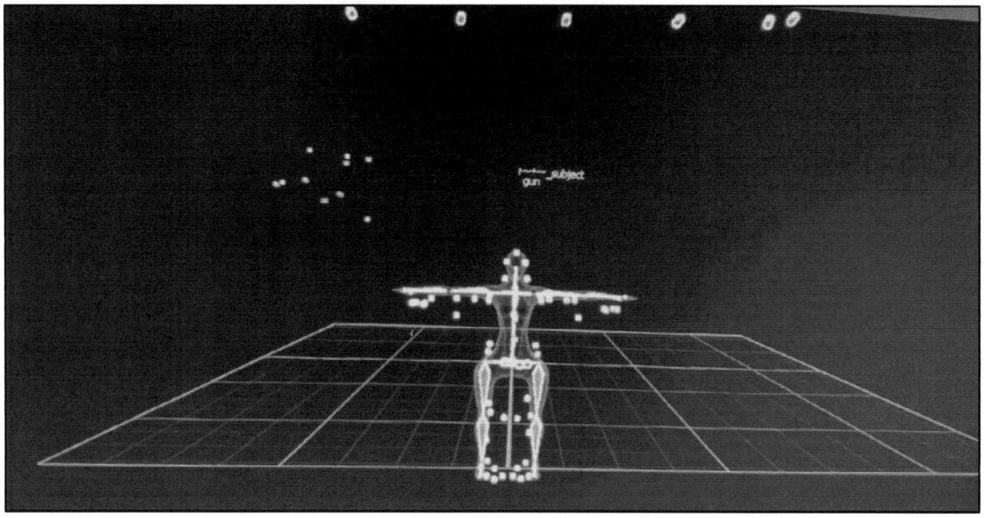

Figure 1 Motion Capture Skeleton

Zero waste

Zero waste is a goal to aspire to in the fashion industry (as it is within any [manufacturing] industry today) mimicking nature where all resources are extracted, used, and returned to the natural environment with zero waste. While looking at motion capture as the possibility of a new approach to experimental pattern cutting the research recalls zero waste and how this concept might also be incorporated. The issue of waste within the fashion industry is a growing problem due to consumer overconsumption of clothing and planned obsolescence within the fast fashion sector. Textiles globally are the fourth highest user of primary raw materials including water (EEA, 2019, p.8) and yet it is estimated that only 1% of textiles are recycled after use into new textiles (EMF, 2017, p.20) despite the claim that all textiles are nearly 100% recyclable and as such should not be wasted (Hawley, 2006). Waste in the fashion and textile industries is generally described as post-consumer waste which is clothing textile waste after the consumer usage phase and pre-consumer waste which is waste produced in the production of textiles and clothing, which might consist of over production, cancelled orders, faulty and over stocks of textiles and dead stocks. "More than USD 500 billions of value is lost every year due to clothing underutilisation and the lack of recycling" (EMF, 2017, p.3)." Added to this waste of course is energy loses due to unrenewable energy usage, transportation impact, water and land pollution through chemicals and waste of skills, time in human labour. Recognised methods of avoiding waste in fashion and textiles both pre-consumer and postconsumer are reuse, recycle, remanufacture and upcycling and all these methods have seen a resurgence in recent years due to concerns for the environment and how much waste textiles are ultimately going to incineration or landfill.

Zero waste fashion design (ZWFD)

Zero waste fashion design refers to the process of producing garments which consider the elimination of waste in all its forms from the outset, including fibre generation, textile production methods, clothing design and manufacture, logistics and distribution, sales, consumer usage and closing the loop at the end of life. To assist in this far-reaching goal of zero-waste, rather than accepting waste as a biproduct of clothing manufacture designers are looking at ways of producing no waste from the design outset (Clark, 2012; Rissanen, 2013; Fletcher, 2013; Niinimaki, 2013; ElShishtawy et al, 2022), while also considering fabric type, longevity of use, durability, and possible reuse qualities in a more circular economy. The process of producing clothing without waste is not a new concept and can be traced back to a time when animal skins were draped around the body. With the introduction of woven cloth, the whole rectangular piece of cloth produced on looms of varying widths was draped around the body, examples of which would be the Indian sari, Roman toga, and Chiton of Ancient Greece. Contemporary examples of designers incorporating zero-waste into their company ethos would be Berlin designer Bojana Draca who produces the brand Farrah Floyd Primarily made from geometric pieces of fabric pieced together so that no waste is produced in the cutting stage, referencing a system of zero waste pattern cutting used by St James (2009) and Hishinuma (1986). The production process has over time been refined to a level where industry practices of out-sourced production and digital pattern making allows for multiple garments to be made cost-effectively, while also producing to a high standard of finish. Another approach to ZWFD has been adopted by the company Doodlage (doodlage.com) based in India who produces all its garments from a mixture of eco-friendly fabrics such as organic cotton, corn fabric and banana fabric, post factory cutting waste and left -over textiles from large manufacturers, which otherwise would have been discarded to landfill. Any unavoidable waste cut off fabric is used to make accessories, soft furnishings and paper for packaging diverting all production waste away from landfill.

Zero waste pattern cutting (ZWPC)

It is widely accepted that within clothing manufacturing the pattern lay plan will 'cut off' 15-20% of the material (Rissanen & Gwilt, 2011; Clark, 2012; Rissanen, 2013; Townsend and Mills, 2013), which is then often sent to landfill or other downgraded usage as 'rags'. Advances in technologies has allowed for the opportunity to explore more efficient lay plans but the close fit and the curved nature of contemporary patterns does not allow for 100% zero waste and as a result designers are looking at new pattern shape and methods of fit to encompass zero-waste. Zero waste pattern cutting is a methodology used to create garments without waste by creating a pattern which fits the whole width and length of the fabric. Historically zero-waste pattern cutting has developed from an appreciation of valuable cloth where the whole loom width of cloth would be used in examples mentioned, such as the sari and toga, to examples such as the Japanese kimono where specific widths and lengths of cloth would be woven to complete the pattern pieces for each kimono made without the need to cut into the fabric, allowing sperate pieces to be replaced when worn or reused when no longer required. Designers such as Vionnet, Gris and Balenciaga based much of their work and experimental pattern cutting on appreciation of cloth and one example of this is the one seam coat produced by Balenciaga. The process of making the coat helped to inform and generate new ideas of working within the concept of zero-waste and how very small adjustments to the pattern could create a whole new silhouette.

Over the last ten years, academic research has grown around zero-waste pattern cutting with researchers such as Rissanen, McQuillan, Telfer, Liu and Roberts for example looking at ways of pattern cutting garments with zero or minimal waste. There is no one method of zero-waste pattern cutting, and this can be seen in some of the examples below, but what is evident is that no one method uses the tradition block pattern cutting method set out as a tailoring matrix by Aldrich and other similar pattern cutters of the time. Some of these methods have a more significant impact than others on how garments are pattern cut and produced within a contemporary fashion industry, in terms of scalability, consistency of construction and quality, and gradeability into multiple sizes.

Practice

The motion capture experiment was carried out by two participants, one who identified as male and a second participant who identified as female. The shirt fitted with motion capture markers was worn by each participant to perform a basic body movement (0. This was then captured and visualised as a MOCAP pattern.

Figure 2 Front, Back and side view of the shirt fitted with motion capture markers.

Also, MOCAP patterns of individual body movements were superimposed to explore new patterns such as with only the colour of the marker points differentiating between each individual pattern (0. Using the Gestalt theory of perception and expression random shapes were then drawn using the marker points as a guide. Developed in the 1920's by three German psychologists, Wertheimer, Koffka and Kohler, Gestalt theory is now used by researchers in many other disciplines as a scientific method of explaining human perception and how we might process visual information by using the laws of perception to explore how we interpret and organize visual information from our surroundings (Koffka, 1935).

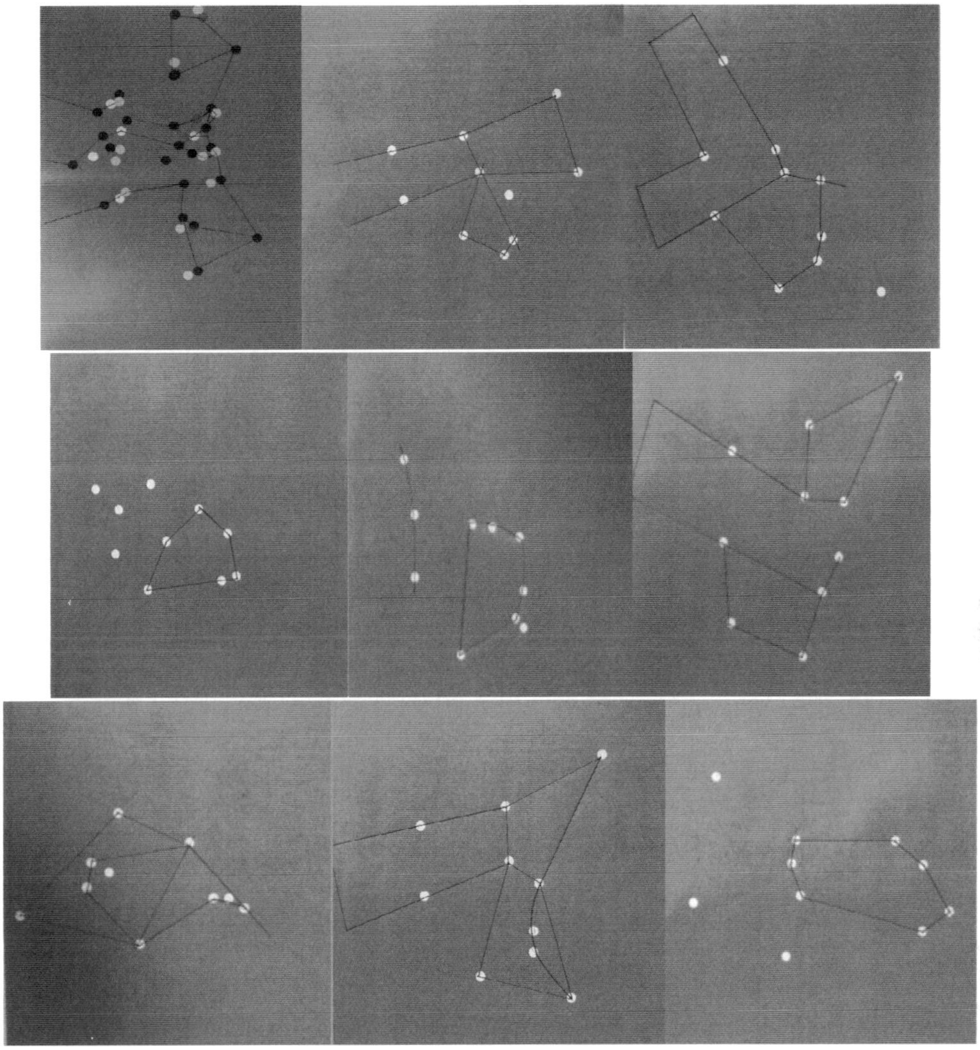

Figure 3 MOCAP patterns of superimposed individual body movements

The Gestalt laws and principles of Figure/Ground (when the eye sees different shapes depending on the background); Proximity, (objects placed together the eye sees them as a group); Closure, (when an object is incomplete the eye/brain will fill the space); Similarity, (objects which look like one another the eye perceives them as a group/pattern of colour shape texture or size) and Continuation, (the eye is drawn to move from one object through another) highlight the idea that the human brain has the capacity to organise and select visual signals to generate design outputs. In this study the visual signals represent the motion capture images resulting from the generated body data during a performance and then manipulated and translated to create patterns for fashion design, where individual components are organised by our perception as a physical whole garment (i.e., a shirt in this case). Using the Gestalt principle to extract pattern from these digital images is the perceptual principle of organising geometrical shapes and patterns based on the loose perception of a shirt and all its component parts (Mwakisunga, 2010). As the researchers have experience in pattern development, pattern cutting and design the knowledge of what the pattern pieces might look like on a shirt and the silhouette of the shirt played some part in influencing the drawn pattern shapes. However, as the objective was to be as free as possible, the drawn shapes were mostly geometric in nature as this would aid in the zero-waste pattern approach due to curved patterns traditionally being more wasteful.

These pattern shapes were then transferred and developed into a zero-waste pattern which could be then toiled and manufactured into a wearable shirt. The zero-waste 'jigsaw' pattern cutting approach (Rissanen, 2013) was used whereby a measured piece of cloth was used to the full to create each individual shirt with no waste or minimal waste. This is an iterative process and initial pattern shapes had to be developed to allow for the successful construction of a shirt with essential elements of the traditional shirt such as the collar and stand, cuffs and button stand all within a zero-waste or minimal waste pattern. While design elements were considered it was important to allow the design to develop freely at this stage to show how using the motion capture technology created the potential to transform the design and pattern cutting methods.

Results

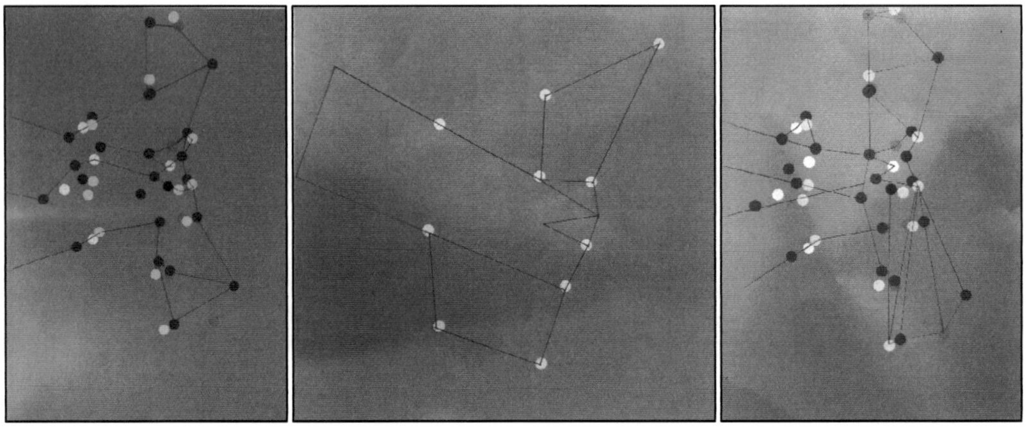

Figure 4 (l-r) MOCAP scan inspiration for Unisex Shirt 1, MOCAP scan inspiration for Unisex Shirt 2, MOCAP scan inspiration for Unisex Shirt 2

The following images show toiles produced in this study from initial inspiration of data collected from MOCAP scans produced by two participants (Fig 4). Rather than using the traditional method of 2D or 3D pattern cutting, the pattern design has been produced within a single piece of cloth with zero or minimal waste with pattern shape inspired by the shapes formed from the MOCAP scans using a gestalt theory of perception and continuation. Although the traditional silhouette of the shirt has been considered in this study, the use of the shapes created by the MOCAP scans has allowed for greater freedom of expression in what is considered pattern cutting for a garment. The construction of the garment (shirt) also needed to be adapted to suit the [sometimes] irregular shapes which would not traditionally be used in shirt production and fit. In addition to presenting alternative methods of approaching fashion design and creative pattern cutting, the resulting garments also questions the traditional use of grainlines and how these support the garment but also respond to the body in terms of comfortable fit. Where the need for a strong grain tends to be placed on the warp grainline and a bias grain at 45 degrees, patterns from the MOCAP scans didn't always accommodate this and as such opened further questions when constructing the garment.

Figure 5 (l-r) Unisex Shirt 1, Unisex Shirt 2, Unisex Shirt 3

The practiced-based research produced three different final outcomes of a unisex shirt which were pattern cut using a zero-waste or minimal waste jigsaw method as well as using inspiration for pattern shape from the MOCAP data (Fig 5). The styled images show the Unisex Shirts 1,2 &3 worn over a fitted body suit designed by one of our graduate student Emily Seabright (2023). Each shirt produced used different MOCAP data as inspiration for pattern design and all three outcomes used more than one scan for shape inspiration. The zero-waste jigsaw pattern approach needed to develop the initial inspiration (while keeping true to the original inspiration) to be able to conform to the necessities of pattern shape which resulted in the essential elements of a shirt i.e., the collar, cuffs, and button placket. Some knowledge and experience of shirt making, and traditional shirt pattern cutting was also needed to aid in sizing and functional silhouette. The three final outcomes were photographed on live models to illustrate the cutting of the fabric, and how the garment responds to movement.

Discussion

This study demonstrates an alternative approach to fashion design and creative pattern cutting for new expressions of garment cut and design using data collected through movement on MOCAP. The MOCAP data has been visualised, manipulated, and translated to create new patterns which were then used to produce a collection of three shirts. This innovative approach to fashion design and pattern cutting opens new expressions in garment design and manufacture as well as the garments remaining functional and produced with sustainable best practice at the forefront. Strong connection to sustainable best practice is shown in the pattern cutting with zero-waste, use of certified organic fabrics sourced in the UK and localised production of the shirts. A shirt was chosen as the garment to work with in this study however the process of using data collected from MOCAP could potentially be used for any garment or a series of garments on a body of any size. Further research has the potential to create more expressive forms by breaking away from the traditional forms of fashion design and pattern cutting. This could be further enhanced by participation of consumer as a as a performer, a method of public engagement. The patterns, toiles and final shirts in this study were produced over a very short period as a means of creating a starting point for further research into both pattern cutting and fashion design inspired by MOCAP, while incorporating sustainable best practice within all aspects of the supply chain and manufacture.

Using MOCAP as a method for collecting body data to inform design and pattern cutting also opens the practice of emphasizing movement of the body and how the garment responds to this.

References

Aldrich, W. (2004). Metric pattern cutting. Oxford: Blackwell publishing.

Aldrich, W., 2013. *Fabrics and pattern cutting*. John Wiley & Sons.

Aldrich, W., 2015. *Metric pattern cutting for women's wear*. John Wiley & Sons.

Almond, K. and Power, J., 2018. Breaking the rules in pattern cutting: an interdisciplinary approach to promote creativity in pedagogy. *Art, Design & Communication in Higher Education*, *17*(1), pp.33-50.

Amaden-Crawford, C., 1989. *The art of fashion draping*. Fairchild Publications.

Brewer, M.K., 2019. Slow fashion in a fast fashion world: Promoting sustainability and responsibility. *Laws*, *8*(4), p.24.

Behrens, R.R., 1998. Art, design and gestalt theory. *Leonardo*, *31*(4), pp.299-303.

Burnham, D. (1973). Cut my cote. Ontario: The Royal Ontario Museum.

Cassidy, T.D., 2017. Exploring the Garment Fit paradigm from a Sustainability Perspective and its Meaning for FirstCycle and Second-Cycle Fashion Retailers. *Current Trends in Fashion Technology & Textile Engineering*, *1*(3).

Chakraborty,M (2021). Applications of Motion Capture Technology in Diverse Industries. Available from: https://www.analyticsinsight.net/applications-of-motion-capture-technology-in-diverse-industries/ [Accessed 20.March.2023]

Cheung, R. and Ng, G., 2007. Efficacy of motion control shoes for reducing excessive rearfoot motion in fatigued runners. Physical Therapy in Sport, 8 (2), 75–81.

Clark, H., 2012. Uncut: the materiality of textiles and the politics of sustainment in fashionable clothing. Published in *Textiles and Politics: Textile Society of America 13th Biennial Symposium Proceedings*, Washington, DC, September 18- September 22, 2012.

Cunnington, C.W. and Cunnington, P., 1992. *The history of underclothes*. Courier Corporation.

Dresp-Langley, B., 2015. Principles of perceptual grouping: Implications for image-guided surgery. *Frontiers in Psychology*, *6*, p.1565.

Duburg, A and vander Tol, R., 2009, 2nd ed. Draping: art and craftsmanship in fashion design. Arnhem, Artezpress

ElShishtawy, N., Sinha, P. and Bennell, J.A., 2022. A comparative review of zero-waste fashion design thinking and operational research on cutting and packing optimisation. *International Journal of Fashion Design, Technology and Education*, *15*(2), pp.187-199.

Ellen MacArthur Foundation. *A new textiles economy: Redesigning fashion's future.* Ellen MacArthur Foundation, 2017.Available from: https://emf.thirdlight.com/link/2axvc7eob8zx-za4ule/@/preview/1?o [Accessed 22. March.2023].

European Environment Agency (EEA) (2019), 'Textiles in Europe's circular economy', Available from: https://www.eea.europa.eu/publications/textiles-in-europes-circular-economy [Accessed 22.March.2023]

Fletcher, K. (2013). Sustainable fashion and textiles: Design journeys. Routledge.

Frayling, Christopher (1993), 'Research in art and design', Royal College of Art Research Papers, 1:1, pp. 1–5.

Graham, L., 2008. Gestalt theory in interactive media design. *Journal of Humanities & Social Sciences*, *2*(1).

Hawley, J.M., 2006. Textile recycling: A systems perspective. In *Recycling in textiles*. Woodhead Publishing Limited, UK.

Koffka, K., Principles of Gestalt Psychology. (London: Routledge & Kegan Paul, 1935).

Kiisel, K., 2020. *Draping*. Hachette UK.

Lindqvist, R., 2014. The transformative cuts: new foundations in pattern cutting and approximations of the body.

Lindqvist, R., 2015. Kinetic garment construction: Remarks on the foundations of pattern cutting (Vol. 13). Rickard Lindqvist.

Luible, C. and Charbonnier, C., 2016. Fashioning Movement A new approach to Fashion Design.

Mwakisunga, H.A., 2010. Gestalt theory for remote sensing image analysis (Master's thesis, University of Twente).

Niinimäki ,K. (2013) A Renaissance in Material Appreciation: Case Study in Zero Waste Fashion, Journal of Textile Design Research and Practice, 1:1, pp.77-92 https://doi.org/10.2752/175183513X13772670831191

Ozdamar Ertekin, Z. and Atik, D., 2015. Sustainable markets: Motivating factors, barriers, and remedies for mobilization of slow fashion. *Journal of Macromarketing*, *35*(1), pp.53-69.

Paramanik, D., 2013. Gestural patterns: a new method of printed textile design using motion capture technology. PhD, Nottingham Trent University, UK

Rissanen, T. and Gwilt, A., 2011. Shaping sustainable fashion: changing the way we make and use clothes. Earthscan, Londo

Rissanen, T., *Zero-waste fashion design: a study at the intersection of cloth, fashion design and pattern cutting [dostopno na daljavo]* (Doctoral dissertation, PhD Thesis Sydney: University of Technology, 2013.

Rissanen, T. I. (2013). Zero-waste fashion design: a study at the intersection of cloth, fashion design and pattern cutting (Doctoral dissertation).

Rissanen, T. and McQuillan, H., 2016. *Zero waste fashion design* (Vol. 57). Bloomsbury Publishing.

Rudofsky, B. (1947). Are Clothes Modern? An essay on contemporary apparel. Chicago: P Theobald.

Sevin-Doering, G. (2007). Itineraire – Du costume de theater à la coupe en un seul morceau. Colombes: Les editions du jongleur.

Townsend, K. and Mills, F., 2013. Mastering zero: how the pursuit of less waste leads to more creative pattern cutting. *International journal of fashion design, technology and education*, *6*(2), pp.104-111.

UCL Petrie Museum's Tarkhan Dress: world's oldest woven garment, 2016. Available from: https://www.ucl.ac.uk/news/2016/feb/ucl-petrie-museums-tarkhan-dress-worlds-oldest-woven-garment#:~:text=The%20Tarkhan%20Dress%2C%20a%20V,the%20late%20fourth%2Dmillennium%20BC. [Accessed 28 April 2023]

Wertheimer, M., 1938. Gestalt theory.

Wertheimer, M. and Riezler, K., 1944. Gestalt theory. *Social Research*, pp.78-99.

Zong, Y. and Lee, Y.A., 2011. An exploratory study of integrative approach between 3D body scanning technology and motion capture systems in the apparel industry. *International Journal of Fashion Design, Technology and Education*, *4*(2), pp.91-101.

Textile Coloration & Processing

SUSTAINABILITY APPROACH OF BANARASI BROCADE

Jasminder Kaur[1]

[1]*Textile Design Section, Faculty of Visual Arts, Banaras Hindu University, Varanasi, Uttar Pradesh, India*

Abstract

Banarasi Brocade is the most popular fabric of India's rich textile tradition. It has a history of several millennia and has been mentioned in many mythological texts. Brocade's importance can be gauged from the fact that almost one-fourth of the Varanasi population is someway associated with its business. *Jamdani, Tanchui, Gyasar,* and *Katan* are the main varieties of this fabric. This handloom-based product has to go through many traditional processes in the making.

Banarasi Brocade witnessed many ups and downs in the era of development and modernization. It faced serious challenges from the power loom, fake *zari* and silk, and substandard products. Many efforts, such as the availability of quality silk, handloom development, technical improvement and efficiency, conservation and promotion of traditional, along with marketing and consumption have been taken in recent years to revive this industry. The artisans, who were leaving their traditional weaving art were directly linked with consumers by providing technical training and support. They were given incentives of health cards, weaver's cards, and subsidies. Solar handlooms and zero-wastage technology were adopted for convenience. The main objective of this research paper is to shed light on the importance of handlooms in Banarasi textiles, and from weaving to marketing, through these factors.

Keywords

Brocade, Traditional Process, Handloom, Training, Revive, Improvement, Marketing.

Introduction

Indian textiles have been famous in the world for their uniqueness and beauty since ancient times. Banarasi brocade is the most popular fabric of this rich Indian textile tradition. Banaras, the cultural and religious capital of India, is known as a seat of knowledge, but its rich brocade fabrics and saris are famous all over the world. Although Banaras produces many things, Brocade is its most precious and proud product. This is the main industry of Banaras, where 70% of its population is directly or indirectly involved in making it. The brocade textile industry employs more than 200000 workers in weaving and allied activities and has an estimated business turnover of $ 500 million annually (Varman and Chakrabarti, 2006, 24).

It is difficult to predict when this rich tradition of brocade weaving came into existence, but the ancient and authentic texts of India give evidence of the very ancientness of this art. *Jataka tales* (Mehta, 1939, 149), and Pali literature related to the life of Gautam Buddha also tell about its antiquity. These beautiful clothes have been mentioned in the *Jataka tales* as *Kashika* or *Kasayani* (Kashi fabric, Motichandra, 1973, 10-11). They have been described *as Kashikuttam* (Jatak, 335) and *Kashiya* (Jatak- 500) in Pali literature. The words *Kashik Vastra*, *Kashi* and *Kashikansu* etc. are mentioned for the textiles made in Banaras in *Divyavadana*, the Buddhist Sanskrit treatise of the Gupta period.

According to the *Mahaparinibbanasukta*, during the Buddha's time, the cloths made in Kashi were so famous that the body of Gautama Buddha was wrapped in a cloth made in Banaras, which had blue, red and yellow shades (Mehta, 1970, 7). Indian ancient texts, *Kautilya's Arthashastra, Panini's Ashtadhyayi, Acharanga sutra, Valmiki's Ramayana* etc. have mentioned the textiles made in Banaras (Krishna and Krishna, 1966, 156).

The costumes depicted in the frescoes of the Gupta period, and the ornamental motifs inscribed on the Dhammekh Stupa at Sarnath present a glimpse of the style of ornamentation made during that period (Krishna and Krishna, 1966, 15). The influence of the Mughal period is also seen in Banarasi textiles when Iranian-inspired motifs were incorporated while maintaining the original style. Some of the European travellers who came to Banaras in the sixteenth-seventeenth century of the Mughal period i.e. Ralph Fitz (1538-91), Peter Mundy (1632, Krishna and Krishna, 32), Tavernier (Tavernier, 1889, 233), have shed more light on Banarasi brocade fabrics in their writings. After the decline of the Mughal rule and the emergence of British rule, the pattern of ornamentation on these clothes changed again, and they were made keeping in mind the European taste, which shows European influence on the clothes made during that period (Watt, 1902-1903, 326-327).

The Official Catalogue of the Craft Exhibition on Indian Art held at Delhi in 1902-1903 by George Watt throws much light on the various aspects of this fabric. He writes that Banaras was the main centre of Brocade Weaving (Watt, 237, 319).

The technical and decorative aspects of the banarasi brocade

Technique: The Banarasi Brocade weave is known for its technical skill and efficiency. There are two weaving techniques of Banarasi Brocade, Which are used to weave the embellishments with the extra weft.

a. Fenkwa- In this technique of weaving, the extra weft is thrown by a shuttle from one end to the other through the gap of the warp threads of the design. When the brocade is weaved through this technique, floats of excess yarn are visible on the back side of the fabric, which is cut and removed later on.

Figure 1 Weavers weaving the pattern in the Kadhwa technique, Source of image –personal

***b. Kadhwa* -** The patterns are created using discontinuous extra weft in this weaving technique. Separate needles (a type of bamboo stick) locally known as *sirki* are used to weave each ornamentation in this technique (Figure 1). Usually, two weavers work together on the loom in this weave. It takes technology and labour both to weave brocade cloths with this technique, resulting in these brocades being much more expensive than those made with the throw shuttle technique.

Types of brocade- Many varieties of brocade are woven in Banaras. Some of these have been developed due to external influences on the industry from time to time. But today they are limited to the Banaras region only. There are different styles of brocade weave in Banaras that are used to weave embellishments from the extra weft. Some of the brocades are,

i.Kimkhab- This was one of the oldest and most sought brocades of Banaras at one time. It is woven with gold or silver threads.

ii. Amaru- This Brocade type is weaved with pure silk threads.

***iii. Baftas* or Pot than**- In this style of weaving, the brocade patterns are weaved with gold or silver threads, whereas the greater part of the fabric is made up of closely woven coloured silk. These are also known as *katan* Brocade.

iv. Abrawan: It is the lightest of the brocades or silk gauzes woven with very few gold and silver threads (Watt, 297).

v. Gyasar (Buddhist Ceremonial Cloth)- *Gyasar* is a Tibetan word for silk brocade. The patterns are weaved with discontinuous supplementary weft in this brocade, with gold, silver and silk threads on a satin base. The fabric is embellished with auspicious Buddhist symbols, such as *Ashtamangala* (eight lucky signs), the Thunderbolt and Bell, Dragons, Clouds, and Lotus Flowers.

vi. Jamdani- It is a transparent cotton (Muslin) or Silk with *zari* weaved with the *kadhwa* technique. It is a very light brocade compared to others.

***vii.* Cutwork-** It is transparent cotton with *zari* brocade, weaved with the throw shuttle technique. Cutwork is the cheapest version of *jamdani* brocades.

viii. Tissue- This is also known as woven water. This brocade is made by single warp and single gold weft threads. It is a thin and transparent lightweight cloth.

Apart from this, the credit goes to the weavers of Banaras who worked hard to save the existence of many other types of extinct or to-be-extinct brocades like *tanchui* (Satin Weave Brocade), *baluchar* etc. and revived them again.

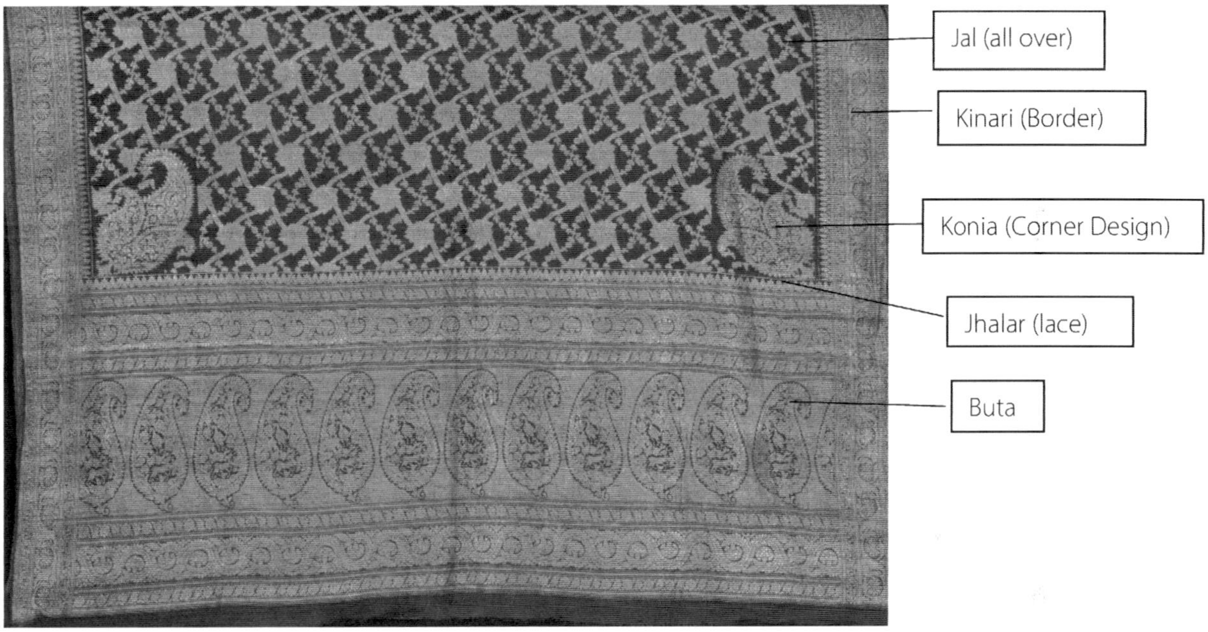

Figure 2, Banarasi Sari, C.20th Cent. C.E., Early, Banaras, U.P. / India, Silk with Golden work, Source of Image: Bharat Kala Bhavan Museum, Banaras Hindu University, Varanasi

Design: Banarasi brocade fabric and saris are the main product of this city. They are made in pure silk, *zari* and cotton threads. Due to the uniqueness of the traditional ornaments inscribed on it, Banarasi Brocade maintains its distinct identity among all other Indian brocades (Plate-2). Its traditional motifs are known as *Buti* and *Buta*, while the small one is called *Buti*, the big size motif is called *Buta*. They are designated after a single flower or figure which are not joined to a connected pattern (Singh, 1979, 22). Each variation of the *Buti* has a different name. *Ashrafi buti, Keri buti*(Mango shape), *Chand Tara buti* (Moon-Star), *Genda buti* (Marigold flower) etc.

Running patterns of leaves and flowers all over the groundwork are known as *Bel* under different names such as *Adibel, Daurbel, Genda Ki bel* etc. Other than these, *Jungla* or woodland design, geometrical patterns in compartments or closed figures are called *Jal, Badrum ka Jal, Kamarakhi jal. Konia, Jhalar, Mothra* are also very popular motifs of Banarasi brocade.

Its traditional ornaments, Buti, Buta, Jhalar, Mothara, Jal, Jangala, Konia, Kangri, etc. are typical. Motifs, such as flowers, leaves, animals and birds, *Keri*, geometric etc. are made in this ornamentation. They were given the names of *Genda* (Marigold flower) *Buti, Latifa Booti*, Shikargah (hunting), *Khajuri Bel, Chakora* etc. for the identification convenience of the weavers, as the majority of them were illiterate.

Challenges- Made entirely on handlooms, Banarasi brocade saris and garments faced stiff competition from other weaving centres which made power-loom products, such as Surat (Gujarat, India) and China. It was the era of the 80s when the handloom industry went into a slump after the introduction of power looms. Saris produced on power looms were very cheap, and the power loom gradually captured the handloom market, and the handloom industry came to the verge of extinction.

The capacity of the power loom was to produce five to six saris in a day, which were also much cheaper, while it took a minimum of two to three days to make a handloom sari with the throw shuttle technique and two to three weeks to make a handloom sari with other techniques, not to mention that they were expensive too. Because of this unequal competition, the demand for handloom saris started decreasing, and many trained weavers were forced to leave their traditional work and started driving rickshaws, becoming potters and working as labourers. The sad part is, that there came a time when weavers started committing suicide due to lack of work.

The struggle for existence led Handloom to make many compromises. In order to compete with the power looms and to remain in the market, the traditional fine embellishments of handlooms were replaced by big and modern motifs. There were other changes too, as the saris started to be made in imitation *zari* and silk, which were difficult to make on power looms. But this necessity of change took a toll on the handloom, and the traditional brocade clothes and saris of Banaras started losing their identity.

The industry went through very rough weather. The brocades like Jamdani, which was made only by the *fekwa* (Throw shuttle) technique, almost went out of weaving. As it was possible to produce fabrics made by the throw shuttle technique on the power looms, but products made by other techniques were not possible to make on it. The unavailability of many products on power looms was disappointing for the buyers. Other than this, the saris, dhotis and other products made on power looms were stiff & heavy, and also lacked the finesse of handloom products, so the people gradually lost interest in them.

Apart from power loom, the changing Fashion trend also affected the brocade industry. The world's changing into a global village and the emergence of social media changed the taste of a larger section. The new generation got interested in trending global fashion, designer's clothing and modern Western wear. The designer's lehengas replaced the traditional red Banarasi saris worn on occasions like marriages, festivals and other rituals.

These changes adversely affected the Banarasi brocade industry. Other than this, because of similar and many other reasons, the Banarasi handloom gradually started dying out. This alarming situation worried the policymakers, and the government and non-government organizations took many important measures to revive this dying industry and to create attraction for handloom among the people. In the first step, the shortcomings and deficiencies were identified, and then effective steps and improvement measures were taken by them in the interest of the industry. The weavers were provided with a set-up and system that catered for their needs starting from the availability of raw materials to the marketing of the finished product. Every possible step was taken to ensure that the ancient fame and form of this age-old industry could be returned, and such a system could be created, which would not only help the handloom industry but would also create an everlasting demand for its product.

The sustainable steps of banarasi brocade.

Raw material

The most important materials for brocade weaving are silk and *zari*, which were formerly procured from Bengal, Kashmir, and imported from Central Asia, China, and even Italy (Muthilal, 1911, 30). Due to the easy availability of silk in China, 50% of the total consumption was imported from there. But because of the high import duty, the procurement of Chinese silk was an expensive affair, resulting in the high cost of the finished product. The Banarasi industry was greatly affected by this anomaly and other products started occupying the market space against it. Intensive efforts were done to overcome the dependency on China, and measures were taken to promote the production of high-quality silk in India. As a result, today 80% of the silk for Banarasi brocade is sourced from Bangalore, a city in India itself.

The procured raw silk is prepared for weaving by twisting. There are three different techniques of twisting. It is followed by many other steps to make a brocade.

Twisting

Silk fibre twisting is also an important step in brocade manufacturing. In this step, the silk fibres used in warp and weft are twisted and different types of brocade are made by them. It can be classified into,

i. *Chinia*- A twisted thread usually composed of four to twelve strands used for warp threads of flowering fabrics.

ii. *Katan*- Two strands lightly twisted together and used for warp thread

iii. *Pat*- Four to twelve strands doubled together, but not twisted and used for the weft threads of thick fabrics (Krishna and Krishna, 1966, 43).

***Zari*-** The second essential material is *Kalabattu*, the gold and silver threads commonly known as *Zari*. It is said that "If silk is the backbone of the industry, design is its structure, and *Kalabattu* or *Zari* is its life and soul." (Krishna and Krishna, 67)

The competition with cheaper power loom saris, led to the use of fake *zari* in handloom saris, which increased with time, and it became difficult to identify between fake and real *zari*. To address this situation, the weavers were motivated through various workshops to use silver-coated thread *zari* and gold-polished copper wire *zari*. Besides this, steps were taken for the easy availability of good quality *zari* to the weavers.

Colour- Dyeing of the yarn is another important part of Banarasi brocade making. The rawness of colours and limitation of shades was a big challenge for Banarasi brocade fabrics. To overcome this, dyers were trained for this purpose, they were also motivated to use vegetable dyes. As a result, the Banarasi weavers are now using threads dyed with botanical and alkaline colours both. Many shades of botanical colours produced from flowers, leaves, barks, roots and minerals etc. are used in Banarasi brocade.

Preparation of the warp thread forms the third step. The thread has to be starched properly to obtain the necessary stiffness and then delivered to the weavers.

Design or pattern-making- Another step constitutes pattern-making. The Designer or *Nakshaaband* sketches the design. Designers play a vital role in making a Brocade. It is very important for the designer to know contemporary market demand along with decorative and technical skills for a successful product. The skill and ability to visualize the effect of embellishment before the weaving is also a major part of this work. Often this was done by weaving a small swatch before weaving the main product to see the possible effect of embellishments and colour combinations. Although labour, time and money, all three were spent in this process, the shortcomings that could be there in the main product were removed.

Over a period of time, a number of specially dedicated software were developed keeping Banarasi brocade in view. This enabled the stakeholders to see the effect, colour and technical effect of the embellishments prior to weaving. It suited the weavers, as they could make changes as needed without any hassle. Since most of the traditional weavers and designers were illiterate and unaware of the new techniques, special training programs were also conducted for designers and weavers to use the software successfully.

Loom setting- This is also an important step in the weaving process. In this step, the new warp threads are set on the loom along with the mounting of the design-punched cards on the jacquard. Weavers of Banaras never remove the warp from the loom, they start the new weaving by tying the new warp threads together with the old warp threads, so that both time and thread are not wasted.

Weaving- The process of weaving was intricate, slow, time-consuming and demanded a great deal of skill in workmanship. Where the weavers worked continuously for 7 to 8 hours sitting in the same position. Therefore, to make their work easy and comfortable, various changes were made to the handloom. Along with this, the basic requirements like comfortable seats, adequate lighting, ventilated rooms and fans were provided on the looms to provide them with a quality life at the workplace.

Weaver- The most important contributor to the Banarasi brocade weaving is its weaver. The slump in this industry caused the migration of many skilled weavers to other states and professions. Therefore, several schemes have also been launched to preserve this traditional workmanship of Banaras and bring back those weavers to this profession, who migrated due to the power loom crisis. Clusters have been formed at weaving centres, where weavers are provided with threads and loom-related parts.

Apart from this, they have also been associated with trained designers to make them aware of information related to colours and new ornamentations. Apart from providing electricity to them at a subsidized rate, health insurance cards have also been given to protect their health. Several State and National level awards have been instituted to honour and encourage their work. The highest of these awards is named after Kabir, a weaver and Great saint of India. This award is given for exceptional weaving.

Finishing- The woven fabric taken out of the loom was starched, pressed and polished for the glaze. Polishing is the final stage of this process.

Packaging- One of the important aspects of the sale of any product is the packaging of the item. Effective work was also done to make the packaging of Banarasi brocade clothes easy and attractive. Many new packaging designs were developed for this purpose.

Marketing- The past practice of marketing Banarasi saris and garments had a middleman between the manufacturers and consumers, making the brocade clothes expensive. To overcome this deficiency, manufacturers and weavers were linked directly to the consumers through online shopping platforms, such as Flipkart, Amazon etc. Along with this, weavers were trained to make and manage their presence on the social media platforms effectively, so that they can market their products through Facebook, Instagram and website. Today Banarasi textile products can be bought online from anywhere. This initiative not only revived this dying art but also helped in sustaining this industry.

Apart from these efforts, to further strengthen the Banarasi handloom sari industry, the products were given **Handloom Mark**, **Silk Mark** and **GI tag**, so that the credibility of its products remains doubtless.

Conclusion

The advantage of all these efforts was that Banarasi Brocade Handloom and Power Loom are no more working as competitors today. Banarasi handloom has its own consumers, whose specific demand has turned it into the most sought-after product today. The rising demand made the big players enter this field, and today the big corporate houses, such as Tata, Reliance, Birla, etc. have engaged the weavers and opened their factories here.

Due to the constantly increasing demand for this product, its weaving has also started in the villages around Banaras. In addition to its efforts, many exhibitions and fairs are being organized in the country and abroad to introduce the people to this unique art and incredible workmanship of Banaras. The teamwork spirit, joint discharge of responsibilities and efforts made at each level has put life in this industry, and today, the Banarasi brocade textile industry is now passing through the golden period.

References

1.Varman and Chakrabarti (2006), Case Studies on Industrial Cluster: A Study of Kanpur Leather and Footwear, Varanasi Silk Sari and Moradabad Brassware Cluster, Department of Scientific and Industrial Research, Government of India, Technical Report.

2. Mehta, N. Ratilal.(1939), Pre-Buddhist India, Examiner Press

3. Motichandra,(1973) Costumes, Textiles, Cosmetics & Coiffure in Ancient and Mediaeval India, Oriental Publishers, India.

4. Jatak (English translation) 6, 47. 151; 1, 335; 6,500

5. Divyavadana, 391, line 6, 17, 23-27

6. Mehta, J. Rustam,(1970) Masterpieces of Indian Textiles, D. B. Taraporevala Sons, Mumbai.

7. Mahaparinibbaṇa Sutta 3, 30-32

8. Krishna Annand and Krishna Vijay, (1966) Banaras Brocades, New Delhi.

9. Watt Sir George, (1902) Indian Art in Delhi: Official catalogue of crafts exhibition, Delhi, 24

10. Tavernier Jean-Baptiste,(1889) Travels in India, London.

11. Singh Chandramani, (1979)Textiles and Costumes from Maharaja Sawai Man Singh II museum, Jaipur.

12.Dovel, Nikita,(2016), The rise and fall of the Banarasi Handloom Sari, livemint.com

13. Muthilal, K.S.,(1911) Smiling Benaras, Madras.

DISPELLING DOUBTS ABOUT TEXTILE PLASMA TREATMENTS

Robert Dewhurst[1] and Robert Mather[2]

[1]Ceretex Ltd. Summerfield's, Church Fenton Lane, Ulleskelf, LS24 9DW

[2]Heriot-Watt University, Riccarton, Edinburgh EH14 4AS

Abstract

The acceptance of plasma technologies over the last ten years by the textile industry worldwide has been remarkably slow. Although many research papers and reviews have been published about plasma treatments of textiles, much of the industry has concerns about adopting the technology, even though it is widely used in other sectors, notably microelectronics.

Chief amongst these concerns are machinery costs, the formulation of process parameters and the durability of plasma treatments. Knowledge of how plasma works (physics), chemical knowledge (chemistry) and the wide variety of machines available (engineering) are all key but are scarcely used in combination. This may also explain why the adoption of the process has been so slow.

The plasma process produces changes to the surface of a textile without altering its bulk properties. It is a dry process, emits no effluent and so possesses good environmental credentials. In this presentation, we discuss machinery costs, noting that the cost can be as low as £150,000.

We also illustrate how a judicious combination of approaches can progress the formulation of processing parameters for a desired plasma treatment in a particular machine. In addition, we outline several strategies for promoting the durability of a treatment.

Introduction

Demands are rapidly growing regarding the environmental friendliness of textile finishing processes. In particular, it is now well recognised that the removal of water is energy intensive and therefore both environmentally harmful and expensive. In textile wet processing, water consumption is far greater than the quantity of textile processed. Rinsing and subsequent washing operations alone consume enormous volumes of water. After textile wet processing, a large amount of effluent will be discharged.

"Pollution produced by the textile industry has a huge impact on the planet, and reasons are quite easy to understand. Clothes are probably the most common items that people buy in today's world, and the average number of clothes that an individual buys every year increased drastically in the recent years.

The textile industry uses millions of gallons of water every day. That's because to produce one kg of fabric, typically, 200 litres of water are consumed: washing the fibre, bleaching, dyeing and then cleaning the finished product. The problem does not rest in the high usage, though, but in the fact that often waste waters are not treated to remove pollutants before they are disposed in the environment. As a consequence, according to some studies 20 percent of all fresh water pollution is made by textile treatment and dyeing."

(Impact of the Textile Industry on the Environment | greenofchange)

A suitable dry process therefore represents an attractive, economical alternative, and there seems little doubt that plasma treatments offer a practical way forward. Plasma treatments play a dominant role in the microelectronics sector for etching and the deposition of thin films. They are also used in the fabrication of semiconductor devices and in the modification of the surfaces of many types of material, including plastics, metals, ceramics and – increasingly – biomaterials (Grace and Gerenser, 2003; Kim *et al.*, 2003; Wu *et al.*, 2015; Chu, 2007). A comprehensive book 'Plasma technologies for textiles' was published over 15 years ago (Shishoo, 2007).

Plasmas are complex mixtures of positive and negative ions, free radicals, excited molecules, photons and electrons. All these species can in principle interact with a textile surface, although it is often the case that only a few of these species are dominant in altering a textile surface in any particular treatment (McCoustra and Mather, 2018). Moreover, plasmas can be maintained at, or slightly above, room temperature if energy is supplied to the plasma from an external source, usually by means of an electrical discharge. Atmospheric pressure devices are generally preferred, as they can often be readily incorporated into continuous textile processing.

Plasma Formulation

It has to be recognised that there are a myriad of reactions and other processes occurring in a plasma chamber during the plasma treatment of a textile. Successful formulation of the optimum conditions for a specific plasma treatment of a specific textile in specific machinery could therefore seem daunting at best. However, in a recent review (McCoustra and Mather, 2018) it was shown that formulation need not in fact be quite so challenging. The review highlighted examples of practical strategies utilising a whole range of disciplines:

a. physics,
b. chemistry,
c. chemical and process engineering,
d. statistical analysis,
e. computing.

This combination of disciplines, often along with art and design, is by no means new to textile practice. The review also noted that there may well be only a few plasma species that are the main reactants with a textile surface and that a major influence on the efficacy of these species will be the distance of the sample from the plasma source.

In addition, we envisage that in the near future formulation will be assisted by the application of artificial intelligence – and in particular machine learning. In machine learning, systems learn from data presented to them and then make a set of predictions. It is enabled by learning algorithms called artificial neural networks that crudely mimic the capabilities of the human brain. In fact, detailed reviews of the potential of artificial neural networks to textile technology have been available for a number of years (Veit, 2012).

The key practical advantages of low temperature plasma treatments are:

a. They produce physical and chemical modifications on the surface of the treated substrate without changing bulk fibre properties and can even confer functionalities not readily achievable with traditional wet processing;
b. There is minimal consumption of chemicals and no drying process required.
c. The process has a high level of environmental compatibility.
d. The process can be applied to almost all kinds of fibre.
e. Machines that are low temperature plasma are commercially viable. Price in region of £150k with a throughput of 3 million meters + per annum.

It should be emphasised that the individual equipment used has a significant bearing on the conditions of a plasma treatment. Important control factors include:

a. The pressure of the gas in the plasma chamber, and any fluctuations in pressure, although it appears that modern machines use air instead of a separate gas chamber. This reduces the cost of installation and as air is composed of the following elements.

Element	Volume by %
Nitrogen	78.08
Oxygen	20.95
Argon	0.93
Carbon Dioxide	0.040
Neon	0.0018
Helium	0.0005
Krypton	0.0001
Hydrogen	0.00005
Xenon	8.7×10^{-6}

b. The power and frequency of the electrical supply.
c. Higher input power increases the number of high-speed electrons in plasma and improves plasma treatment effect (Wang and Qiu, 2007)
d. The duration of the treatment. Generally, the longer duration of treatment the more severe the modification of substrate surface.
e. The rate of flow of gas into the reactor. Higher input power increases the number of high-speed electrons in plasma and improves plasma treatment effect (Wang and Qiu, 2007).
f. The nature of the substrate.
g. Discharge power, which in practice is generally the combined factor of pressure and discharge power (T. Yasuda, Gazicki, and Yasuda 1994)

h. Ageing of plasma treated polymer surfaces can be minimised in a number of ways. An increase in the crystallinity and orientation of a polymer surface enhances the degree order and thus reduces the mobility of polymer chains, resulting in slower ageing. Also, a highly cross-linked surface restricts the mobility of polymer chains and helps reduce the rate of ageing (Brennan et al. 1991).
i. Sample distance is the distance between the sample and the plasma source. Zhang and Fang (2009) studied the relationship between the distance of the electrodes and the fabric and *K/S* value of inkjet-printed fabric under the effect of plasma treatment. The highest *K/S* value of the samples was obtained when Zhang and Fang set the distance between the electrodes at 3 mm for air. Conversely, the field intensity becomes weaker with the increase of the gap, which weakens the effect of plasma discharge.

A treatment formulation that works well for one system may have to be adjusted for the same treatment in another system. Nevertheless, it has been demonstrated that a combination of experimental techniques, the application of statistical experimental design and the use of computer simulation and modelling can be invaluable in formulating the key processing parameters for a particular system (McCoustra and Mather, 2018).

Yet despite its uses in many other industrial sectors, the impact of plasma technologies over the last 10-15 years on the textile industry worldwide has been slow. We believe that there are a number of contributory factors, including:

a. Lack of clear information and direction on achieving practical formulations for economic treatments.
b. The perceived high cost of plasma machinery. This seems to have developed from the introduction of high temperature plasma machines which have a cost of £1.5 million each. Now low temperature plasma machines do not rely on such expensive lasers or installation and are at a commercial price of around £150,000.
c. Doubts about the durability of plasma treatments. Here we gauge durability by:
   i. Stability, which indicates no change over 7 days without further treatment.
   ii. Washability, which indicates no change over a prescribed number of washes.

In this paper, we seek to quell concerns about the commercial application and involve interested parties of plasma treatments and results to remove the mystique that is commonly associated with them. The added properties that we shall discuss here are:

a. Flame retardancy, assessed by horizontal and vertical flame tests.
b. We recognise there are many tests for textile flame retardancy but have limited ourselves to the vertical flame test for drapes (BS5867 part 2 type B) and the horizontal flame test for domestic upholstery (Furniture and Furnishings (Fire Safety) Regulations 1988 (Match & Cigarette test). The significant difference between these two tests is that BS5867 part 2 type B holds a piece of fabric of the prescribed size vertically and presents a butane flame into the centre of the fabric sample for a prescribed time. If the fabric continues to burn after the flame is removed or the fabric burns to the edge of the sample the result is fail. By comparison the flame test for domestic upholstery (the Match & Cigarette test) has a butane flame presented to fabric draped over foam. If the fabric draws away from the flame it exposes the foam and fails, the test. Often fabrics that pass one test fail the other and vice versa.
c. Water repellency, assessed by contact angle measurement and spray testing.
d. Durability of treatment, assessed by no. of wash cycles (see above).

e. Breathability is the ability of a fabric to allow air and moisture (or sweat) to pass through it. It is an essential contributor to the properties of thermal and anatomical comfort in clothing and is vital in filtration and medical textiles.

We will carry out the tests on a range of substrates.

Ref.	Table 1: substrates
1.1	Woven polyester
1.2	Warp knitted polyester
1.3	100% woven cotton
1.4	Non-Woven Polypropylene.
1.5	Polyester/cotton 70/30

Flame Retardancy

The need to render textile products flame retardant has been well recognised for a long time. Indeed, more than two centuries ago, the Montgolfier brothers coated their balloons with alum to reduce fire hazards. In the second half of the twentieth century, a wide range of treatments was developed, many of them based on phosphorus compounds or highly halogenated aromatic hydrocarbons. However, many of these compounds were later recognised to be toxic, e.g., through the release of formaldehyde or gaseous halogen compounds, and potentially harmful to the environment.

Lots of attention has also been given to intumescent flame retardants that promote char formation, for example in fibres that can melt, such as polyester and nylon. Whilst some have been shown to be effective, again they may impair fibre formation and properties.

The successful conferment of flame retardancy is clearly then highly complex, and alternative approaches to achieve it are now being attempted. For instance, it is evident that an important first barrier to the ignition of any material, including textile fibres, is the surface. It therefore seems appropriate to find out what can be done to modify fibre surfaces to improve flame retardancy. The focus of attention has now very much shifted to the application of nanotechnology – in particular, the creation of very thin layers on fibre surfaces. In this respect, plasma deposition techniques could well offer a practical way forward, provided the plasma treatment is sufficiently durable to the subsequent processing and eventual application destined for the particular textile. Some results have already been reported (Horrocks *et* al, 2011; Li *et al.*, 2019; Qi *et al.*, 2022).

Domestic drapes are not required to be flame retardant in the UK, however drapes act like a wick up the side of high-rise buildings, and there have been several disasters where non-FR drapes would have contributed to the calamity. Of the little work published on using atmospheric plasma, work has shown that deposition of silicon-based chemicals on textile surfaces can significantly improve their flame retardancy defined in terms of improved fire resistance.

Ref.	Table 3: Tests for flame retardancy	Standard description
3.1	Domestic upholstery FR (horizontal)	BS5852: Match & Cigarette test.
3.2	Domestic drapes FR (vertical)	BS5867: Part 2: type B

Water Repellency

Water repellency is required for a wide variety of textile applications. Textile products manufactured for outdoor use, such as outdoor clothing, tents and tarpaulins, are obvious examples. Other examples include seat covers in road vehicles, trains and aircraft, and bandages to curtail the wicking of blood and the entry of water from the outside.

Traditionally water repellent finishes can be applied by a variety of means, such as dip processing and spray coating, but all are applied from the liquid state. Moreover, some finishes have, at best, limited durability to washing, whilst with others (notably, treatments with fluorochemical compounds, which also confer oil repellency) there are concerns about toxicity and adverse environmental impact. Resistance of carpets and furnishings to the spillage of beverages is also conferred by water repellent finishes.

The conferment of water repellency by plasma treatment offers an approach that by-passes wet processing and overcomes toxicity and environmental concerns. A number of papers and reviews have been written about this application of plasmas, but most of them refer to the fluorination of textile fibre surfaces. An approach that is now being seriously investigated is plasma treatment involving hexamethyldisilane (HMDSO). Although a number of papers on this topic describe the use of low-pressure plasma equipment, it has nevertheless been shown that successful treatment on polyester fibres can be achieved in atmospheric plasma equipment (Ref: Ji et al. 2008). HMDSO is polymerised at the fibre surfaces to give a thin continuous water repellent coating, although it can also be hard and brittle (Ref: Keller et al, 2007).

There is an important distinction between "waterproofing" and "water repellency". The former prevents water passing through the fabric unless a pressure is applied. Waterproof fabrics are usually coated and offer no breathability. Water repellent fabrics on the other hand are breathable and repel water which is measured by the contact angle of the droplets and a spray rating.

We intend to use the following chemicals for water repellency.

Ref.	Table 5: Tests for water repellency	Standard description
5.4	Contact angle	ASTM D5946 test
5.5	Spray rating	EN 24920 - Textiles. Determination of resistance to surface wetting (spray test) of fabrics

Durability

As already highlighted, a prominent concern in the textile community is the durability of textile plasma treatments, especially as the depth of treatment at the fibre surface is only of the order of 10 nm, and hence far smaller than for more conventional coatings. How stable is a textile fibre surface, even a few days or weeks after the treatment? How durable is the treated fibre surface to the subsequent steps involved in conversion to the end-product, and also how stable to washing and wear-and-tear during the lifetime of the product? There are many ways in which treatment stability and durability can be assessed, and these assessments are often governed by the intended application of the textile. For our work we have chosen to consider a plasma treated textile to be stable if no change is detected seven days after treatment. We deem durability to be good if no change is detected after a prescribed number of washes.

Instability of the treated fibre surface often arises from reorientation of surface polymer chains. In addition, any volatile compounds that were formed during the treatment are liable to migrate to the fibre surface and then be released from it. Synthetic fibres with low glass transition temperatures, such as polypropylene fibres, are particularly susceptible to these ageing processes.

Several approaches are available for improving stability. For instance, the reorientation of surface polymer chains is hindered if they are present mostly in crystalline domains. Another approach is to graft a radical species in the plasma onto the fibre surface. The radical species can be a monomer, which can then itself be polymerised. Again, the reorientation of the surface polymer chains becomes restricted. Still another approach is plasma polymerisation, whereby a very thin layer of cross-linked polymer can be formed from species in the plasma, which in effect encapsulates the fibre. One prominent example of such a cross-linked polymer has already been noted above. A coating of the cross-linked polymer of HMDSO not only confers hydrophobicity but also in general renders a textile more durable due to the robust nature of the coating.

We have chosen to measure durability in 3 ways (see Table 3). First after 7 days we retest the fabric without any further processing. This measure is sometimes called ageing. Second, we will test the fabric after 5 washes, and third after 50 washes. These tests will give us a guide onto how the combined treatment and chemicals perform, and we will assume if it fails after a week, it will not pass any further tests.

We plan to test treated samples to the following standards.

Ref.	Table 6: Tests for durability	Standard description
6.1	Durability: stability/ageing	Retest after 7 days without further processing.
6.2	Durability: washability 3 washes at 40 degrees	Retest after 5 washes at 40 degrees
6.3	Durability: washability 50 washes at 40 degrees	Retest after 50 washes at 40 degrees

Breathability

Breathability is the ability of a fabric to allow moisture vapour to be transmitted through the material.

It is important especially for the comfort of the wearer, and is a combination of the structure of the fabric and the coating applied. We are working at nano level with plasma which does not affect the structure of the fabric.

Conclusion

We hope this will be the beginning of providing a basis for future development of chemicals used in plasma treatment. Already we have carried out research with bacteria repellent chemicals on cotton that have proved successful. There remains a host of other chemicals that can be treated this way e.g., insect repellent.

References

Brennan W J, Feast W J, Munro H S & Walker S A (1991), *'Investigation of the ageing of plasma oxidised PEEK'*, Polymer 32, 1527-1530.

Chu P K (2007), *'Plasma-treated biomaterials'*, IEEE Trans. Plasma Sci. 35, 181-187.

Grace J M & Gerenser L J (2003), *'Plasma treatment of polymers'*, J. Disp. Sci. Technol. 24, 305-341.

Horrocks A R, Nazaré S, Masood R, Kandola B & Rice D ((2011), 'Surface modification of fabrics for improved flash-fire resistance using atmospheric pressure plasma in the presence of a functionalized clay and polysiloxane', Polym. Adv. Technol. 22, 22-29.

Ji Y-Y, Hong Y-C, Lee S-H, Kim S-D & Kim S-S (2008), 'Formation of superhydrophobic and water-repellency surface with hexamethyldisiloxane (HMDSO) coating on polyethyleneteraphtalate fiber by atmospheric pressure plasma polymerization', Surf. Coatings Technol. 202, 5663-5667.

Keller M A, Fortunato G, Körner E & Hegemann D (2007), *'Continuous coating of synthetic fibers using hexamethyldisiloxane'*, Plasma Process. Polym. 4, S1063-1067.

Kim M C, Yang S H, Boo J-H & Han J G (2003), 'Surface treatment of metals using an atmospheric pressure plasma jet and their surface characteristics', Surf. Coatings Technol. 174-175, 839-844.

Li J, Tong W & Yi L (2019), 'Flame-retardant composite coatings for cotton fabrics fabricated by using oxygen plasma-induced polymerization of vinyl phosphoric acid/cyclotetrasiloxane', Text. Res. J. 89, 5053-5066.

McCoustra M R S & Mather R R (2018), 'Plasma modification of textiles: understanding the mechanisms involved', Textile Progress 50, 185-229.

Qi L, Wang B, Zhang W, Yu B, Zhou M, Hu Y & Xing W (2022), 'Durable flame retardant and dip-resistant coating of polyester fabrics by plasma surface treatment and UV-curing', Prog. Org. Coatings 172, 107066.

Shishoo R (Ed) (2007), *Plasma technologies for textiles*, Woodhead Publishing Limited, Cambridge, UK.

Veit D (2012), 'Neural networks and their application to textile technology' in Simulation in textile technology, Veit D (Ed), Woodhead Publishing Limited, Cambridge, UK, pp 9-71.

Wang C X & Qiu Y P (2007), 'Two sided modification of wool fabrics by atmospheric pressure plasma jet: In fluence of processing parameters on plasma penetration', Surf. Coatings Technol. 201, 6273-6277.

Wu C-C, Wei C-K, Ho C-C & Ding S-J (2015), Enhanced hydrophilicity and biocompatibility of dental zirconia ceramics by oxygen plasma treatment', Materials 8, 684-699.

Yasuda T, Gazicki M & Yasuda H (1984), *'Effects of glow discharges on fibers and fabrics'*, J. Appl. Polym. Sci. 38, 210-204.

Zhang C & Fang K (2009), 'Surface modification of polyester fabrics for inkjet printing with atmospheric-pressure air/Ar plasma', Surf. Coatings Technol. 203, 2058-2063.

The Circular Economy & LCA

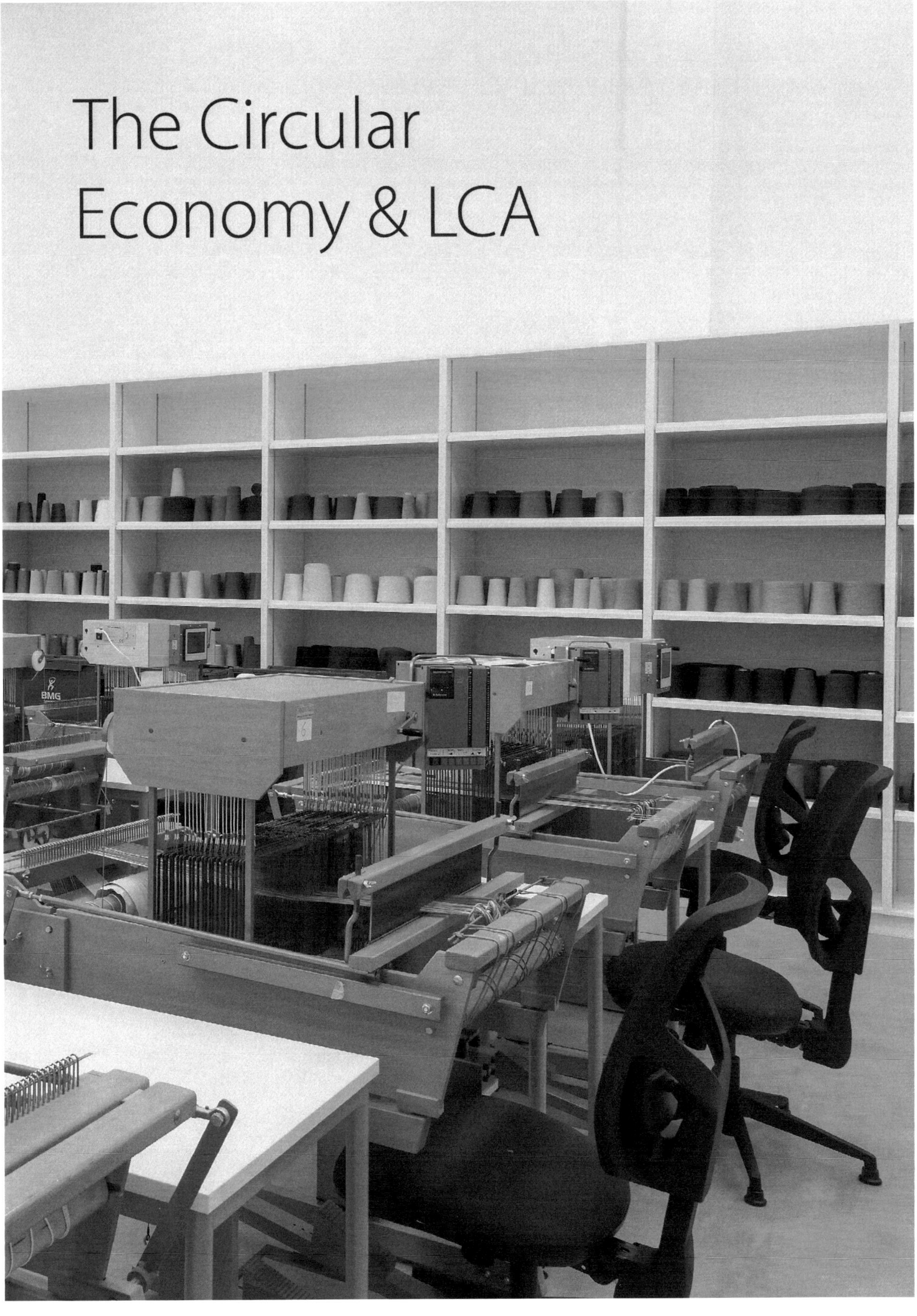

POST-CONSUMER POLYESTER WASTE RECYCLING IN EUROPE; CHALLENGES AND OPPORTUNITIES OF CLOSED-LOOP PRACTICES.

Teena Antony[1]

[1]*London College of Fashion, University of the Arts, London, 272 High Holborn, WC1V 7EY*

Abstract

Increasing textile waste is a significant environmental problem. The limitations of the current 'take-make-waste' linear economic paradigm are clear. Recycling textile waste has been the preferred approach in proactive attempts to create a more sustainable process. The circular economy moves the recycling paradigm forward by emphasising waste reduction at source and keeping high-value materials in circulation for longer. The research highlighted here bridges academia and industry to provide a broad overview of the scope and possibilities of implementing closed-loop practices in post-consumer polyester textile recycling systems across the European market whilst identifying existing constraints. The study adopts an inductive approach and has drawn on insights from textile recycling experts to provide actionable recommendations to specific stakeholders. The potential of implementing closed-loop practices within textile recycling was evaluated and opportunities were identified across the value chain – from securing investments necessary for commercial scale to drafting EPR policy and incorporating design for circularity into current production processes. Investors, policymakers, designers, manufacturers, and retailers can use these insights to enhance future strategies as they move toward a closed-loop system in the circular economy.

Keywords

Closed-loop practices; textile recycling; post-consumer; polyester; textile waste; circular economy.

Introduction

The global textile industry is well-criticised for its unsustainable use of fossil fuels and its various manufacturing processes that employ hazardous chemical usage contributing to environmental pollution (Sandin and Peters, 2018). The current fashion and textile industries heavily rely on "non-renewable resources across" various stages of their value chain because the current economy is inefficient and wasteful in its value creation model (Ellen MacArthur Foundation, 2017). Polyester, derived from petroleum, is the dominant material in clothing manufacturing, representing 60% of the global market and 16% in Europe (Sajn, 2019, Circle Economy, 2020). According to WRAP's (2019) report, polyester material saw a quick rise in popularity on a global level, due to its advantage of low water consumption during the production process.

With more than 15 kg of textile waste per person produced annually within Europe, the post-consumer clothes and home textiles discarded by the consumer stand as the largest source of textile waste accounting for 85% of the total waste (McKinsey, 2022). Landfill and incineration in and around Europe and UK have become the primary end destination of discarded textiles, creating detrimental consequences for the people and the environment. While ethical concerns regarding the disposal of end-of-life clothing are still evident, the EU's waste prevention report considers recycling an important strategy for recovery (Piribauer and Bartl, 2019). In recent years, the separation of post-consumer textile fibre using chemical recycling has sparked a lot of interest. With fast-paced innovation happening around recycling technologies, there is a race for scale. Although well-established technologies for the mechanical recycling of cotton are common, developments in the chemical recycling of polyester textiles are subject to extensive research and development and need to be commercialised. If developed well, it is estimated that 70% of the textile waste could be recycled using fibre-to-fibre recycling technology (McKinsey, 2022).

The production of new clothing items has experienced a significant increase in the last decade, resulting in more than 150 billion items of clothing manufactured each year (Bedat, 2016). The amount of clothing manufactured and disposed of outpaced the capacity of recycling facilities, leading to negative environmental consequences. Furthermore, the lack of demand for recycled materials and limitations in technology which enable recycling clothes into virgin fibres resulted in the UK and European markets having low recycling rates (Jaamaa and Kaipia, 2022).

Closed-loop recycling is considered a major opportunity for reducing textile waste and extracting virgin raw materials. Creating 'closed-loop material recovery' in recycling is the sole way to extract material value over multiple lifetimes (Goldsworthy, 2014). Fabric made from PET (polyethylene terephthalate) constitutes most of the recycled material available in the market, which is a meagre 10% of the total textile market. It is important to accelerate textile-to-textile recycling innovation to move beyond the recycling of single-use plastic (Prahl, 2020). Therefore, verifying the recyclability of polyester and polyester blends is vital to provide recommendations to brands, manufacturers, and textile recycling units.

Falling far short of the 1.5C threshold set forth in the United Nations Paris Agreement, by a startling 50 per cent, the fashion industry is on a 3C trajectory. Swift changes along with integrating circular practices into the value chain have now become the need of the hour for the fashion industry to survive and thrive. Significant and swift action is needed from all stakeholders to accelerate the transition of the current linear industry to a circular economy. As outlined in Figure 1, from the design and development, through manufacturing and waste management, the current industry has limited infrastructure to support textile waste management to achieve a closed-loop lifecycle (McKinsey, 2020). This highlights the importance of investigating the existing barriers, feasibility, and opportunity of closed-loop practices within post-consumer polyester textile waste recycling, as it is no longer about whether the fashion industry will adopt circular practices, but rather how.

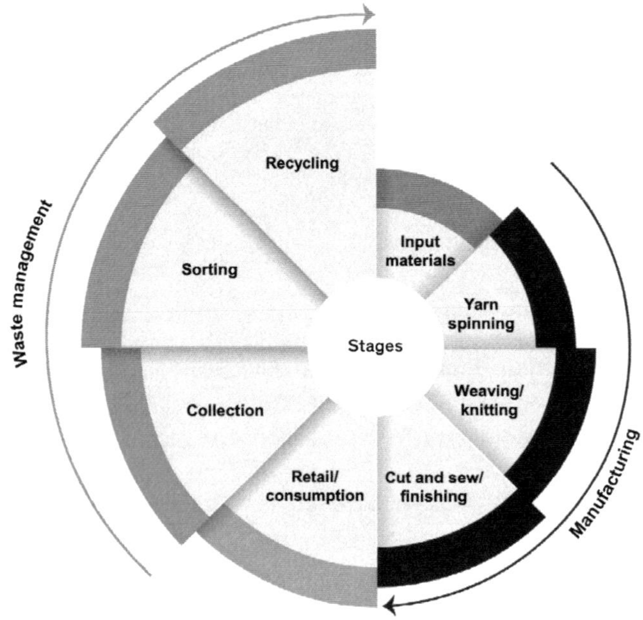

Figure 1 Closed-loop fashion lifecycle stages (McKinsey,2020)

The academic journals to date investigate the strong claims of the impact of textile recycling, however, research regarding post-consumer polyester waste is limited (Sandin and Peters, 2018). Therefore, this study specifically considers the opportunity for post-consumer polyester textile waste recycling in the European fashion industry and how it facilitates bridging the gap by investigating closed-loop circular economy practices. Although the focus of this study is on post-consumer polyester textile waste recycling across the UK and Europe, this serves as a guide for assessing the existing situation and possible advantages of closing the loop for various textile waste on a global level, proving that further research and analysis into the textile-to-textile recyclability of polyester is key to achieve a closed-loop system.

This research aims to explore the recycling opportunities for post-consumer polyester textile waste in Europe to analyse how it contributes to the fashion industry and the environment in developing a closed-loop circular system in the current fashion industry.

Literature Review

The textile and fashion industry's significant contribution to the aggravation of environmental problems increases the need for closed-loop practices and circular business models and exposes the limitation of the current linear model in which the industry operates. With less than 1 per cent of clothing recycled into new clothes, a majority of the textile recycling capacity worldwide is still made up of downcycling textiles into low-value products (Europa, 2021). Although tons of textiles within the UK and Europe are taken into charities, the current fashion system is mainly linear (take-make-waste) and is disconnected from the post-purchase phase (Ellen MacArthur Foundation,2021).

Post-Consumer Textile Waste Market Analysis

The number of textiles made from fibre blends has greatly increased over the past few decades, making polyester blend textiles a highly accessible product on the market (Eileen Fisher Foundation,2022). Due to Western Europe's fast consumption and disposal patterns the textiles that are introduced to the market are rapidly approaching their life span. According to the report by Circle Economy (2020), each person in the Netherlands discards an average of 40 articles of clothes per year. This rising mountain of post-consumer textiles waste offers diverse opportunities for possible applications for market reinsertion. With approx. 64% of the garments collected from Western Europe are reusable, just a small part of the textiles that are collected are currently recycled into new ones (Circle Economy, 2020). Whereas the majority of used textiles are currently being downcycled, burned, or dumped (Ellen MacArthur Foundation, 2017). As a result of the saturation that the used clothes market is currently experiencing, these textiles are often considered unwearable (Ljungkvist et.al, 2018).

Textile waste is very fragmented. Over 7 million tons of textile waste are produced annually in Europe, which is estimated to be at least fifteen kilograms per person (Eileen Fisher Foundation,2022). Given the essentially local nature of the collection stage of the lifecycle of textiles, achieving full circularity necessitates a local, as opposed to a global model (Niinimaki, 2019). The majority of this textile waste, or around 85% of it, is made up of domestic textiles and clothes that are thrown out and typically end up being dumped in landfills or incinerated around the world with a detrimental effect on the ecosystems, whereas the post-industrial, pre-consumer and post-consumer commercial waste make up the remaining trash (Green peace, 2022; McKinsey, 2022). Table 1 gives an overview of the expected growth of textile waste.

Table 1 Gross textile waste in Europe (McKinsey,2022

GROSS TEXTILE WASTE IN THE EUROPE (IN MILLION TONS)		
WASTE STREAMS	**Year 2020**	**Year 2030 (Expected)**
Production (Post-industrial Waste)	0.5	0.6
Retail (Pre- consumer Waste)	0.2	0.3
Private (Post-consumer Household Waste)	6.0	7.3
Commercial (Post-consumer commercial Waste)	0.4	0.5
TOTAL	7.1 million Ton	8.7 million Ton

Although the consumer segment is scattered, 85 per cent of the waste is produced from post-consumer household waste. Even with the lack of transparency in the value chain, brands and retailers account for a small portion of overall waste volume (McKinsey,2022). This highlights the lack of collection of PCTW as a major barrier to widespread textile recycling across Europe.

Post-Consumer Textile Waste (PCTW)

The fashion and textile industry has been highly polluting over the last few decades with growing concern revolving around the environmental impacts of clothing. Although the production of fashion and textile products impacts the environment due to the vital use of resources and fossil fuels, the waste produced as the used and damaged textile is equally to blame (Kumar et al., 2020). Contributing majorly to the landfills due to lack of over-production and extreme consumption patterns, textile waste is classified according to its origin. With the potential to be recycled into new raw materials, pre-consumer textile waste which is bio-degradable, consisting of fibre, yarn waste, scraps, and fibre lint, is widely recycled within the industry (Wanassi et al., 2016). With a mixture of various biodegradable and non-biodegradable substances, consisting of various types of textile materials such as cotton, wool, polyester and other trims and accessories like buttons, zippers, hooks, and other plastic materials the recyclability of PCTW stands as a major concern. Coskun and Basaran state that its almost 100% possible to recycle PCTW (Coskun & Basaran, 2019), whereas Rotimi argues that the lack of advancement in technology to assist in the process of separation of blended fibres makes recycling a complex process (Rotimi, Topple, and Hopkins, 2021). Whilst the convenience of sorting makes pre-consumer textile waste easier to recycle, the complexity of components involved in PCTW makes it a harder choice (Coskun & Basaran, 2019). This variability in clothing composition is a major challenge for recyclers (Goldsworthy, 2012). Figure 2 depicts the textile waste recycling routes.

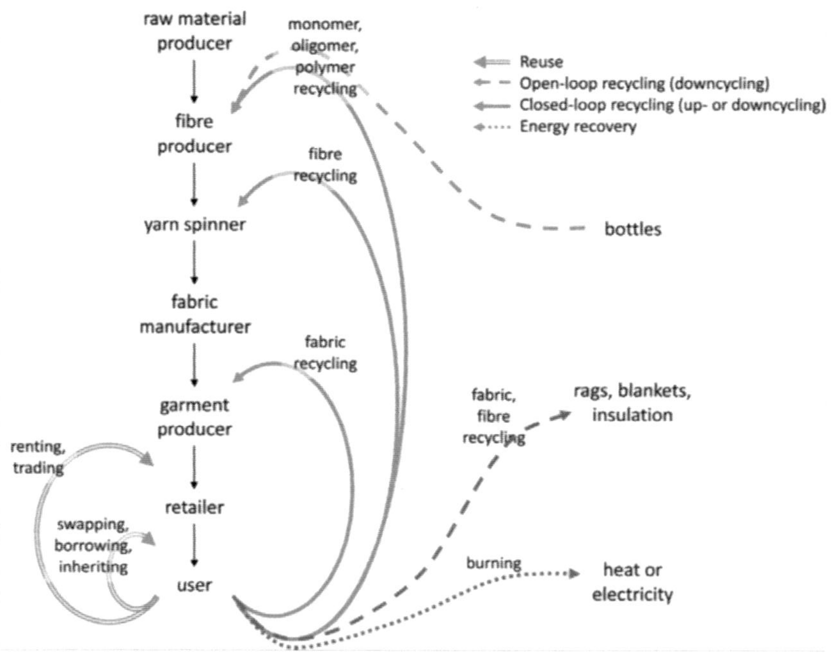

Figure 2 Recycling and reuse routes for Textile waste (Sandin & Peters, 2018)

Piribauer and Bartl argue that the handling of PCTW is an inefficient use of resources as the waste is often used for energy recovery (Piribauer & Bartl, 2019). Globalisation and an increase in the consumption drive of consumers made the production of clothing increasingly inexpensive, treating low-priced garments as disposable. With a major portion of post-consumer waste derived from discarded garments and home textiles, the transition of textile products from a basic human need to a fashion item has facilitated the steady decline in the disposal of products before the end of life of the product (Piribauer & Bartl, 2019). With over 75% of the pre-consumer waste generated during the production cycle being recycled, the rate of post-consumer waste recycling has been at a meagre of 15% (LeBlanc, 2014).

PCTW Challenges and Environmental Impacts

Since the introduction of synthetic materials in the middle of the 20th century, the range of textile materials has significantly increased. It is anticipated that within the next ten years, fibre production will expand by 34% on a global level, driven predominantly by the growth of fast fashion, estimated to double the use of polyester as seen in Figure 3 (Textile Exchange, 2021).

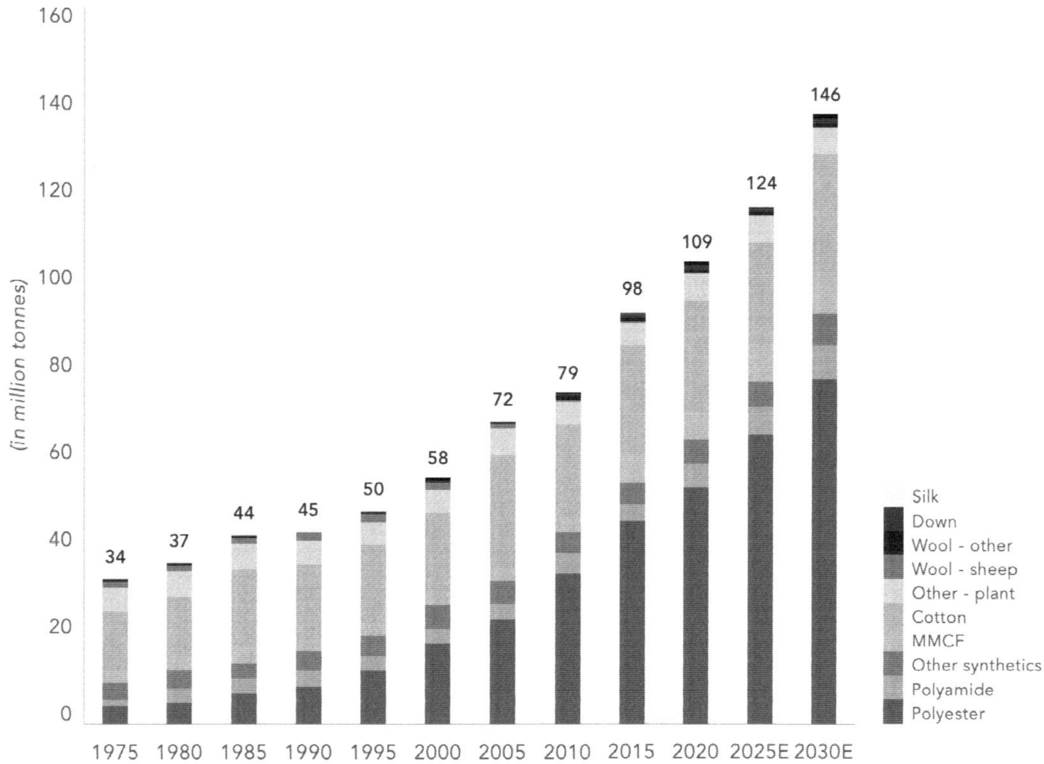

Figure 3 Projected fibre production of Polyester (Textile Exchange, 2021)

Park and Kim (2014) claim that the widespread use of PET (Polyethylene Terephthalate) has increased waste production and boosted the demand for developing recycling systems for these materials of which the most popular commercial method is making textiles by recycling PET plastic bottles, mechanically by melting the bottles, and chemically by adding chemicals.

The production and consumption of textiles impact the environment through both pre-consumer and post-consumer waste. According to Goldsworthy (2012), a piece of textile made from cotton, printed using natural and biologically safe dyes has far more recyclability than the same piece overlaid with a metallic or glossy finish. On average, 80% of a product's environmental impact is decided during the design phase of the garment. Potential barriers to recycling should be 'designed out' and identified at the production stage to create textile products that are totally recyclable (Goldsworthy,2014). Nevertheless, the majority of the products produced within the textile industry are designed without considering the end-of-life stage of the product (Euric, 2022). Hence it is vital to consider the utmost importance of 'Design for Circularity', to achieve a circular economy within the textile industry. Although, 'Design for Circularity' is in the beginning stage, the need for 'design for recyclability' has not just only been under-explored within the industry, but on a wider spectrum, several brands have been greenwashing about using recycled materials within their products and making impactful claims about the environmental credentials without any evidence to back up their claims (Eppinger, 2022).

Polyester Recycling and Existing Technology

The most popular multicomponent garments on the market, cellulose polyester blends strive to adjust material qualities in relation to moisture absorption, wrinkle resistance, and wearing comfort. It is challenging to recycle them using a single procedure because cellulose and polyester behave very differently in terms of hydrophilicity and thermal stability (Haslinger et.al, 2019). Koszewska (2018) argues that the solutions and advanced recycling technologies which could possibly eliminate the barriers to the establishment of a global closed loop in the textiles sector are still very limited. As a result, low-grade materials and blends predominate in the recycling market, placing pressure on technologies that can be used to recycle low-grade textiles and blends while still being profitable. In order to avoid incineration or landfilling, textile recycling might range from recovering fabric or fibres to degrading the material to recover polymers or even monomers (Palacios, Meer & Seide, 2021). Despite these advantages, virgin material is frequently selected over recycled material, which remains unpopular. Only 13% of the market for recycled polyester was present in 2018 (Textile Exchange, 2019).

Although cutting-edge technologies are setting new aesthetic and functional standards to convert discarded plastics from marine into novel raw materials and high-performance fabrics, the recyclability of the end-of-life of products made from these materials is still a question (Koszewska, 2018). However, these technologies are helpful in extending the life of plastics which otherwise would have been an eco-concern. Whilst these advancements are helpful in one way, the application of these discarded plastics in textiles is still a dilemma as the lack of technology to recycle the polyester textile is still in its infancy.

The recycling end markets are dominated by blended and low-grade materials. The lack of expertise in upscaled processes of fibre separation from mixed blends, the commercial viability of recycling technology for low-value textile fractions, the high expenditure involved in the recovery process, the high cost of logistics and the limited availability of textile recycling plants on both local and regional makes it harder for the existing recycling technologies (Damayanti et.al, 2021). Even with such concerns revolving around recycling textiles, especially polyester textiles, a few brands have started collaborating with businesses to create a closed-loop ecosystem for polyester. The most well-known advocates of recycling polyester are American Retailer Patagonia who collaborated with Teijin, a Japanese business that has created its own closed-loop polyester recycling technology in developing an alternative recycling channel for polyester clothes that cannot be worn otherwise (Patagonia, 2022).

Other methods to separate fibre blends from polyester have been under development by Worn Again Technologies (Worn Again Technologies, 2018). Furthermore, the consistency of the feedstock is important for several sophisticated recycling systems since it affects output quality, quantity, and process efficiency (Peng et.al, 2022). However, these technologies are not adopted on a large-scale commercial basis, but rather on a Pilot, Lab, and Demo scale. Table 2 gives an overview of the existing chemical recycling technologies along with their capacity.

Table 2 Overview of the existing chemical recycling technologies

FACILITY	TECHNOLOGY	INPUT	OUTPUT	STAGE & CAPACITY
Evrnu, US	Chemical polyester recycling	Pure cotton, pure polyester, blended cotton-polyester	Recycled polyester Recycled man-made cellulosic fibres	Current pilot (live) Demo plant (planned for 2022/2023): 19,000 tonnes / year
Ambercycle, US	Chemical polyester recycling	Pure and blended polyester	Recycled polyester	Current pilot: 10-100 KG/day Demo plant (in planning): 1 tonne/day
Circ, US	Chemical polyester and cellulosic recycling	Pure cotton, pure polyester, blended cotton-polyester	Recycled polyester; recycled man-made cellulosic fibres	Current demo plant: 4 tonnes / day
Pheonxt, Australia, Europe, US	Chemical recycling, Circular recycling technology Solvent-Free Technology	Polyester, Blended Poly-cotton, plant based natural cellulosic fibres	Recycled polyester, Cellulosic fibres	Pilot (live) Demo plant (in planning)
Worn Again, UK	Polymer recycling technology Solvent-based technology	Pure polyester, blended cotton-polyester	PET and cellulose outputs. Recycled Polyester	Demo plant (in planning): 1,000 tonnes / year

An increase in demand for the need to adopt new technologies in polyester recycling calls for the need of integrating circular economy practices starting from the polyester manufacturing at the raw material stage (Palacios, Meer & Seide, 2021) Current finishing techniques and chemicals used in the polyester fabric to increase the functionality affects in processing, as it makes it irreversible, creating barriers in adopting closed-loop chemical recycling for polyester PCTW (Goldsworthy 2012). For polyester, it is important to recycle it for perpetuity. To work within the system, adopting a mono-material concept will help in transforming the industry's sustainable future by creating a closed-loop (Wang, 2006). Furthermore, as the mono-materials are constructed of a single type of fibre, mono-material fabrics are simple to recycle after their end of life. However, achieving the levels of comfort, functionality, aesthetics, and performance demanded by consumers in mono-material fabrics has proven to be challenging, as it fails to satisfy the majority of the qualities that people need in their products. Furthermore, advances in mono-material clothing rely primarily on polyester for its performance qualities, but as most of the polyester produced worldwide is made from petroleum, it is not thought to be environmentally sustainable (Bridge, 2022).

Recyclability of post-consumer textile waste which is composed of synthetics and blended fibres is undoubtedly one of the most difficult problems the fashion industry has faced to date, highlighting the imperative need to transition toward a circular economy. Addressing the bottlenecks preventing the transition from the linear model, the adoption of closed-loop practices within post-consumer polyester textile waste recycling is important to effectively implement a textile-to-textile recycling system that keeps materials in circulation materials considering the increasing demand for diminishing resources (McKinsey,2022). Analysis of key theories and the PCTW recycling landscape concerning textile recycling and closed-loop practices within the industry and academia was analysed, considering how it can be utilised to achieve a closed-loop ecosystem for polyester PCTW recycling. Research questions are then derived from these key aspects.

Methodology

The methodological choice employed for this study was mono-method qualitative, which equips a 'single data collection technique' (Saunders et. al, 2019). A purposive sampling strategy was adopted for this study which equips the data collection method of semi-structured interviews. It was also vital to ensure the data collected was aligned with the goals of this research study (Bell et al.,2019). In order to enable participants to be carefully chosen and best support comprehension of the topic, this sampling approach was used (Creswell, 2018). As a result, a sample of eleven participants who were experts, and professionals from the textile recycling sector with knowledge of polyester textile waste recycling were selected and interviewed online. Participants were invited through email, and an information sheet and consent forms were sent prior to the interview.

Upon consent from each participant, the interviews were recorded on Teams call and voice recorded, to mitigate any risk that acts as a barrier to the data collection process. The collected data was then stored safely on the computer and backed up, adhering to ethical procedures. Furthermore, the data collected was assessed and validated by the participants in the form of transcripts and recordings to ensure the data collected is accurate and the interpretations were correctly made (Creswell, 2018, Saunders et. al, 2019). Following the data collection through semi-structured interviews, thematic analysis of the interview transcripts was done in the order that they took place, which was helpful in identifying the themes 'told' by the participants (Riessman, 2014). A qualitative dataset was developed, analysed, and patterns were interpreted using the thematic analysis approach, which involved systematic data coding to derive themes (Braun and Clarke, 2021). The inductive approach utilised to iteratively categorise the raw data collected as codes and subcodes were helpful for further analysis. (Creswell, 2018).

As polyester textile waste recycling is still in its infancy, it should be considered relatively a new field of research. No academic study has been done to date focusing specifically on post-consumer polyester textile waste recycling, making this study more descriptive and exploratory, than deriving conclusions and finalised answers. Furthermore, the time horizon is also a limitation. Hence, the study focuses on the early stage of polyester textile waste recycling and has limited evidence on the feasibility of implementing closed-loop practices. Although purposive sampling was useful in deriving robust data (Creswell, 2018), the limited sample size of the participants limits the scope for generalising findings. However, it would be more challenging to have a larger sample size as it would affect the depth of the findings.

Findings

The study explored the scope, challenges, and barriers of post-consumer polyester textile waste recycling in Europe to analyse the feasibility of closed-loop practices. The existing recycling technologies and their adaptability is analysed along with critical success factors to implement closed-loop practices. Through interviews and thematic analysis, key factors that emerged from the study are outlined in the following section which highlights the key findings and addresses the research questions that were derived from the literature review.

RQ1: What are the existing barriers and challenges in recycling the post-consumer polyester textile waste in the UK and Europe?

Financial Incentive

Lack of financial aid in the field of scaling up textile recycling technologies was identified as a major barrier. The significant benefits that could be derived from scaling up recycling technologies, which in turn help in striving towards a circular economy are limited due to the lack of funding available. Technologies that are at the pilot, research and development stage are not commercialised on a massive scale due to the barrier of lack of funding available for the proven technology. The interest among the participants to scale their business depended highly upon the investments received.

In addition, a consistent stream of supply of good quality material available to feed into the system is crucial as the quality of feedstock determines the quality of the end product. In order to supply sufficient quality feedstock to facilitate recycling, the system needs to be scaled, which is prevented by the lack of financial incentives available. The need for investments in this sector was undeniably a major barrier. Finally, the entire value chain of PCTW recycling which includes, the collection of old and worn-out garments, sorting for reuse and recycling requires financial incentives and investments to scale for development. Hence, the participants highlighted the need for investment in critical scaling and commercialisation of recycling technologies.

Buying & Disposal Habits of Consumers

Buying & Disposal habits of consumers were identified as a factor affecting the recyclability of post-consumer polyester textile waste. The key factors contributing to this were the low price of oil which is holding back from offering cost-competitive alternatives to virgin polyester. With a large proportion of the textiles consumed ending up in the municipal bin of the country where it is purchased, this disposal habit of the consumer is giving it no opportunity at even a hint of a second life. However, with the recent amends in legislation adopted by EU Waste Framework Directive, pressure from the EU to mandate textile waste collection in all EU member states by 2025, has the potential to create economic benefits and opportunities for textile recycling, with the possibility to influence consumer's textile buying and disposal habit.

Product & Design Innovation

The longevity of the material used in designing the products was pertinent. Apparel reuse and second-hand purchases have the potential to reduce the horrendous environmental cost of producing excessive garments. Effective design innovation is crucial and textile products should be designed and developed after considering the end-of-life path of the product, with limited complex materials involved. The need for timeless designs to limit seasonal drops and overconsumption was emphasised, and the material choice of each product should be reflected with authentic labelling. Additionally, mono-materiality was a key factor stated by the participants. Although mono materiality could facilitate polyester recycling and ease the process, it has unintended consequences, such as significant push-ups in the price and large-scale disruption of markets. This demonstrates the significant complexities involved in design and material consideration as the majority of polyester textile products in the market are blended, making it hard to recycle, and serving as a key strategic barrier to PCTW recycling.

Government Policies & EPR Implementation

Government policies and EPR implementation were identified as crucial factors in achieving textile circularity in the EU and the UK. Participants stressed the necessity for government liaisons to engage closely with recyclers to facilitate chemical recycling as the EU is mandating textile waste collection and sorting by 2025. To encourage the development of textile-to-textile recycling infrastructure, there is a need for government policies to specify that the source needs to be recycled textiles. Participants emphasised that the right policies could act as a catalyst to ensure the brands meet their commitments to the people and the environment. The absence of policies, notably the adoption of EPR, was highlighted as a major barrier in the UK.

RQ 2: What is the potential of Textile recycling to adapt to the existing and emerging polyester recycling technologies?

Scaling of Recycling Technologies

There was agreement among the participants that the advanced recycling technologies at a commercial scale were very limited. Participants expressed interest in shifting to new technologies and innovations if available at scale. Whilst evaluating the polyester recycling scenario, the analysis revealed that commercial-scale infrastructure for polyester textile waste recycling is non-existent, and the existing ones are in the pilot stage with ongoing research and developments which is expected to consume a few years to be scaled. Fibre to Fibre (F2F) recycling was noted as a promising approach for the recovery and separation of blended fibres, with surging and access being the key requirements. Additionally, demand concerns for recycled polyester materials were raised, with the expense of scaling technologies and the mindset of businesses identified as potential barriers.

Cost Efficiency and Commercial Viability

The majority of the participants agreed that there is a need for investment to develop and scale the recycling technology, with very limited scaling occurring due to the high cost involved.

Whilst few participants expressed their openness to emerging technologies, some highlighted that a few of their clients, who runs large-scale businesses were reluctant to work with start-ups due to the concern over the newness of start-up companies and the risk factor involved. Additionally, commercial viability emerged as a factor affecting the potential of adopting new and emerging technologies emphasising the need for widespread infrastructure throughout textile recycling.

Waste minimization

Optimising the manufacturing processes to minimise the waste being produced was of utmost importance to the participants. Maximising the material resource efficiency was a key factor along with emphasis given to increasing remanufacturing. Additionally, improving the recyclability of the apparel and textile products was key in optimising the manufacturing processes. Prevention of landfill and incineration was highlighted, with most participants emphasising the need for reusing being considered the best option before assessing the recyclability.

However, the need for addressing the pre-existing textile waste in the system was stressed, which gains paramount importance in textile recycling. This highlights the criticality of considering the design for recycling from the sourcing and manufacturing stage to minimise waste to avoid disposal, landfilling and incineration of unwanted and unsold textile products.

RQ 3: How might implementing closed-loop circular economy practices within polyester textile recycling create opportunities for development?

Encouraging Circularity

Upon evaluating the opportunities for the development of closed-loop practices within polyester textile waste recycling, encouraging circularity was formulated as a key category. Most participants highlighted the need for implementing closed-loop practices within polyester recycling to enable resource efficiency through sustainable production and consumption of the resources. The shift from the current linear model to the circular business model was the most resounding theme cited by most of the participants. A shift from the current linear business model is imperative in the textile and fashion industry whilst integrating recycled material is crucial.

Reducing the barriers affecting implementation.

Reducing barriers affecting implementation emerged as a crucial factor, with investments and price parity of recycled polyester with virgin polyester being the key drivers to facilitate and promote circularity and uptake of best practices.

Another resounding factor emphasised by the participants was the lack of transparency and education in the highly fragmented supply chain, along with the technical aspect of polyester for producing recyclable material from old and worn-out clothing being crucial in determining the recyclability.

Category 3: Opportunity for Fibre-to-Fibre (F2F) Recycling

Participants expressed that although significant investments are required, textile-to-textile recycling is a chance for brands to manage their waste and utilise it as a resource to make a profit out of waste. With strict feedstock input requirements for polyester fibre-to-fibre recycling must increase their capacity to handle blends and decrease their price to increase the demand. This limitation is currently prohibiting the circular textile economy from scaling.

There was general agreement among participants that textile-to-textile polyester recycling would undoubtedly be an achievement. Although immature, F2F recycling has the potential to turn textile waste into value, demonstrating the opportunity to bring significant change within recycling by scaling F2F recycling. Overall, the findings suggest that implementing closed-loop circular economy practices can create opportunities for development, but critical scaling across the value chain is required to provide adequate quality feedstock.

Discussion and Conclusions

Critical analysis of the literature review and primary research revealed that, with less than 1 per cent of the discarded clothing recycled back into clothing, the current fashion and textile industry doesn't have a closed–loop recycling system and needs a systemic change. The analysis also revealed that the recyclability of the garment needs to be identified and 'designed out' at the design and product development stage to facilitate recycling, emphasising the need for design for circularity (Goldsworthy,2014).

The existing barriers to recycling polyester outweigh the technological development and scale-up of polyester recycling, however, the very recent research by McKinsey (2022) preludes fibre-to-fibre recycling as a potential path toward achieving circularity. Although fibre-to-fibre is immature at its current stage, it indicates a latent opportunity for the textile and fashion industry to minimise waste and recycle products at their end of life. In this study, the potential of implementing closed-loop practices within textile recycling was evaluated, and opportunities were identified across securing investments, commercial scaling, EPR policy, and design innovation, as seen in Figure 4.

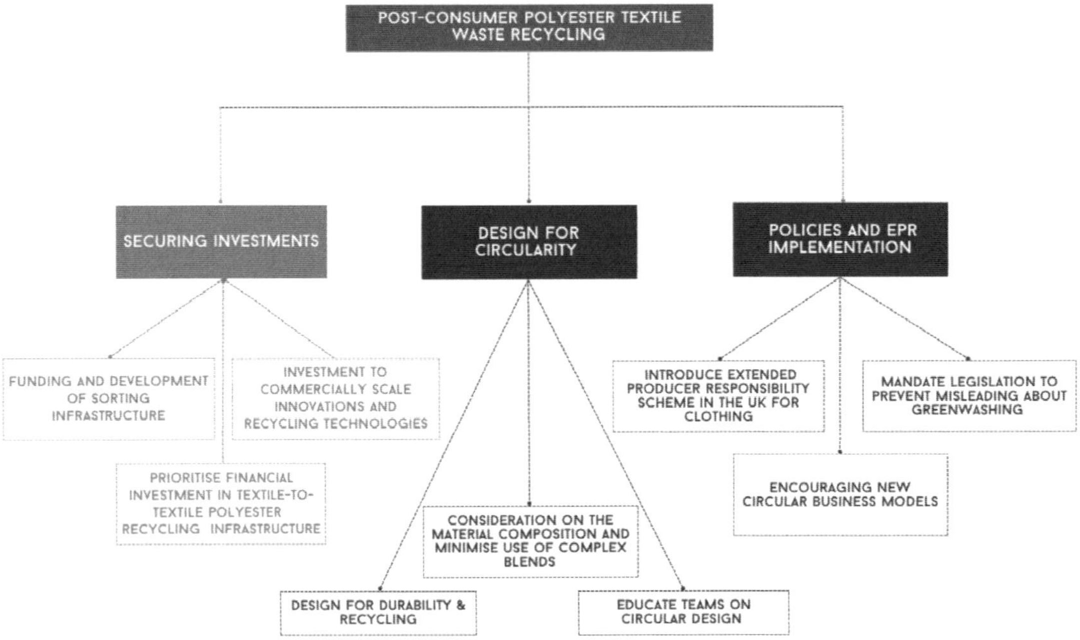

Figure 4 Breakdown of recommendations

Lack of financial incentives, policies and design innovations majorly affect the recyclability of post-consumer textile waste. Buying and disposal habits and barriers within the sorting infrastructure contributed to challenges. However, improving design and product innovation and consumer buying and disposal behaviour exclusively cannot solve the barriers holding back the recyclability of the garments. Despite the need for design and product innovation, the need for incentives to fund the process is crucial, for the development of the textile recycling value chain, which calls for substantial capital investments.

Many of the major issues that lie ahead as a barrier to the scaling of recycling technologies can best be resolved through intense collaboration. To engage in an operational effort to overcome the barriers to scale, business executives from throughout the textile value chain, and private and public investors, need to collaborate in an unprecedented manner. Although the analysis reveals that textile recycling could become matured, profitable, and scaled commercially with the help of financial incentives, there exists the need for transition funding such as Extended Producer's Responsibility (EPR) funds to cater for future developments within the textile recycling sector. Therefore, it can be inferred that the major barriers and challenges such as financial incentives, sorting infrastructure, and lack of policies from the government and the stakeholders that emerged consensually as the key factors are preventing the opportunity for further development within post-consumer polyester textile waste recycling.

Considering the existing investment landscape, recycling technologies must be scaled up considerably for implementation. Regulatory mechanisms like EPR schemes and targets for recycled content must be used to support these advancements. Bold, directive policies from governments are necessary to ensure quick action across all levels of the fashion sector to advance the implementation of closed-loop practices.

Designing for Circularity whilst reinventing the material choice and design & development process along with redefining what is achievable through recycling will help to accelerate the transition to a resource-efficient circular economy. It is highly recommended that textile recycling units and emerging recycling technologies secure investments to develop the sorting infrastructure, facilitate textile-to-textile recycling for post-consumer textile waste, and commercially scale the technologies. A more balanced distribution of investments will guarantee a versatile array of material options to advance circularity with recovered materials essential to reduce textile waste.

The aid of effective regulation that provides both businesses and individuals with clear economic incentives, will help in the transition from the current linear business model to the circular business model. Introducing the EPR scheme in the UK will facilitate shifting the textile waste collection responsibility from municipalities to the brands and prevents the unsold inventories from being incinerated and discarded, making brands and manufacturers hold the responsibility for the products they create. Additionally, the EPR scheme can force brands to consider the end-of-life phase of their products which reduces greenwashing and encourage durable products, whilst developing an improved system for disposal. On a business ecosystem level, legislation, and policy to prevent waste generation act as key catalysts. EPR will also elevate the need for a circular business model within the fashion industry which has the potential to increase the material provenance and transparency within the supply chain. Moreover, advocating for stringent legislation, and implementing EPR can be advantageous in ensuring that all stakeholders within the textile industry work collaboratively to actualize circular practices. This will potentially enable the transition to circular systems whilst meeting the climate target pathway of 1.5°C outlined in the UN Paris Agreement.

Limitations and Future Research

Although this study gives a broad overview of the scope and possibilities of implementing closed-loop practices in textile recycling systems, it focuses solely on European regions, limiting the opportunity for the data to be generalised and scaled across the globe. Furthermore, there is a need for research on advanced polyester textile waste recycling, particularly for the separation of polyester blended fibres. Such studies could provide potential guidance for recycling systems within the polyester supply chain and help fuel developments in Europe's textile industry towards a circular economy.

References

Bédat, M. (2016) *Our love of cheap clothing has a hidden cost – it's time for a fashion revolution, World Economic Forum*. Available at: https://www.weforum.org/agenda/2016/04/our-love-of-cheap-clothing-has-a- hidden-cost-it-s-time-the-fashion-industry-changed/ (Accessed: 22 August 2022).

Bell, E., Harley, B. and Bryman, A. (2019) *Business Research Methods*. 5th edn. London: Oxford University Press.

Braun, V. and Clarke, V. (2021) *Thematic analysis: A practical guide*. London: SAGE Publications.

Bridge, A. (2022) *Textile manufacturers use mono materials to improve recyclability, PCIAW®*. Available at: https://pciaw.org/textile-manufacturers-turn-to- mono-material-fabrics-and-cellulosic-fibres-as-pressure-mounts-to-improve-the- recyclability-of-textile-products/ (Accessed: 28 August 2022).

Circle Economy (2020) *Recycled post-consumer textiles - an industry perspective*. Interreg NWE Fibersort project. Available at: https://circulareconomy.europa.eu/platform/en/knowledge/recycled-post-consumer-textiles-industry-perspective.

Coskun G. & Basaran, F. N. (2019) 'Post-Consumer Textile Waste Minimization: A Review', *Journal of Strategic Research in Social Science*, 5 (1), pp. 1–18. doi: 10.26579/josress-5.1.1.

Creswell, J. W. and Creswell, J. D. (2018) *Research design: Qualitative, quantitative, and mixed methods approaches*. Thousand Oaks, CA, USA: SAGE Publications.

Damayanti, D. *et al.* (2021) 'Possibility routes for textile recycling technology', *Polymers*, 13(21), p. 3834. doi: 10.3390/polym13213834.

Eileen Fisher Foundation (2022) HEY FASHION! — Fashion's waste crisis and how to solve it - prepared for Eileen fisher foundation by pentatonic®, HEY
FASHION! Available at: https://www.heyfashion.org/report (Accessed: 22 August 2022).

Ellen MacArthur Foundation (2017) *A New Textiles Economy: Redesigning fashion's future, Ellenmacarthurfoundation.org*. Available at: https://ellenmacarthurfoundation.org/a-new-textiles-economy (Accessed: 20 June 2022).

Ellen MacArthur Foundation (2021) *The trends and trailblazers creating a circular economy for fashion, Ellenmacarthurfoundation.org*. Available at: https://ellenmacarthurfoundation.org/articles/the-trends-and-trailblazers-creating-a- circular-economy-for-fashion (Accessed: 7 March 2022).

Eppinger, E. (2022) 'Recycling technologies for enabling sustainability transitions of the fashion industry: status quo and avenues for increasing post-consumer waste recycling', *Sustainability: Science Practice and Policy*, 18(1), pp. 114–128. doi: 10.1080/15487733.2022.2027122.

Euric (2022) The textiles re-use & recycling industry welcomes the EU Strategy for Sustainable and Circular Textiles, Euric-aisbl.eu. Available at: https://www.euric- aisbl.eu/position-papers (Accessed: 12 August 2022).

Europa (2020) *The impact of textile production and waste on the environment (infographic), Europa.eu*. Available at: https://www.europarl.europa.eu/news/en/headlines/society/20201208STO93327/th e-impact-of-textile-production-and-waste-on-the-environment-infographic (Accessed: 7 August 2022).

Europa (2021) *Latest trend keeps clothes out of landfill, Europa*. Available at: https://ec.europa.eu/research-and-innovation/en/horizon-magazine/latest-trend- keeps-clothes-out-landfill (Accessed: 25 July 2022).

Europa (2022) *EU Strategy for Sustainable and Circular Textiles, Europa.eu*. Available at: https://eur-lex.europa.eu/legal- content/EN/TXT/HTML/?uri=CELEX:52022DC0141 (Accessed: 7 August 2022).

Goldsworthy, K. (2012) 'Laser-finishing: a new process for designing recyclability in synthetic textiles'. Available at: https://ethos.bl.uk/OrderDetails.do?uin=uk.bl.ethos.606520.

Goldsworthy, K. (2014) 'Design for Cyclability: Pro-active approaches for maximising material recovery', *Making Futures Journal*, 3. doi: ISSN2042-1664.

Green peace (2022) *From donations to the dumpsite: textiles waste disguised as second-hand clothes exported to East Africa, Greenpeace.org*. Available at: https://www.greenpeace.org/international/publication/53355/poisoned-gifts-report- fast-fashion-textile-waste-disguised-as-second-hand-clothes-exported-to-east-africa/ (Accessed: 25 June 2022).

Haslinger, S. *et al.* (2019) 'Solid-state NMR method for the quantification of cellulose and polyester in textile blends', *Carbohydrate polymers*, 207, pp. 11–16. doi: 10.1016/j.carbpol.2018.11.052.

Jäämaa, L. and Kaipia, R. (2022) 'The first-mile problem in the circular economy supply chains - Collecting recyclable textiles from consumers', *Waste management (New York, N.Y.)*, 141, pp. 173–182. doi: 10.1016/j.wasman.2022.01.012.

Koszewska, M. (2018) 'Circular economy — challenges for the textile and clothing industry', *AUTEX research journal*, 18(4), pp. 337–347. doi: 10.1515/aut-2018-0023.

Kumar, J. *et al.* (2020) 'Challenges for waste in fashion and textile industry', in *Recycling from Waste in Fashion and Textiles*. Wiley, pp. 19–32. doi: 10.1002/9781119620532.ch2.

LeBlanc, R. (2014) *Textile and garment recycling facts and figures, The Balance Small Business*. Available at: https://www.thebalancesmb.com/textile-recycling- facts-and-figures-2878122 (Accessed: 7 July 2022).

Ljungkvist, H., Watson, D., Elander, M. (2018) *Developments in global markets for used textiles and implications for reuse and recycling*. Available at: http://mistrafuturefashion. com/wp-content/uploads/2018/09/Mistra-FutureFashion- 2018.04-M-Elander-D.3.3.4.1.pdf.

McKinsey (2020) *Closing the loop: Increasing fashion circularity in California*. Available at: https://www.mckinsey.com (Accessed: 25 June 2022).

McKinsey (2022) *Scaling textile recycling in Europe turning waste into value, Mckinsey.com*. Available at: https://www.mckinsey.com/~/media/mckinsey/industries/retail (Accessed: 26 July 2022).

Niinimaki, K. (2019) Sustainable fashion in A circular economy. Aalto-yliopisto.

Palacios-Mateo, C., van der Meer, Y. and Seide, G. (2021) 'Analysis of the polyester clothing value chain to identify key intervention points for sustainability', *Environmental sciences Europe*, 33(1), p. 2. doi: 10.1186/s12302-020- 00447-x.

Park, S. H. and Kim, S. H. (2014) 'Poly (ethylene terephthalate) recycling for high value-added textiles', *Fashion and textiles*, 1(1). doi: 10.1186/s40691-014-0001-x.

Patagonia (2022) *Recycled polyester, Patagonia.com*. Available at: https://eu.patagonia.com/gb/en/our-footprint/recycled-polyester.html (Accessed: 28 June 2022).

Peng, Z. et al. (2022) Advanced recycling: Opportunities for growth as interest in the circular economy grows, emerging recycling technologies that are complementary with mechanical recycling are accelerating, Mckinsey.com. Available at: https://www.mckinsey.com/~/media/mckinsey/industries/chemicals/our%20insights/advanced%20recycling%20opportunities%20for%20growth/advanced-recycling- opportunities-for-growth.pdf?shouldIndex=false (Accessed: 27 August 2022).

Piribauer, B. and Bartl, A. (2019) 'Textile recycling processes, state of the art and current developments: A mini review', *Waste management & research: the journal of the International Solid Wastes and Public Cleansing Association, ISWA*, 37(2), pp. 112–119. doi: 10.1177/0734242X18819277.

Prahl, A. (2020) *Performance Days: Closed loop functional fabrics still some way off, Innovationintextiles.com*. Available at: https://www.innovationintextiles.com/performance-days-closed-loop-functional- fabrics-still-some-way-off/ (Accessed: 12 June 2022).

Riessman, C. K. (2014) *Narrative methods for the human sciences*. Thousand Oaks, CA, USA: SAGE Publications.

Sajn, N. (2019) *Environmental impact of the textile and clothing industry, Europa.eu*. Available at: https://www.europarl.europa.eu/RegData/etudes/BRIE/2019/633143/EPRS_BRI(201 9) 633143_EN.pdf (Accessed: 28 July 2022).

Sandin, G. and Peters, G. M. (2018) 'Environmental impact of textile reuse and recycling – A review', *Journal of cleaner production*, 184, pp. 353–365. doi: 10.1016/j.jclepro.2018.02.266.

Saunders, M., Lewis, P. and Thornhill, A. (2019) *Research methods for business students*. 8th edn. London: Pearson Education.

Textile Exchange (2019) *Preferred Fiber & Materials Market Report*. Available at: https://textileexchange.org/wp-content/uploads/2019/11/Textile- Exchange_Preferred-Fiber-Material-Market-Report_2019.pdf.

TextileExchange (2021) *Preferred fiber & materials market report 2021*, *Textileexchange.org*. Available at: https://textileexchange.org/wp-content/uploads/2021/08/Textile-Exchange_Preferred-Fiber-and-Materials-Market- Report_2021.pdf (Accessed: 9 July 2022).

Wanassi, B., Azzouz, B. and Hassen, M. B. (2016) 'Value-added waste cotton yarn: Optimization of recycling process and spinning of reclaimed fibers', *Industrial crops and products*, 87, pp. 27–32. doi: 10.1016/j.indcrop.2016.04.020.

Wang, Y. (2006) *Recycling in Textiles*. 1st edn. Cambridge, England: Woodhead Publishing. Available at: https://www.elsevier.com/books/recycling-in- textiles/wang/978-1-85573-952-9 (Accessed: 5 May 2022).

Worn Again Technologies (2018) *Polyester & cotton fibre recycling technology*, *Worn Again Technologies*. Available at: https://wornagain.co.uk/polyester-cotton-fibre-recycling-technology/ (Accessed: 18 June 2022).

WRAP (2019) *Textiles market situation report 2019*, *WRAP*. Available at: https://wrap.org.uk/resources/market-situation-reports/textiles-2019 (Accessed: 28 July 2022).

EDUCATING FOR CIRCULARITY – ESTABLISHING A FRAMEWORK TO ADDRESS THE PRIORITY KNOWLEDGE GAPS IN FASHION HIGHER EDUCATION

Jo Conlon[1] and Jane Wood[1]

[1]Fashion Business and Technology, Department of Materials, The University of Manchester, Oxford Road, Manchester, M13 9PL

Abstract

As WRAP launches Textiles 2030, academia has a key role in ensuring curricula are addressing industry knowledge gaps in achieving global climate goals for fashion to net zero, and through graduates with the knowledge and agency to reduce the impact of clothing around the world. This pilot investigation focussed on the knowledge and practice of a team of academics in one department as the source of influence and responsibility to embed circular principles in the curriculum.

The wider project aims to support the transition to a sustainable circular fashion system through the identification of key priorities in the technical understanding that underpins sustainability, thereby addressing the skills gap of graduates entering the industry in both general and technical roles. The project is informed by the theoretical framework of Rogers' diffusion of innovation theory to investigate the required changes in higher education to foster innovative thinking and the knowledge, skills and abilities that enable graduates to enter business as competent change agents to advance circular fashion. A key sustainability competency (KSC) framework was employed to underpin a pilot survey to capture understanding and experiences of circular economy (CE) in higher education from academics based in a UK HEI fashion department.

The pilot highlighted that whilst academic staff recognise the importance of embedding CE in teaching practice, CE is not always explicitly included in programme learning outcomes. Additionally, there may be some gaps in practical experience of CE practice and thus wider collaborations across HE institutions and industry are critical to transform the educational landscape.

Keywords

Circularity, Fashion, Education, TIWC 2023, Sustainability, key sustainability competency (KSC)

Introduction

Sustainable development can be seen as a process in which the aspirations for the future are described through the UN's Sustainable Development Goals (SDGs) and Agenda 2030 (United Nations, 2015). Education plays a key role in achieving these aspirations. Quality education is promoted through goal 4 within which Education for Sustainability target 4.7 states the specific aim that "by 2030, ensure all learners acquire the knowledge and skills needed to promote sustainable development…" Education for Sustainable Development (ESD) is considered to represent a goal for holistic and transformational education rather than a prescribed body of information and knowledge (Giangrande, *et al.*, 2019). As a key educational sector, universities are responsible to educate future decision makers in ways that foster the knowledge and competencies to enable them to contribute to achieving this goal (Barth and Rieckmann, 2012). Further, with recognition of textiles and fashion as a key industrial sector with significant impacts across many SDGs, the imperative to reform industry practices becomes evident.

The circular economy (CE) model supports the SDGs for people, prosperity, and the planet (Weetman, 2020) and provides an agenda and framework for implementing education for sustainable development. Businesses are also exploring ways to restructure from linear to CE models and the increasing importance of the CE will have a significant impact on the shape of future jobs and competencies for future decision makers. With the CE model there is a focus on materials and how to preserve their value. Extending the service life of products and recovering materials for re-use requires products to be designed with this specifically in mind. These challenges will require education on materials and CE principles to foster innovative thinking and the knowledge, skills and abilities that enable graduates to enter business as competent change agents to advance circular fashion. The ambition to support these competencies and orientations requires innovative and appropriate teaching, pedagogical approaches and thorough evaluations concerning their actual impact (Brandt *et al.,* 2019). This research acknowledges the significance of academics as an important starting point for change (Moore, 2005) and the need to address importance of "educating the educator" (Lozana, 2006, p.793) by working collaboratively as a hub to support change. It acknowledges the obligation of higher education to prepare younger generations with an agentic mindset for the challenge to come (Moore, 2005). As educators we must urgently foster in our students the knowledge, skills, and abilities to act as agents for change and advance circular fashion.

This research began with the university becoming an affiliate partner WRAP's Textiles 2030 initiative. As an institution, the university is strongly committed to taking action and the Textiles 2030 programme will sit alongside the institutional goals for sustainability, involving all students across the university with Textiles 2030 activities. This article represents a first step towards embedding the guiding principles and strategies for CE in the curriculum across our fashion, business and technology courses, creating the industry professionals of the future armed with all the knowledge and tools to implement change. This paper draws on extant literature of embedding sustainability in higher education to provide the context from which to advance education for the circular economy in the fashion system.

As a first step, this paper aims to identify the knowledge, skills, abilities and effective pedagogies to foster circular thinking within higher education fashion programmes and advance circular fashion systems. Although this initial research is UK HEI focussed, we situate it in a global context given the global nature of higher education and the challenge of radical change in textiles and fashion to achieve the UN SDGs. The starting point of this investigation is to focus on the knowledge and practice of the academics in one team as the source of influence and responsibility to embed circular principles in the curriculum.

Literature Review

Defining circular fashion.

From the 1970's the term 'sustainability' has been part of social awareness and has since gathered momentum socially, industrially and throughout academia (Gamage 2022). Therefore, education for sustainable development (ESD) is now a mature field from which education for the circular economy can draw (Kircherr and Piscicelli, 2019). Sustainable development (SD) is based on three fundamental pillars: social, economic, and environmental. The circular economy offered an alternative to the take-make-waste traditional way of doing business and the promise of significant environmental, social, and economic benefits (Ellen MacArthur Foundation, 2013) but from this time there has been a drift towards a focus on the economic dimension. The result of this is that the relationship and contribution of CE to SD is currently under discussion (Schöggl et al., 2020) particularly as the historic over-emphasis on environmental and economic issues lacks sufficient social sustainability focus. Acknowledging this oversight, the amended CE now more deliberately brings people and society into an equity-oriented circular economy framework (CEF) through a renewed focus on restorative or regenerative innovations in the transition to CE (Morseletto, 2020). The UK Parliament report, Fixing Fashion (2019), clearly identifies some of the major issues faced in the UK in this arena and recognises the hurdles faced in addressing these. CE is highlighted as a critical issue to be addressed within education, starting as early as Key stage 2 in the broader curriculum (UK Parliament, 2019) and therefore crucial in specialist HE subjects, such as fashion programs.

This study adopts the definition: A fashion system that moves towards a regenerative model with an improved use of sustainable and renewable resources, reduction of non-renewable inputs, pollution, and waste generation, while facilitating long product life and material circulation via sustainable fashion design strategies and effective reverse logistic processes. (Dissanayake and Weerasinghe, 2021).

This model for CEF offers a roadmap for the transition to a sustainable fashion industry by demonstrating the interconnectedness of the solutions (Figure1). However, the current position within industry is one of strong desire but only low-level understanding of concepts and applications (Dissanayake and Weerasinghe, 2021). This highlights the obligation for universities to be leaders and more actively involved in this area (Moore, 2005), and the urgent need for knowledgeable graduates to enter the industry.

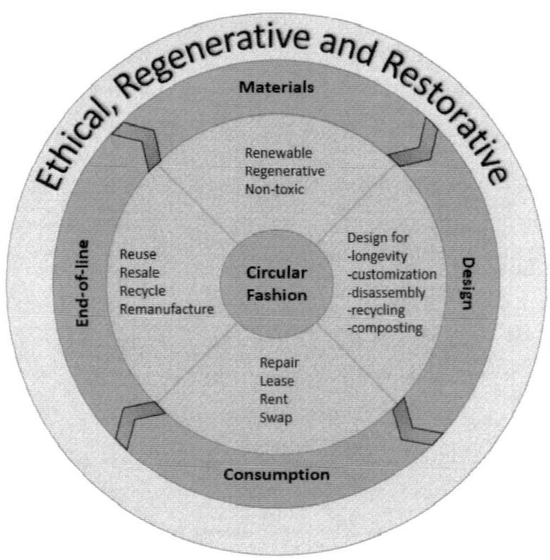

Figure 1. Circular Fashion (Adapted from Dissanayake and Weerasinghe, 2021).

Translating knowledge into actions.

To establish sustainable and circular education in the curriculum requires a transformative approach to the way subjects are taught and learnt. Curriculum change processes are complex resulting in different levels of response and change (Figure 2). Therefore, although higher education institutions (HEIs) will be at different stages in engaging with ESD they share a common task. The challenge for educators is how to translate knowledge of core concepts and aspirations for a sustainable future (as expressed through the UN SDGs) into teaching and learning activities.

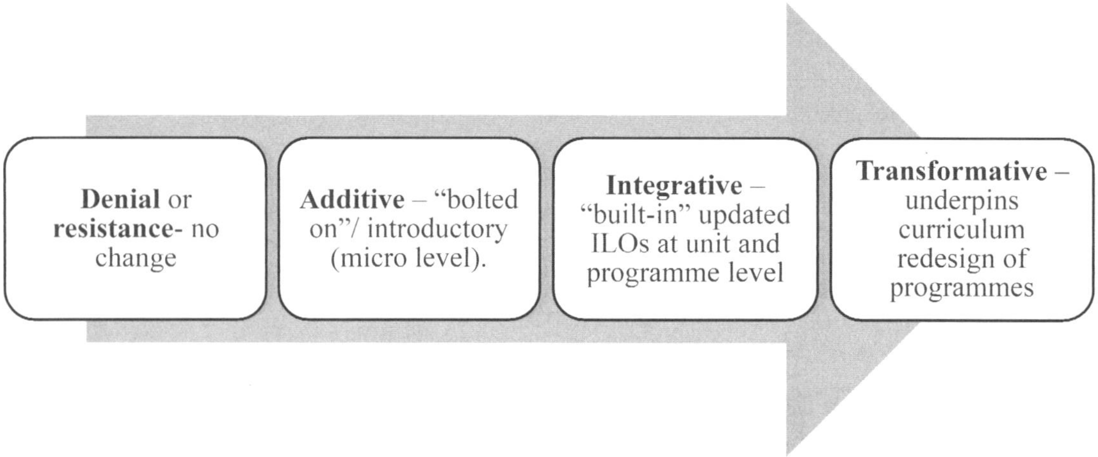

Figure 2. Levels of curriculum change

The three areas of Awareness, Willingness and Ability in relation to sustainability must be developed through the curriculum if students graduating from university are expected to contribute to sustainable development (Lidgren et al. 2006). Although students arrive with a broad awareness of sustainability when they enter HE, this does not necessarily give them the ability to evaluate information, prioritise issues or take action (Thomas, 2004). More information is not the solution. Students need support to learn how to think critically, find and evaluate information, mature their decision making and collaborate towards innovative solutions. Consequently, education should not restrict itself to giving students information but provide them with an experience of exploring contemporary sustainability challenges that prompt creative thinking and innovative solutions. Therefore, it is imperative to identify and teach the knowledge, skills and abilities that graduates will most need to advance sustainability though the CE while confronting unprecedented change.

There is increasing adoption key sustainability competency (KSC) frameworks as a heuristic for discussing and disrupting the business-as-usual way of teaching and learning and support education to lead and promote meaningful change (Pacis and Van Wynsberghe, 2020). Much of the recent work (Ceulemans and Severijns, 2019; Evans, 2019; Pacis and Van Wynsberghe, 2020; QAA and Advance HE, 2021; Sumter et al., 2021) on KSCs builds on the work of Weik et al., (2011) for a large-scale educational transformation to equip and empower a new generation of professionals (Table 1). Such KSC frameworks can be aligned with learning outcomes to reimagine teaching and learning strategies to build the capacity in learners to think critically and have agency in problem-solving in this area. As with any framework there are acknowledged limitations in that they will continue to extend / are not fixed and weaknesses if applied in an overly mechanistic or prescriptive way.

Table 1. Key sustainability competency (KSC) framework (adapted from Weik et al, 2011, Wiek et al. 2016 and Pacis and Van Wynsberghe, 2020)

System thinking	Systems thinking in practice encourages us to explore inter-relationships (context and connections), perspectives (each actor has their own unique perception of the situation) and boundaries (agreeing on scope, scale and what might constitute an improvement).
Anticipatory (future) thinking	Anticipatory (future) thinking in practice encourages us collectively develop future scenarios related to sustainability and sustainability problem-solving frameworks.
Strategic thinking	Strategic thinking in practice encourages us to collectively design and implement relevant and often alternative decision-making strategies towards sustainability.
Normative competency	Normative competency describes the ability to collectively elicit, map, specify, apply, reconcile, and negotiate divergent sustainability values, principles, goals, and targets.
Interpersonal competency	The ability to motivate, enable, and facilitate collaborative and participatory sustainability research and problem-solving.
Integrative problem solving	The ability to apply different problem-solving frameworks to address complex sustainability problems and develop viable solution options.

Effective pedagogies that foster key sustainable competencies.

CE demands a reconfiguration in how we produce and consume products. In turn, this demands new curricula and pedagogic approaches that foster innovative thinking and the knowledge, skills and abilities that enable graduates to enter business as competent change agents to advance circular fashion. The extant literature recommends an initial focus on the knowledge and practice of the academic staff to enable transformative change (Barth and Rieckmann, 2012) as the source of influence and responsibility to embed circular principles in the curriculum. In a multiple case study of teacher education, Brandt et al (2019) found that educators need three components: knowledge of sustainability (content knowledge), knowledge of effective teaching and learning strategies (pedagogical content knowledge, (PCK)), and the motivation to act (attitude) to actively support and implement ESD. This approach could be used to advance circular fashion.

It is important to disseminate applicable methodologies to raise awareness, prompt adoption and thereby support the development of pedagogical content knowledge within academic staff. Nine pedagogical approaches (Table 2) were identified as effective pedagogical approaches to support key sustainable competencies (Evans, 2019).

Table 2. Top nine pedagogical approaches (adapted from Evans, 2019)

	Approach
1	Organisational or community project (problem-based learning, PBL) note: case studies were seen as supportive of generating ideas for these projects.
2	Integrative learning (inter- and transdisciplinary)
3	Project or class based PBL
4	Active learning
5	Service learning
6	Research-based learning
7	Critical text / information analysis / interpretation
8	Reflexive learning
9	Collaborative learning

Towards Future Fit Fashion Education

Embracing and advancing circular fashion systems requires engagement and empowerment of all stakeholders and higher education has an opportunity to provide knowledge support to industry and foster the knowledge and tools to implement change the future industry professionals. There is a need to rebalance curriculum beyond keeping students abreast of the current trends to emphasise holistic learning towards sustainability by including eco-design, awareness of material impact, alternative business models of consumption, transparency and EPR responsibilities. Significantly, after graduation, students of UK HEI's will enter decision making roles in global brands where they can lead change by applying knowledge of the environmental impact of these decisions, confidence to make sustainable design decisions, e.g., through the selection of recycled materials and engaging consumers with new business models (Dissanayake and Weerasinghe, 2021). The circular economy requires closing the loops on knowledge and material streams (Ellen MacArthur Foundation, 2013). Currently in the industry there is a focus on end of life (EOL) waste streams (Textiles 2030, 2023), whereas within education the curriculum reflects business-as-usual or designing for circular economy and the role of the consumer.

AMFI's Reality School provides an example on embedding a CE agenda as a core pedagogic approach in its curriculum design and dynamic learning environment (Figure 3). A problem-based approach to student learning combines phased opportunities for mastery (through guided learning) and discovery (independent learning with personal meaning) to encourage deep learning and evolve competencies (Hall & Velez-Colby, 2018).

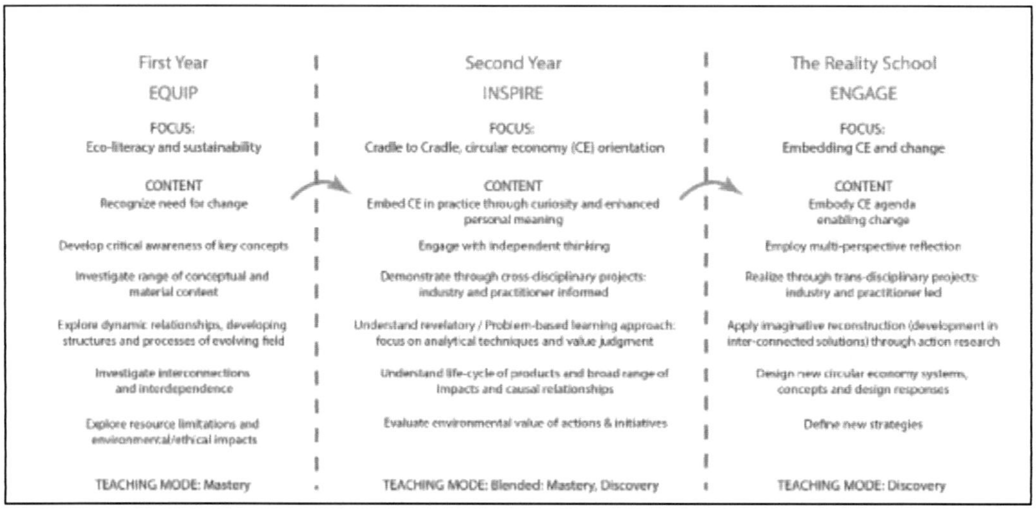

Figure 3. Model of expected student learning progression (Hall & Velez-Colby, 2018)

Methodology

Theoretical approach

The theoretical framework of Rogers' diffusion of innovation theory has been employed to investigate change in organisations and services. To establish sustainable and circular education in the curriculum requires a transformative approach to the way subjects are taught and learnt. Educators as change agents in education are important to the diffusion of this approach which will in turn foster the knowledge and skills in graduates able to support change in the industry. Additionally, Dahle and Neumayer (2001) and Agarwal (2021) suggest a 'top down' approach can accelerate change due to academics 'unique power to inspire students to follow their example' (Thomas, 2004, pp42)

Data collection device

Sumter et al (2021) is an adapted form of Weik et al's (2016) KCF for the circular economy. The five competencies from Weik are sub-divided to are expanded to a suite of nine key competencies to guide the development of courses. As with all KCF this is expected to evolve over time but provides a useful starting point.

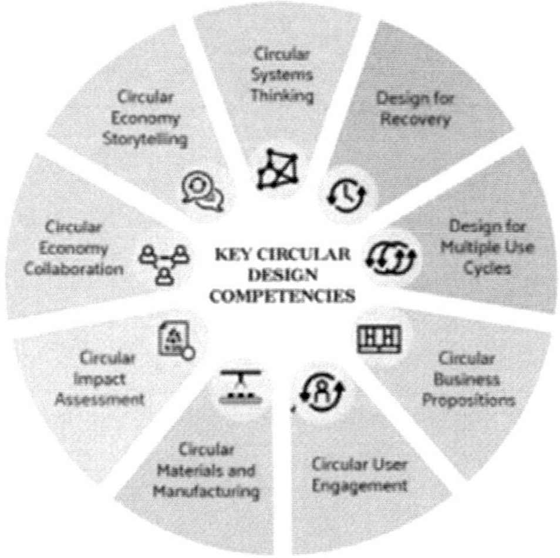

The competencies that reflect the main qualifications

Systems Thinking

Anticipatory (Design) Thinking

Strategic Thinking

Normative thinking

Interpersonal ability

Figure 5. Key circular design competencies (Sumter et al.2021, Weik et al., 2016)

Survey and data collection method.

A pilot survey was conducted over 4 weeks to capture understanding and experiences of CE in higher education. It was distributed to academics based in a UK HEI fashion department.

The survey was designed, built and responses anonymously recorded using the Qualtrics XM platform (www.Qualtrics.com). A total of 10 questions were posed; 9 were closed answer multiple choice and 1 was open ended free text.

Questions were developed to ascertain the amount of time academics had taught in HE and their involvement in specific disciplines within fashion (e.g., marketing, buying, product development, design). In conjunction with the 'Key Circular Design Competencies' model (Figure 5), questions were then designed to ascertain individuals' views on CE in their own teaching, unit design and involvement in CE outside academia.

As a precursor to the questions, the definition of CE was adapted from the Common Objective (2016) to give context to the survey (Figure 6).

Please note that for the purposes of this survey, circular economy is defined through a fashion lens. It is a considered model of production in which waste and pollution are designed out and products and materials are kept in use for as long as possible.

Figure 6. CE definition statement (Common Objective, 2019)

Findings & discussion

Respondents and background

Anonymous responses (n = 21) revealed an average (mode) of 11 – 20 years of involvement in higher education (figure 7).

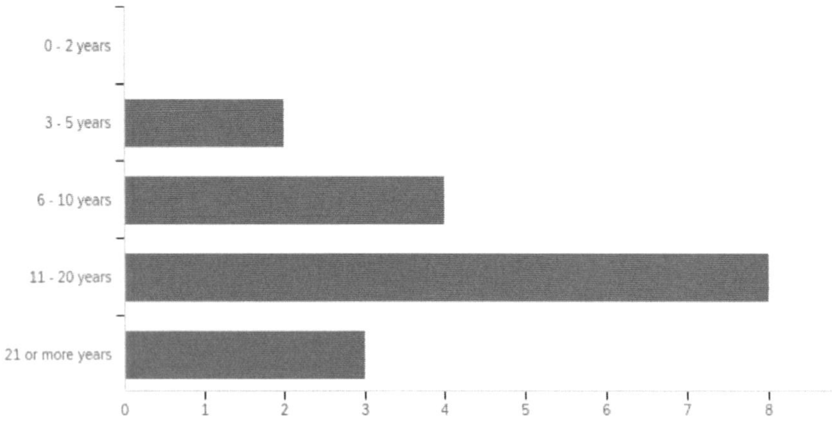

Figure 7. Number of years involved in HE.

Disciplines taught within fashion education were skewed towards 'business' subjects (marketing (21%), management (16%), buying (11%), merchandising (5%) and business (18%)) making up 66% of the responses (Figure 8). Whilst this may be indicative of the expertise of the small number of respondents, it also illustrates the importance of our study; much research in the field CE and sustainability in fashion education to date has focussed on design practice which is only part of the spectrum in fashion education (Agarwal, 2021; Sumter et al., 2021). This further concurs with the issues raised in the literature, where is it stated that effective CE education is a product of a holistic approach (Lee and Regni, 2009; Hall & Velez-Colby, 2018).

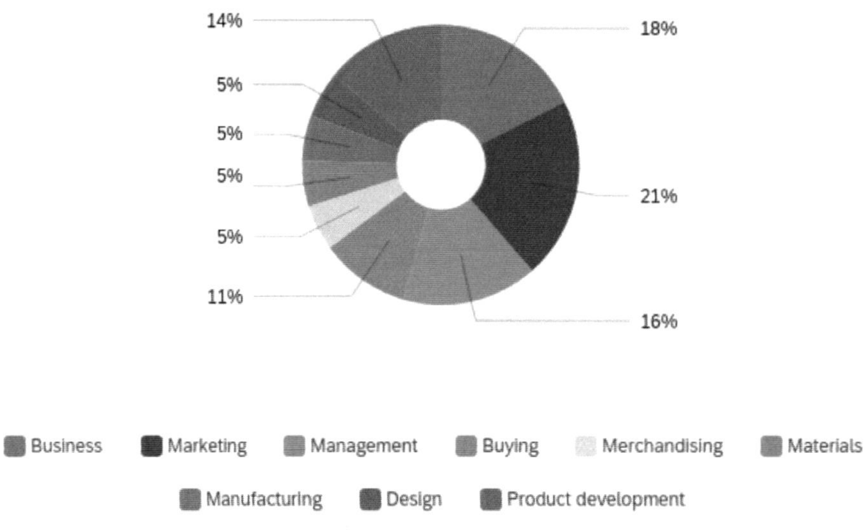

Figure 8: Disciplines taught by respondents in HE.

Approach to CE in personal teaching practice

These questions were developed to gain insight into respondent's own views on their approach to embedding CE in their teaching, regardless of the constraints of the curriculum.

Most respondents felt they embedded CE into their teaching, but not intensely, with only 2 respondents (13%) claiming CE did not form part of their teaching at all (Figure 9).

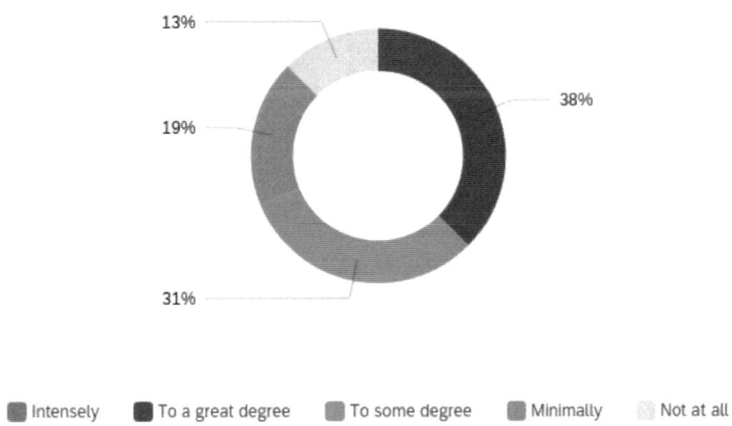

Figure 9: Degree to which CE is embedded into teaching practice

When questioned on the mode of delivery of CE concepts, those respondents embedding CE into their teaching practice used a combination of case studies and live briefs to support theoretical approaches (figure 10). However, literature suggests that effective CE teaching in HE needs a wider pedagogical approach to enhance the learner's experience (Evans, 2019). Whilst table 1 provides a diverse overview of pedagogical approaches for CE education, the literature identifies the need for students to be able to explore issues pertaining to CE for themselves, suggesting that problem-based learning (PBL) provides the most appropriate vehicle to facilitate this (Hall and Velez-Colby, 2018; Agarwal, 2021). In our study, a PBL approach did not explicitly feature in the survey questions. However, respondents could have classified PBL as a 'case study', although it is acknowledged that these 2 approaches are very different.

It is encouraging that none of the respondents relied on 'theoretical principles only' in their practice, further agreeing with literature that CE education needs to be firmly grounded in contemporary global practice (Moore, 2005, QAA and Advance HE, 2021).

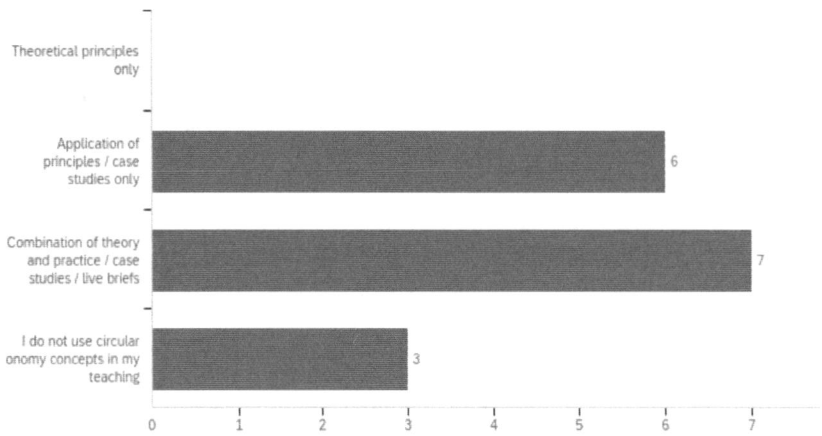

Figure 10: Mode of delivery of CE principles

CE as part of unit design

Respondents were asked to consider how CE was embedded into the design of the units on which they taught, in terms of CE as part of unit learning outcomes and unit assessment. The results to this section of the survey were in stark contrast to those in teaching practice. 94% respondents agreed that CE was not part of the learning outcomes (figure 11), whilst 69% also acknowledged the lack of assessment of CE in these units (figure 12). If CE is not part of learning outcomes, this raises the issue of the need for appropriate assessment of CE in these units. Interestingly, despite the high proportion of units not explicitly mentioning CE in learning outcomes, 32% respondents still felt understanding of its concepts were measured in some way, either formatively or summatively.

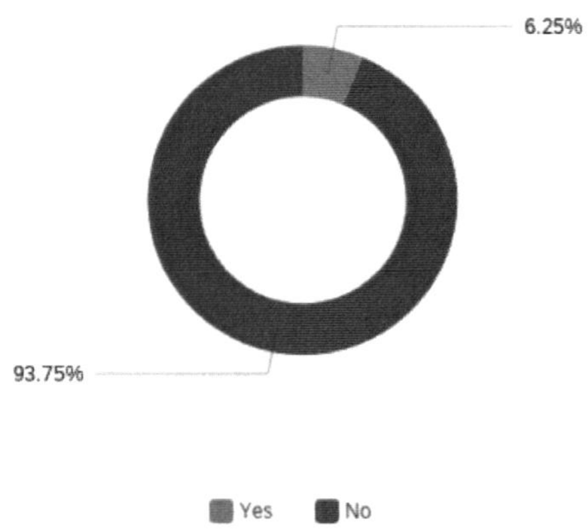

Figure 11: Is CE embedded into unit learning outcomes?

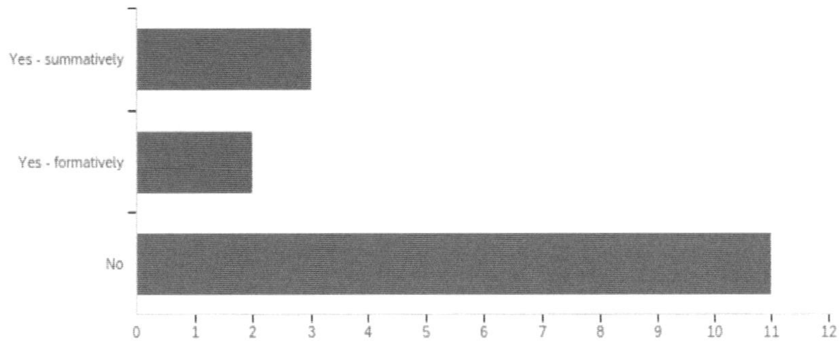

Figure 12: The measurement of CE in assessment

This part of the survey starkly highlighted another issue of CE education in HE. Staff in this pilot study indicated the importance of CE in fashion education, yet survey results suggest modern units and programmes are still lacking explicit learning outcomes and assessment in this field. However, encouragingly, these issues have already been addressed from several angles. The adoption and adaptation of KSC frameworks (Evans, 2019; Sumter et al., 2021) by educational bodies such as the QAA will promote transformational changes in this field and assist in the development of future programmes and units, enabling the clear linking of CE with learning outcomes and assessments.

Personal expertise

The survey then asked respondents about their experiences in CE, from a perceived knowledge and expertise viewpoint and from personal experience outside academia.

The 'Key Circular Design Competencies' model (Sumter et al, 2021), was critical at this stage in the development of these questions and the analysis of the responses in order to address the 'gaps' in CE expertise. However, whilst a breadth of knowledge was reported, unsurprisingly, given the background of the respondents, the greatest level of skill was noted in 'Circular business Model Integration (25%)' and Circular Economy Communication (25%)'. 'Circular Impact Recovery' and areas related to design concepts were noted as the least amount of knowledge within the respondent group (Figure 13).

The recommendations in the literature are that holistic approaches are required for effective student development in the field of CE. In our survey, the focus of expertise fell in the 'business' category of CE, with less expertise reported in design or social fields. This highlights the importance of continuous staff development, even in those teams that are considered 'experienced' due to the ever-changing nature of CE concepts and practice.

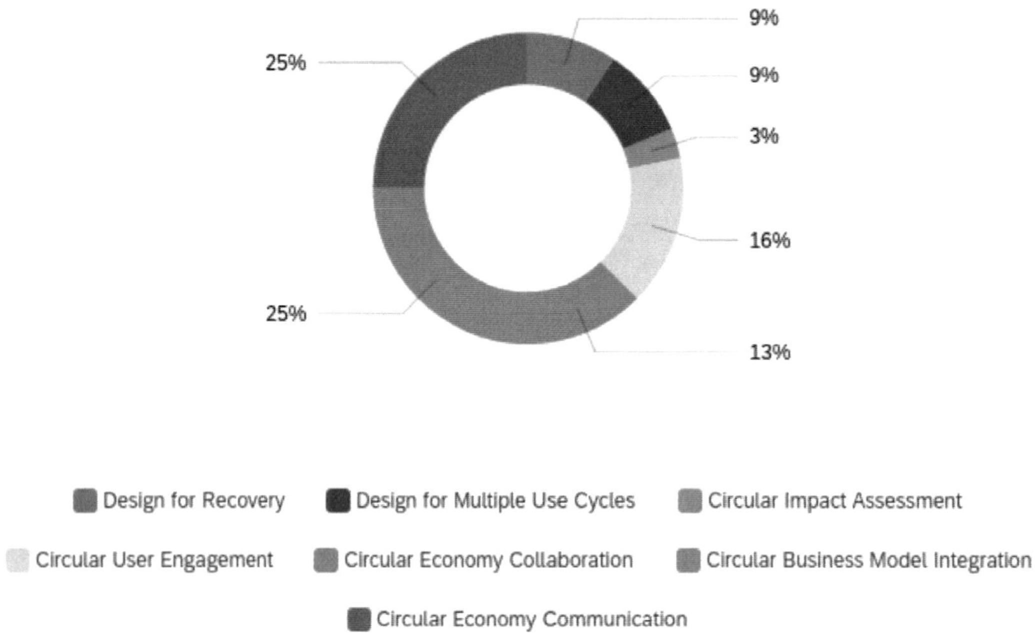

Figure 13: Reported areas of most knowledge in CE

Respondents we also asked if they had ever worked on a CE project outside HE or academia; 63% reported they had not. When rating the level of their own expertise in CE concepts, 80% reported knowledge at an intermediate or advanced level (Figure 14). None of the respondents claimed their level of knowledge of CE to be 'expert'.

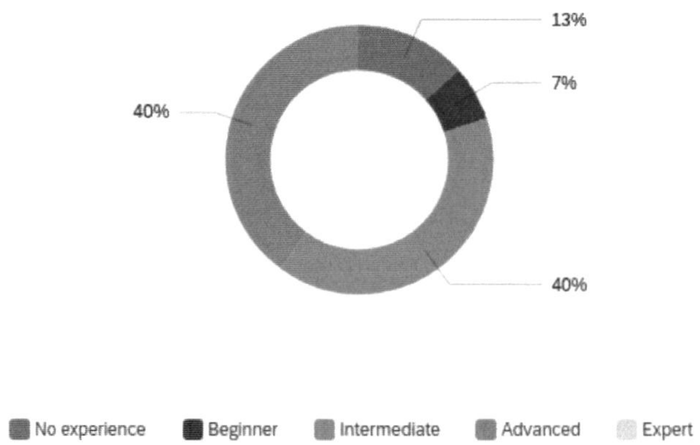

Figure 14. Self rated level of CE expertise

This result raises some interesting questions when considered alongside the whole survey. Whilst respondents felt they had a proficient level of expertise in CE and used case studies predominantly in their communication of concepts, over half of the respondents reported not having first-hand experience of CE outside the confines of HE. This indicates that some educators are gaining CE knowledge second hand, without industry experience. Whilst this may not be an issue when addressing theoretical concepts, it could raise some problems when trying to convey 'real world' situations. As previously mentioned, PBL is suggested as the most effective model of developing students thinking for CE (Agarwal, 2021), which could feasibly be conveyed without first hand experience. However, other authors suggest that it is live projects that have the greatest impact, and it is those academics with industry experience or strong industry links that will be able to facilitate this approach most effectively.

Conclusion

The overall aim of this research project is to produce a collective framework of knowledge and skills to support further debate, collaboration and curriculum development thereby accelerating this agenda within HE. As a pilot study, this first step in the project surveyed a group of academics in a UK HE fashion department to attempt to identify key priorities in the technical understanding that underpins sustainability and thus address the skills gap of graduates entering the industry in both general and technical roles.

Using predominantly closed questions developed from a key circular design competency model, the initial survey revealed that the respondents had the most expertise in fashion business and management fields. This raised the first tension noted in the literature; effective CE education should be delivered holistically, covering the range of fields described in CE competency models.

The survey revealed that whilst fashion staff appreciate the importance of embedding CE in teaching practice, this is not always reflected in the curriculum in terms of inclusion in programme and unit design from a learning outcome and formal assessment perspective. However, literature does suggest that developments are taking place in this field, with bodies such as the QAA engaging with key competency frameworks. Additionally, the nature of embedding CE principles in teaching practice raised another tension with the literature. Whilst literature suggests PBL is the best approach, and industry-based briefs most impactful, over half of the respondents had no practical experience of CE outside academia.

To address some of the issues raised in this pilot study, the authors suggest wider collaboration across industry and HE institutions are required to allow interdisciplinary thinking to develop in the fashion arena, and to expose students to the reality and challenges of CE in the global industry.

Whilst this pilot survey highlighted many of the issues raised in literature with addressing CE in the HE fashion curriculum, it also provided the foundation for a deeper and broader study across academic and industrial partners. It also further underpins the commonly held thought that transformative change is required in order to effectively embed CE practice for students in fashion HE.

References

Agarwal, S., (2021), Integration of Sustainable Practices in Fashion Design Education: An Experimental Study Based on Experiential Learning, The International Journal of Design Education, Vol 15, Iss 2, pp 153-166.

Barth, M. and Rieckmann, M., (2012), Academic staff development as a catalyst for curriculum change towards education for sustainable development: an output perspective, Journal of Cleaner production, 26, pp.28-36.

Brandt, J., Burgener, L., Barth, M., Redman, A., (2019), Becoming a competent teacher in education for sustainable development; Learning outcomes and processes in teacher education, International Journal of Sustainability in Higher Education, pp. 630-653.

Ceulemans, G. and Severijns, N., 2019. Challenges and benefits of student sustainability research projects in view of education for sustainability, International Journal of Sustainability in Higher Education.

Dahle, M. and Neumayer, E. (2001), Overcoming barriers to campus greening: a survey among higher educational institutions in London, UK, International Journal of Sustainability in Higher Education, Vol. 2 No. 2, pp. 139-60.

Dissanayake, D.G.K. and Weerasinghe, D., (2021), Towards circular economy in fashion: Review of strategies, barriers and enablers, Circular Economy and Sustainability, pp.1-21.

Ellen MacArthur Foundation, (2013), Towards the circular economy Vol. 1: an economic and business rationale for an accelerated transition Ellen MacArthur Foundation https://ellenmacarthurfoundation.org/towards-the-circular-economy-vol-1-an-economic-and-business-rationale-for-an [accessed 15th May 2023].

Ellen MacArthur Foundation, (2017), A new textiles economy: Redesigning fashion's future, Ellen MacArthur Foundation.

Evans, T. L., (2019), Competencies and pedagogies for sustainability education: A roadmap for sustainability studies program development in colleges and universities, Sustainability, 11(19), 5526.

Hall, N., & Velez-Colby, F., (2018), AMFI's reality school: a circular economy agenda for fashion education, Art, Design & Communication in Higher Education, 17(1), 11-24.

Geissdoerfer, M., Savaget, P., Bocken, N.M. and Hultink, E.J., (2017), The Circular Economy–A new sustainability paradigm?, Journal of cleaner production, 143, pp.757-768.

Giangrande, N., White, R.M., East, M., Jackson, R., Clarke, T., Saloff Coste, M. and Penha-Lopes, G., (2019), A competency framework to assess and activate education for sustainable development: Addressing the UN sustainable development goals 4.7 challenge, Sustainability, 11(10), p.2832.

Ioannou, O. and Klein, T., (2022), Closing the Knowledge Gap on Circularity: the CBE Hub Lifelong Education Programmes, In CLIMA 2022: Towards digitalized, healthy, circular and energy efficient HVAC (pp. 1-7).

Ki, C.W., Chong, S.M. and Ha-Brookshire, J.E., (2020), How fashion can achieve sustainable development through a circular economy and stakeholder engagement: A systematic literature review, Corporate Social Responsibility and Environmental Management, 27(6), pp.2401-2424.

King, L. G., McKim, A. J., Raven, M. R., & Pauley, C. M., (2019). New and Emerging Technologies: Teacher Needs, Adoption, Methods, and Student Engagement, Journal of Agricultural Education, 60(3), 277-290.

Kirchherr, J. and Piscicelli, L., 2019. Towards an education for the circular economy (ECE): five teaching principles and a case study, Resources, Conservation and Recycling, 150, p.104406.

Lidgren, A., Rodhe, H. and Huisingh, D., (2006), A systemic approach to incorporate sustainability into university courses and curricula, Journal of cleaner production, 14(9-11), pp.797-809.

Lozano, R., (2006), Incorporation and institutionalization of SD into universities: breaking through barriers to change, Journal of cleaner production, 14(9-11), 787-796.

Moore, J., (2005), Seven recommendations for creating sustainability education at the university level: A guide for change agents, International Journal of Sustainability in Higher Education.

Morseletto, P., (2020), Restorative and regenerative: Exploring the concepts in the circular economy, Journal of Industrial Ecology, 24(4), pp.763-773.

Orr, D., (1991), What is education for?, Context, 27(53), pp.52-58.

QAA and Advance HE, (2021) Education for Sustainable Development Guidance https://www.advance-he.ac.uk/teaching-and-learning/education-sustainable-development-higher-education, [accessed 15th May 2023].

Schöggl, J.P., Stumpf, L. and Baumgartner, R.J., (2020), The narrative of sustainability and circular economy-A longitudinal review of two decades of research, Resources, Conservation and Recycling, 163, p.105073.

Sumter, D., de Koning, J., Bakker, C. and Balkenende, R., (2021), Key competencies for design in a circular economy: Exploring gaps in design knowledge and skills for a circular economy, Sustainability, 13(2), p.776.

Textiles 2030 (2023), Textiles 2030 Roadmap, available at https://wrap.org.uk/resources/guide/textiles-2030-roadmap [accessed 15th May 2023].

Thomas, I., (2004), Sustainability in tertiary curricula: what is stopping it happening? International Journal of Sustainability in Higher Education.

UK Parliament (2019), Fixing fashion: clothing consumption and sustainability, available at https://publications.parliament.uk/pa/cm201719/cmselect/cmenvaud/1952/report-summary.html, [accessed 15th May, 2023].

UN General Assembly, (2015), Transforming our world: the 2030 Agenda for Sustainable Development, 21 October 2015, A/RES/70/1, available at: https://sdgs.un.org/2030agenda [accessed 25 March 2023]

Van den Berg, B., Poldner, K.A. and den Hoedt, T., (2020), Learning Towards a Circular Economy: A New Research Direction for Circular Business, In: Proceedings of the 5th International Conference on New Business Models, Nijmegen, The Netherlands (pp. 1-2).

Weetman, C., 2020. A circular economy handbook: How to build a more resilient, competitive and sustainable business. Kogan Page Publishers.

Wiek, A., Bernstein, M., Foley, R.W., Cohen, M., Forrest, N., Kuzdas, C., Kay, B. and Keeler, L.W., (2015), Operationalising competencies in higher education for sustainable development. In Routledge handbook of higher education for sustainable development, (pp. 241-260). Taylor and Francis.

Weiss, M. and Barth, M., (2019), Global research landscape of sustainability curricula implementation in higher education, International Journal of Sustainability in Higher Education, 20(4), pp.570-589.

LIFECYCLES OF BIO-BASED MATERIALS: INSIGHTS FROM THE HEREWEAR SAMPLE COLLECTION

Laetitia Forst[1] *and Rebecca Earley*[1]

[1]*Chelsea College of Arts, University of the Arts London, 16 John Islip St, London SW1P 4JU*

Abstract

The role of designers is pivotal in a shift to circular fashion systems, but often material knowledge is difficult to access and innovation is hard to rationalise or contextualise. We therefore need to create tools to better understand the field of emergent biobased materials and their potential for a decarbonised, circular fashion sector.

The EU-funded HEREWEAR project explores systemic innovation for biobased, local, circular fashion systems as an alternative to current global and fossil-based approaches. This paper presents insights from the creation of a biobased material collection in the early stages of the project. Analysing the collection alongside literature and interviews with experts, produced a map representing the range of options for feedstocks, transformation processes, and end of life for biobased materials.

The biobased material lifecycle map aims to support designers in choosing the appropriate materials for their designs and then implementing adequate circular design strategies for this material. It is already being used with stakeholders inside and outside the project to produce scenarios for biobased, local, and circular fashion that represent systemic understanding of the potential of these materials.

Keywords

biobased, circular, local, information visualisation, systemic design innovation

Introduction

Bio-based materials offer promising perspectives for the fashion industry. They can support a shift away from the high environmental impacts associated with the use of conventional cotton and polyester. They also present opportunities for locally sourced and circular fashion systems. However, this category of materials is often misunderstood and miscommunicated. New tools to describe and classify biobased materials are needed to support a shift to sustainable practices.

This paper presents the review and analysis of the field of innovative biobased materials performed in the early stages of the EU Horizon 2020 funded HEREWEAR project. To scope the field of bio-based materials, a collection of samples was gathered to understand and communicate challenges and opportunities for a transition to decarbonised and circular fashion systems. The collection of over 90 samples enabled a review of the current various resources, processes, and end-of-life trajectories relevant to these materials. The research was conducted from a design perspective, with an emphasis on making the technical information concerning material innovation available to design decision-makers.

The European Committee for Standardization (2021) puts forward the following definition for bio-based: "The term bio-based product refers to products wholly or partly derived from biomass, such as plants, trees or animals (the biomass can have undergone physical, chemical or biological treatment)." This definition is very broad, providing little guidance for designers intending to use biobased materials in their products; this further definition was needed. There is also entrenched confusion concerning the biodegradability of biobased materials and their relative environmental benefits compared to other resources.

In the context of a climate emergency, best practice approaches for making, using, and recovering products must function within planetary boundaries and enable social justice. The fashion industry, notorious for its impact on people and the environment, is slowly coming to terms with the transformation ahead (Amed *et al.*, 2022). Within design for sustainability principles - which includes many strategies aimed at reducing impacts - circular design focuses on the circulation of resources within fair and energy efficient systems (McDonough and Braungart, 2002; Ellen MacArthur Foundation, 2017). This can be associated with a shift away from carbon-heavy, fossil-based resources (Biomimicry Institute, 2020; WBCSD, 2020). Fashion and textiles are a key challenge for circularity as the complexity of the supply systems and (blended) materials involved make it difficult to adhere to the biological and technical cradle-to-cradle 'loops'. However, many researchers have been rethinking textile design, considering ways of revaluing waste materials and pioneering circular solutions in this field.

Design is changing, becoming a practice that is more able to concretely plan for a better future. Traditionally seen largely as a solely creative and aesthetic act in the fashion and textiles industry, the widespread realisation that design makes a significant contribution to global impacts has generated an urgency to use design differently. Researchers and practitioners are realising that it can be a systemic influencer, reaching beyond product conception and production, in both directions along the extended value chain. Thus, tools to enable designers to work in new, more considered ways, are urgently needed.

This paper presents the review of a collection of bio-based materials that was interpreted as a novel lifecycle map showing the feedstocks, transformation processes, and end-of-life trajectories for these materials. Presenting this information in a visual form, and as a tool useable in online and in-person settings to connect technical knowledge with creativity is key to the effective use of these materials for a shift to a decarbonised and circular fashion system.

Context

The work presented in this paper is set in the context of the HEREWEAR project which aims to explore innovation for bio-based, local, and circular textiles. The broader context for the research is the field of biomaterials innovation, populated by a range of innovators and designers at all scales of activity. This work aims to provide insights for this community.

The HEREWEAR project

The HEREWEAR project is based on three pillars which act as a compass for the project aims: local, circular, bio-based. The ambition is to create an approach which enables local eco-systems of garment production, using locally abundant resources such as straw waste, local 'micro-factories' and considers the local capacity for collection, sorting and processing. This approach relies on circular design to ensure that the materials are recoverable and can be cascaded through the waste hierarchy, from repair to re-use to remanufacturing, mechanical recycling and finally to chemical recycling. Circularity demands that material loss from the system is minimised – or at the very least – managed. Therefore, in the HEREWEAR system, production waste is recovered, and micro-fibre loss is prevented, as far as possible.

The HEREWEAR project aims at 'systemic materials innovation', meaning that the development of technologies for new materials is located within a framework which actively considers the wider context of the materials: the interrelation of the new materials with people and planet. In line with this, alongside the development of new fibres from agricultural waste, an important aspect of the project is to provide guidelines for designers to use bio-based materials in a circular way in their local context. To support this, a collection of bio-based materials was gathered to better understand this category of materials and what challenges and opportunities exist for their adequate use. This paper focusses on this initial part of the research aimed at understanding bio-based materials.

Bio-based materials

Interest in the potential of biomaterials for a sustainable transition has brought together a global community of design stakeholders, makers, industry specialists and consumers around an enthusiasm for the potential of these materials to address environmental issues. The words have been used extensively in the last few years as a proxy for "non-oil-derived". Biomaterials are put forward as a part of circular bio economies which acknowledge the incompatibility of oil-based materials with natural systems in which products and fibres inevitably disperse (Biomimicry Institute, 2020). The focus on biomaterials follows the imperative to decarbonise fibre production (Textile Exchange, 2022). However, a common misconception is that the term "bio" means that it is automatically better in terms of environmental impacts, and biodegradable. According to the definition put forward by the European Committee for Standardization, conventional materials such as a poly-cotton blend could claim the bio-based label if they are 50% cotton. This is contentious as it means that materials with a high proportion of fossil-based content could claim the benefits associated with bio-based materials in the eyes of the industry and thus the consumers.

To narrow the definition used in HEREWEAR and focus on innovative fibres that offer an alternative to current extractive practices, the term 'biosynthetic' is helpful. Biosynthetic - or regenerated materials - offers the opportunity to extract compounds that can be transformed into textile materials from renewable resources when the fibre isn't present or usable is the plant or animal itself, as it is for cotton and wool. Textile Exchange (2017) proposes the following definition: "a biosynthetic fibre consists of polymers made from renewable resources, either wholly or partly". This category includes fibres like man-made cellulose from wood pulp or other sources, as well as other processes which transform plant materials into fibres. Rayon, or viscose is a conventional biosynthetic material in which the cellulose contained in wood is transformed to produce a fibre. This process was historically a way to make textile fibres in a very inexpensive way, but recent regulation on air and water pollution have shifted the balance (Changing Markets Foundation, 2020). It is also relevant to note here that the production of regenerated fibres therefore isn't systematically environmentally sustainable, in particular if transformed through the viscose process, and only by controlling all the steps along the supply chain can this category of materials provide a vision for building resilience and accelerating regeneration (Forum for the Future and Textile Exchange, 2020).

The Textile Exchange (2017) defines bio-synthetic feedstocks as corresponding to three "generations" which are classified in terms of their availability to market. As the industry strives to improve performance and reduce impacts for these materials, efforts are made to optimise the use of biomass. First generation feedstocks are based on crops which provide stable and scaled input but can be in competition with land allocated for food. The second generation are feedstock taken from waste sources such as agricultural by-products. This offers the potential to reduce the land needed to produce fibres by using waste from other industries, however this type of feedstock requires additional research as it is currently only available for regenerated fibres at the pilot scale. The third generation defined by Textile Exchange is those that do not enter in competition at all with food as they rely on other types of agriculture which can be off land such as with seaweed or even using bacteria in lab settings.

One of the challenges here, is to effectively communicate that 'natural' materials aren't necessarily better in environment impact terms than bio-synthetic ones, and that with appropriate control over the production processes, these can deliver reduced impacts. Therefore, communicating clearly on the differences in feedstocks and processes is key to handing adequate information on to key decision-makers.

Method

The sample collection of bio-based materials gathered in the early stages of the HEREWEAR project has provided the basis for the development of a detailed description of the lifecycle of these materials. The analysis of the collection is core to the methodology presented here. The samples provide a basis for mapping the field and discussing more technical aspects of bio-based lifecycles with stakeholders. Other roles of the designer besides creating or making – such as a facilitator and a communicator - are also drawn upon in this work. Alongside academic research approaches which provide the rigour needed to support the insights and contextualise the work in current literature, the methods demonstrate the central role of design in making the complex issues of circular bio-based and local systems easier to understand and include in fashion design.

Sample collection

The sample collection was gathered from medium to large material manufacturers, smaller makers, and various distributors across the world, to show the current range of bio-based materials. This included conventional 'natural' textiles, as well as more innovative bio-synthetic examples.

A call for interest was launched across the HEREWEAR networks in January 2021, and specific suppliers were also approached by email. The suppliers were identified through online searches as pioneers in the field of biomaterials, providing options that can replace current 'problematic' materials. We request included fibres and materials made from agricultural bi-products rather than the crops themselves; or fibres and materials that push the boundaries of production with new, lower impact processes and finishes. Contributors were asked to send a physical sample and fill out an online information sheet providing details on the feedstocks and transformation processes used. The form allowed for some sections to be left blank, acknowledging the discrepancies in the availability of information for novel bio-based materials, as well as information that might be restricted by intellectual property regulation.

As the collection grew, the samples were reviewed through various criteria, such as by raw material or type of applications. Other material libraries were reviewed to compare how the industry is currently encountering these materials. Through this review of material libraries, it was found that a resource showing how materials are made could fill a gap in the current offer. The HEREWEAR sample collection therefore shows resources, or feedstocks used, types of transformation and processes, and the end-of-life options to recover and regenerate the resource. The information is presented as a circular lifecycle map, aligned to the HEREWEAR project approach (active areas for research), and in order to support designers in thinking about the whole lifecycle of a product when using these materials. This full lifecycle approach is widely known to be a key component of design for circularity (Sumter *et al.*, 2021).

Expert interviews

The information concerning biomaterials innovation is diffuse and sometimes not readily available, as it is subject to pending patents or confidentiality agreements. To overcome the complexity of the field and gain insights from current research and innovation, parallel and complementary information was sought through a series of informal interviews. These discussions with experts and material providers act as a first step to the translation of the technical information relating to biomaterials in terms that are useful to informing design decisions (Baule and Caratti, 2016).

The interviews took place both before and after suppliers and innovators sending the samples. This intended to establish trust between the two parties, clarifying the aims and conditions of material sharing. This approach also enabled a review of the classification and the representation of a given material in the collection. These interviews proved to be invaluable to getting first-hand information from the provider concerning the transformation processes and the market-readiness of the materials.

Interviews were also arranged with the HEREWEAR partners particularly involved in technical textile transformation processes from fractionation, to design for end-of-life. These sessions acted as a collaborative approach to building a common glossary and classification of biomaterials for the project, which fuses the expertise of partners from multi-disciplinary perspectives.

All interviews were carried out remotely using an online whiteboard space to present the iterations on the material classification while taking live notes. The interview structure was guided by the visual representation of the material classification and allowed time to discuss the different stages of a material's lifecycle in detail.

The HEREWEAR bio-based materials collection

Wilkes (2011) argues that material libraries are a key tool for knowledge transfer. The HEREWEAR collection aims to represent a range of samples from external partners alongside materials produced by members of the HEREWEAR project. It shows a snapshot of the field of biomaterials, identifying key players and innovative processes across the material lifecycle. It is also meant to help position the HEREWEAR materials within this field, by demonstrating some important similarities and differences when compared to other materials.

Internal materials

The HEREWEAR partners were first asked to contribute materials which had been made by their organisation in previous projects (figure 1). This provides a basis for collaboration and a reference point for what can be achieved in the project. As suggested by Hornbuckle (2020), using samples alongside dialogue can help avoid misunderstandings inside a collaborative project. This category also includes the samples that were produced in the project since its beginning to demonstrate the technology developed by various partners. In this case too, the samples are a key enabler of efficient communication by providing a concrete example of what can – and can't - be achieved.

Figure 1. Samples of naturally dyed regenerated cellulose and PLA produced inside the project

External materials

External materials were gathered from businesses and innovators not included in the project as partners, but which demonstrate similar interests in transforming materials used for fashion (figure 2). This section of the collection aimed to show the state of the art in the field of bio-based materials, contextualising the innovation provided by the HEREWEAR technologies. To provide this overview, some materials which are somewhat outside of the scope of this research - such as conventional 'natural' bio-based materials like cotton or hemp - were included. On the other side of the spectrum, other materials which are not yet available commercially or even in pilot stage - such as bio-iridescent sequins or bacterial cellulose - were also considered to show the range of possibilities in the future which can inspire design ideation now.

Figure 2. A selection of external materials from the HEREWEAR collection

Reviewing these materials with HEREWEAR expertise, and complemented with interviews with the providers, supported the development of a visualisation of the lifecycle of bio-based materials which encompasses multiple types of innovation across the field.

The biobased material lifecycle map

Figure 3. The Biobased Material Lifecycle Map

The sample collection provides the basis for a design-led understanding of the challenges and opportunities for bio-based materials in a circular and local context, grounding the research in the experience of interacting with the materials. The biobased material lifecycle map interprets the knowledge from the collection into a visual diagram that can support designers in better understanding how biobased materials are made and how they can be used. (Figure 3)

Lifecycle categories for biobased materials

Using the samples to support conversation with material providers and HEREWEAR experts, a categorisation emerged that could describe the bio-material life cycle. The visual representation shown in figure 3 provides a description of the different options of feedstocks and processing available for bio-based materials. The categories presented in the materials mapping were iteratively reviewed through interviews. The aim was to agree on a representation that could balance the level of detail and of scientific accuracy to best inform designers. Each segment of the material lifecycle map represents what options are available in material selection, production and use.

Feedstocks

From a design stakeholder's perspective, understanding the type of raw resource that a textile is made from, not only gives clues as to how it will perform in use, but also points to the end-of-life trajectory and the way in which this can be considered early in the design phase. It is also a critical decision point, as the industry aims to diversify and decarbonise its fibre input, by shifting away from the ubiquitous use of cotton and polyester (Textile Exchange, 2022).

Here the main distinction is between feedstocks in which bio-based fibres are naturally occurring; and feedstocks which can be used for the production of biosynthetic fibres after processing to extract the useful compounds. The former type includes naturally occurring cellulose fibres such as cotton or kapok, lignocellulose bast fibres like hemp and flax, and protein fibres in animal hairs such as wool. These materials are described as 'conventional' biomaterials and are included at the margin of this review.

The other overarching category presented in the lifecycle map is that of feedstocks which are in the form of biomass rather than fibres. These include crops such as sugar cane or maise, and forestry, such as eucalyptus trees from which Lyocell is often made. These feedstocks are grown specifically to make fibres, potentially in competition with food crops. Other feedstocks can be used that are the bi-products or waste of other industries - such as straw or orange peel - contained in the category of 'agricultural biproducts' in the lifecycle map. It is also possible to consider textile waste as a feedstock for new textiles when it is mechanically or chemically recycled into new fibres. Finally, as a third-generation feedstock, algae can be grown offshore and contribute the regeneration of marine biotopes as well as carbon capture. The map still accounts for the presence of some fossil-based feedstocks in the case of blends.

Feedstock preparation

The feedstock is rarely ready to use as it is found, various processes are needed to purify it and make it ready to be made into fibres. For instance, polymers naturally available as fibres like bast fibres or cellulosic fibres like cotton, may need a mechanical cleaning and carding process before they can be spun into yarn. Whereas polymers that need to be extracted and transformed, like the cellulose present in trees, first need to be fractionated and then dissolved before they can be made into a filament yarn. In a bio-polymer process, it is possible that the biomass will be fermented to produce the right type of molecules that can then be made to polymerise.

Filament products

This process is only relevant for fibres that are not naturally occurring, and which need to be produced from a pulp or a master batch. With melt spinning, the polymer pellets are melted and extruded to create the filament. Wet spinning uses a solvent bath from which the fibre is drawn, and dry spinning is a similar process to wet spinning with the solvent being removed from the fibre through evaporation rather than a different bath. In all three cases, this step can be the opportunity to engineer filament fibres and give them specific properties - through texturization for example.

Yarn production

Two main types of yarn production are considered here. On the one hand yarns are considered that are produced by assembling multiple short fibres under the heading 'staple fibre spinning'. On the other hand are 'hybrid yarns' made from multi-filament yarns. Staple fibre yarn-making processes can include conventional ring spinning, open end spinning or rotor spinning, and friction spinning, to name but a few. It involves the twisting of multiple fibres into a yarn - these can be natural fibres, synthetic fibres, or a mix of both. Hybrid yarns can be produced from a range of different yarns including staple fibre yarns. There are many ways to spin fibres together by twisting or wrapping them together. It is important to note how the way in which the fibres are spun into yarns can have a large impact on the performance and look and feel of the final textile.

Textile construction

The lifecycle map defines three different types of textile construction: either woven, knitted or non-woven. In the textile construction phase, 'first order fabrics' - made from only one re-source - can be differentiated from blends which combine different materials. With the fibres assembled as a textile it becomes possible to their behaviour in use, whether this concerns their look and feel, or the way they might fibrillate, or how they behave when wet. In this review, materials which are assembled to form alternative leathers fall in the category of non-woven materials as they are first assembled as a mat, either by mechanical felting or by using a binder, before the finishing stage gives them their final look and properties.

After treatments

This section includes a non-exhaustive listing of possible treatments and finishes that can be applied to materials. These treatments can be applied in any order relevant to their aims and intention. For example, a technical finish can be aimed at improving the dye uptake of a fabric, therefore occurring before the dyeing. These are relevant as some fabrics or yarns will react in different ways to finishing processes; or need an extra step to match the performance afforded by some fossil-based synthetic fibres, or even conventional biomaterials such as cotton. In the HEREWEAR project there are research efforts around developing bio-based finishes and embellishments.

Garment design

Garment design must consider the qualities of the materials afforded by feedstock and the various transformation and finishing processes, to enable the most adequate use of a textile for comfort and longevity, and also address the later stages and end-of-life to enable circular practices and recycling. Here the focus is on the choice of designing garments as mono-material pieces, either for one recycling cycle or for several, or design for disassembly. These decisions are key to enabling the long life and smooth recycling of resources in the form of garments.

Use / reuse

Use is a difficult lifecycle stage to design for, as once a product is in the hands of the user post retail, it is almost impossible to control what happens next. The use and reuse phases are left blank in the lifecycle map to acknowledge that some instructions can be given, and that these phases must be considered as a valid part of the full lifecycle, but that clear paths will be difficult to forecast. Clear product labelling, and digital product passports are areas of opportunity to enhance use and reuse.

Lifecycle extension

Several strategies for lifecycle extension are included in the guidelines produced in other parts of the HEREWEAR project. Some of these guidelines would include actions that can be taken at the start of the design process, and some would cover processes to apply to the garment once it has reached the end of its first use cycle, such as redyeing or other embellishments. These approaches can be considered in relation to the selected bio-based material, how it will age and wear, and how it will react to such processes.

End-of-life

When considering biomaterials in circular systems, the end-of-life stage is of great im-portance; it is the potential starting point to a new material cycle.

In a biodegradation end-of-life trajectory, textiles can be safely added to home or industrial compost so that they break down into nutrients under the action of micro-organisms (Elsasser, 2016). A common misconception with bio-based materials, is that they are de facto biodegradable, or indeed nutrients. While being made from bio-based materials can in some cases be an element that contributes to biodegradability, many bio-based materials will not decompose safely. Even a material labelled biodegradable might only be so in certain conditions. For instance, some materials can biodegrade in home compost, whereas others need the controlled conditions (such as heat) of industrial composting facilities.

With mechanical recycling waste textiles are shredded for fibres to be recovered for a new spinning process. This shortens the length of the fibres which means the addition of virgin fibres is often needed to strengthen the recycled yarn (Elsasser, 2016). Mechanical recycling can be a trajectory for both bio-based and fossil-based materials. Effective mechanical recycling is very dependent on a thorough sorting process (Hall, 2018).

Thermal recycling is an option for thermoplastic materials which can be melted down to reform pellets and spun through an extrusion process. This is a common route for polyester materials; at the moment more commonly in clear packaging form, rather than in textiles, but it is also applicable to well-sorted textiles. Bio-based materials which are polymerised to become similar in molecular structure to conventional polyester or nylon can be thermally recycled in the same way as their petroleum-based counterparts.

In the case of man-made cellulosic fibres, end-of-life textiles can be dissolved with chemicals which will extract the cellulose to recover it for wet spinning processes (Elsasser, 2016). Materials can be chemically broken down further into their polymers or even monomers. Chemicals or enzymes are used to break down the polymer chains so that the result can be polymerised and melt-spun once more (Elsasser, 2016). This reverse chemical reaction can enable the removal of impurities from the sludge, and hence produce quality polymers that can then be made into new regenerated textiles.

Using the Biobased material lifecycle map

The collection of samples, and the lifecycle map that came from it, were then used to support designers in taking bio-based, local, and circular principles into their work. These resources form the basis of new tools for discussion and design, for use in workshops (Figure 4).

Common language for biobased materials

The material collection, presented to the workshop participants through the lens of the lifecycle map to provide a circular view of the landscape of bio-based materials, offered a starting point for discussion on the qualities and challenges of these materials. With the physical material in hand, participants could discuss their ideas for its use in a concrete way, aligning descriptors such as 'soft', or 'luxurious' with other designers. The material postcard format (Figure 5) was devised to gather the questions designers ask about the use of these materials in fashion applications and their effect on the lifecycle of the product - from durability to recyclability.

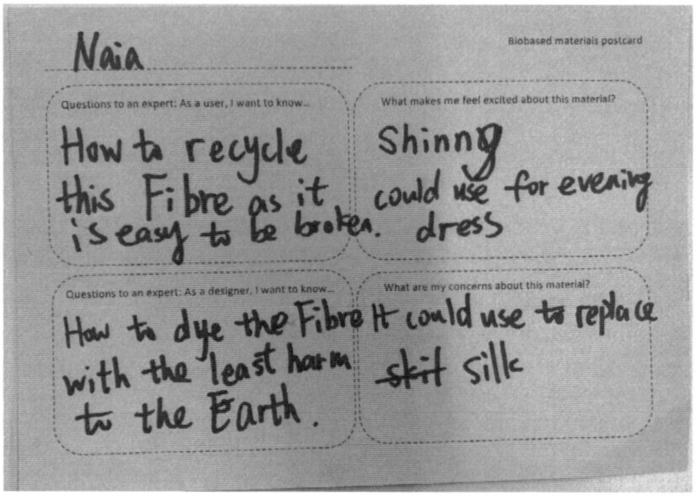

Figure 4 (left). A designer participating in a workshop using a selection of materials from the HEREWEAR collection.

Figure 5 (right). A material postcard reviewing a reference from the HEREWEAR collection.

Material portraits

The material lifecycle map, built from the review of materials from the HEREWEAR collection, was later used to visually represent the 'profile' or 'portrait' for a given material from the collection. These 'material portraits' were first built by interviewing the innovators or suppliers who provided the material for the collection, in order to test the portrait-making process. Further material portraits were then made with the information given for each fabric reference. These portraits show how materials in the same broad category of bio-based materials can differ widely in terms of their feedstocks or processes. The way that the lifecycle map is used accounts for the fact that often, all the information is not available for a given fabric, but some reasonable assumptions can be made regarding its composition, structure, or how to design with it based on labels or touch. This visualisation aims to communicate the diversity of biomaterials to designers in a clear and visual way.

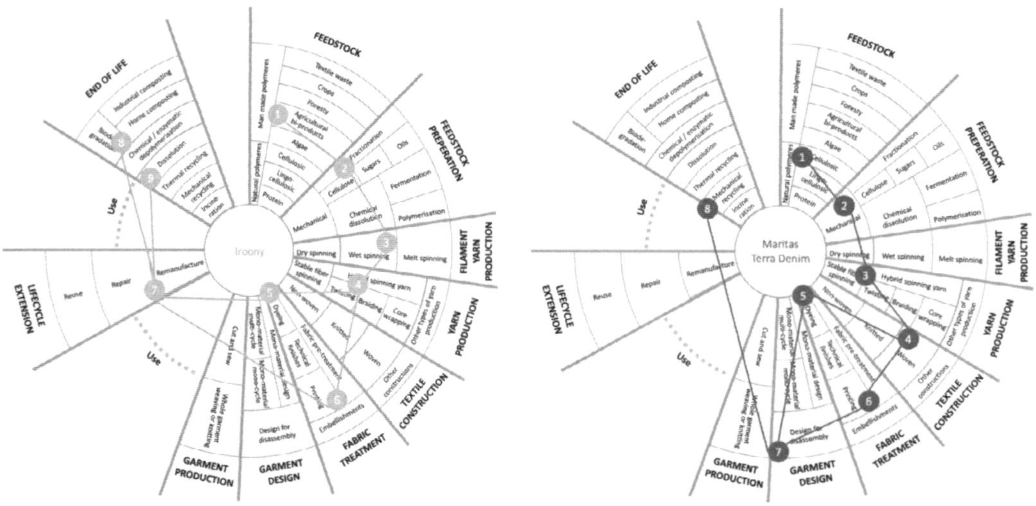

Figure 6. Two material portraits showing a visualisation of the lifecycle of materials from the HEREWEAR collection.

Supporting key design decisions

The aim of developing a better understanding of bio-based materials is for designers to be able to implement novel and reduced-impact fabrics into their work. By segmenting the information related to bio-based materials into stages of a lifecycle, discussion on the impacts of the choices made at each stage can take place between designers and experts. In particular, the material lifecycle map was used with Life Cycle Assessment experts inside the HEREWEAR project to discuss how a change in feedstock or process might affect the product's environmental score.

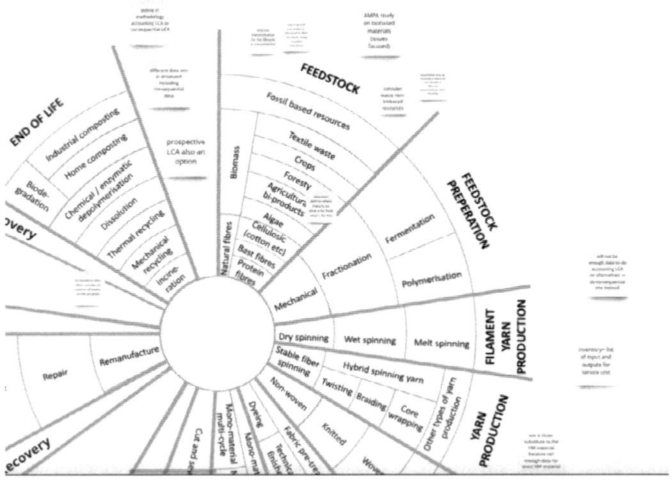

Figure 7. A screenshot from an online collaborative workshop using the Biobased Material Lifecycle Map to highlight key decision points based on Life Cycle Assessment knowledge.

Discussion

This review of bio-based materials through the HEREWEAR sample collection and its interpretation into the bio-based material lifecycle map aims to support designers in furthering their understanding of biomaterials through a visual tool that can inform design decision-making. Some key elements of the diagram contribute to this aim.

Translating experiential knowledge

As the knowledge relating to textiles ranges from the highly technical, to the deeply tacit (Igoe, 2010), it is often difficult to communicate effectively around innovation in this area (Tubito *et al.*, 2018). In particular, with the challenges of a shift to circularity, the scale that is scrutinised can go from the micro and chemical, to the macro and meta. Thus, finding tools and processes that can facilitate effective communication between different experts in this area is key to innovation for sustainability. There is most noticeably a gap between designers' expectations and what science can provide, which is important to bridge through transparent information sharing (Lee *et al.*, 2020).

As designers generally respond more easily to visual material (Brand, 2017), it was therefore relevant to represent the technical information for different feedstocks and processes in a way that could give an overview of the main samples in the collection in term of their trajectory from raw material to end-of-life. The original aim of the representation was to demonstrate how materials made from the same feedstock could be processed in different ways, drawing different paths to end-of-life recycling. It quickly became apparent that this type of visualisation could be useful as a way of gaining a broader understanding of lifecycles and design decisions for bio-based materials. While the lifecycle map cannot hold a large amount of technical detail about a material, it can quickly convey the main characteristics concerning its production, use and recovery trajectories. This can support a joining-up of the conversation between designers who require guidelines on how to use a material, and technical experts, who will understand this information in relation to the production and finishing processes.

Tools for designers

To evolve fashion and textiles within the circular economy an interdisciplinary approach encompassing a range of lifecycle perspectives is required. Facilitating this collaboration requires new thinking and tools. Lifecycle thinking is at the heart of this bio-based material lifecycle map - aiming to help designers 'do' more circular and sustainable design.

Forms of collaborative 'lifecycle thinking' (LCT) tools have been developed to integrate the essential, multiple-expert perspectives within the design process. This is intended to place material, environmental and social scientists, technologists, industry stakeholders and users all together at the core of an iterative design process. The value in the use of these interdisciplinary LCT methods, have been established through multiple research projects (Goldsworthy, Peters and Sandin, 2017; Tubito *et al.*, 2018). The sharing of knowledge in this context provides an important basis for understanding how to bring the theory of circularity into an established industry practice. The work presented here draws from these previous projects to offer a model for lifecycle thinking that is particularly relevant to bio-based materials. This is relevant as a step to overcome the misconceptions around the benefits and challenges pertaining to novel bio-based materials and their perceived role in a circular economy.

The practice of textile and fashion design can no longer be limited to the creation of a product for pleasure. Design stakeholders need to hold on to their ability to create joy and beauty for the wearer, whilst also becoming embedded in a process as systems-thinkers. Knowing the loops and processes of materials and products will be an essential part of their practice, enabled by better communication between technical and creative disciplines.

Conclusion

The understanding of biomaterials and the different processes and end-of-life trajectories that relate to them has been gathered through a sample collection of bio-based textiles. In this work the sample collection has been instrumental in many ways: it forms the basis of a field review, highlighting key areas for innovation; it has engaged networks of biomaterials experts and makers; and it has provided the basis for a lifecycle map relevant to bio-based materials.

Whilst the HEREWEAR sample collection isn't a repository or catalogue that designers can choose from - many other material libraries are currently available for this - it has produced insights useful to designers. The collection has been used to identify the lifecycle stages that are common to bio-based materials, enabling designers to learn about and to better understand these materials for their potential use in circular fashion products and systems. It was also found that the tool functions as a communication device between technical knowledge and creative approaches. The material portrait format offers a full lifecycle view of a range of materials, at a glance. This format can be used in future developments to gather more references of novel bio-based materials for the HEREWEAR collection and other uses.

Acknowledgements

This project has received funding from the European Union's Horizon 2020 research and innovation programme under the European Union's Horizon 2020 research and innovation programme under grant agreement No.101000632.

References

Amed, I. *et al.* (2022) *The State of Fashion 2022*. Business of Fashion and McKinsey and Company, p. 47.

Biomimicry Institute (2020) *The Nature of Fashion: Moving towards a regenerative system*. Available at: https://24rdmo160xr11sgco31bxj30-wpengine.netdna-ssl.com/wp-content/uploads/2020/08/TheNatureofFashion.pdf (Accessed: 29 September 2020).

Changing Markets Foundation (2020) *Dirty fashion: Crunch time, where does the industry stand on stamping out dirty viscose*. Available at: http://changingmarkets.org/wp-content/uploads/2020/12/CM-WEB-DIRTY-FASHION-CRUNCH-TIME-DEC-2020-1.pdf (Accessed: 18 January 2021).

Ellen MacArthur Foundation (2017) *A New Textiles Economy: Redesigning fashion's future*. Available at: http://www.ellenmacarthurfoundation.org/publications (Accessed: 19 March 2019).

Elsasser, V.H. (2016) *Textiles: concepts and principles* [electronic resource]. Third edition. New York: Bloomsbury (Bloomsbury academic collections). Available at: https://arts.idm.oclc.org/login?url=https://doi.org/10.5040/9781501304248?locatt=label:secondary_bloomsburyFashionCentral (Accessed: 16 March 2021).

Goldsworthy, K., Peters, G. and Sandin, G. (2017) 'Towards a Quantified Design Process: bridging design and life cycle assessment', in *Conference proceedings. Circular Transitions 2016*, p. 14.

Hall, C. (2018) 'Mixing it Up in Prato: identifying innovation hotspots within mechanical textile recycling', in *Proceeding of the Global fashion Conference 2018. What's Going On?*, London. Available at: http://gfc-conference.eu/wp-content/uploads/2018/12/HALL_MIXING-IT-UP-IN-PRATO-identifying-innovation-hotspots-within-mechanical-textile-recycling.pdf (Accessed: 9 April 2019).

Hornbuckle, R. (2020) 'Mobilizing Materials Knowledge: exploring the Role of Samples for Supporting Multidisciplinary Collaborative Design for Materials Development', *Design Journal, The* [Preprint]. Available at: https://doi.org/10.1080/14606925.2020.1849963.

Igoe, E. (2010) 'The tacit-turn: textile design in design research', *DUCK Journal for Research in Textiles and Textile Design*, p. 11.

Lee, S. *et al.* (2020) *Understanding 'Bio' Material Innovations Report*. Available at: https://app.box.com/s/amjq9anszv8hvwdexoxg6wubes4aaxqa (Accessed: 18 January 2021).

McDonough, W. and Braungart, M. (2002) *Cradle to cradle: remaking the way we make things*. New York: North Point Press.

Sumter, D. *et al.* (2021) 'Key Competencies for Design in a Circular Economy: Exploring Gaps in Design Knowledge and Skills for a Circular Economy', *Sustainability*, 13(2), p. 776.

Textile Exchange (2017) *About biosynthetics, About biosynthetic fibers*. Available at: https://aboutbiosynthetics.org/ (Accessed: 8 April 2021).

Textile Exchange (2022) The Sustainability of Biosynthetics: How biosynthetics can be part of the fashion and textile industry's journey towards a regenerative and circular future. Available at: https://textileexchange.org/sustainability-of-biosynthetics/.

Tubito, C. *et al.* (2018) *Applied DDMI: A white paper on how Design Driven Material Innovation methodology was applied in the Trash-2-Cash project*. Available at: https://static1.squarespace.com/static/5891ce37d2b857f0c58457c1/t/5c74266be2c4830ed71fb71e/1551115936465/D_1_7-White+Paper-MCI-T2C.pdf (Accessed: 23 December 2019).

WBCSD (2020) *The circular bioeconomy: A business opportunity contributing to a sustainable world.* Available at: https://www.wbcsd.org/contentwbc/download/10806/159810/1 (Accessed: 24 February 2021).

Wilkes, S.E. (2011) 'Materials Libraries as Vehicles for Knowledge Transfer', *Anthropology Matters*, 13(1). Available at: https://doi.org/10.22582/am.v13i1.233.

Defined Applications

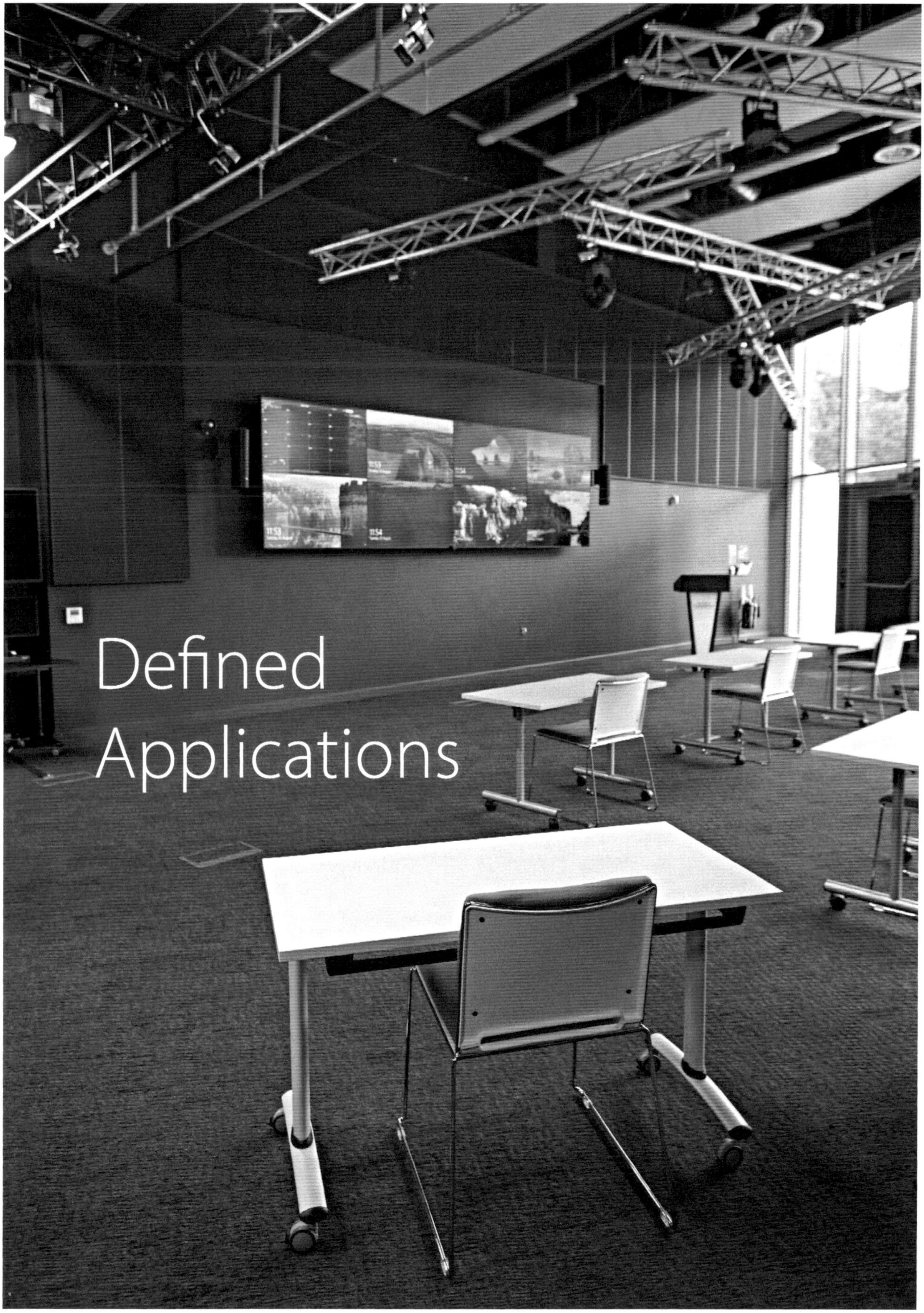

MODULAR KNITTED ARCHITECTURE: COLUMN

Elizabeth Gaston[1,3], *Jane Scott*[2] *and Armand Agraviador*[4]

[1]School of Design, Northumbria University, Newcastle Upon Tyne, NE1 8ST
[2]CCE, Northumbria University, Newcastle Upon Tyne, NE1 8ST,
[3]HBBE, Newcastle University, Newcastle Upon Tyne, NE1 7RU,
[4]Independent researcher

Abstract

The transition to net zero energy use in the building sector requires a fundamental rethink of material strategies. Utilising the zero-waste construction capability of knit, this project used craft practice and computational modelling to produce knitted branching structures that demonstrated structural integrity through their complex geometry at an architectural scale. This offers a potentially transformative approach for construction in two ways, firstly by reducing conventional material requirements and secondly by enabling the transition to low-impact biomaterial alternatives.

One challenge of working with knit at an architectural scale is the dimensional limitations of fabrics manufactured using industrial machines with a needle bed width of 1.85m. Large-scale work can be assembled from multiple pieces, however reconfiguring knitted fabric by joining fabric panels compromises the performance of the material, limiting multidirectional extensibility and the inherent strength the material. Significantly, the challenge of computational modelling of knitted fabric is exacerbated when fabric characteristics are transformed at the joins.

This project found solutions using traditional grafted knit seams that retain the structural performance of the knitted fabric and enable more accurate computational models. This was tested at three scales using hand and digital processes leading to further development with three-dimensional CNC production.

Keywords

Three-dimensional knit, Architectural scale, Computational modelling, Modular formwork, Digital/craft process.

Introduction

The use of weft knit in composite material systems for construction is gaining momentum. The unique structure and production of knitted fabric have contributed to its successful application in construction research by utilising both its formability and tensioning capabilities. This paper will discuss the design challenges of working with knit in the construction industry in the context of the representation of the mechanical properties of knitted fabrics in computational modelling. It builds on previous work by the research team investigating the effect of traditional joining technologies on the accuracy of computational form (Scott et al, 2021), extending the knowledge on the use of joining technologies and complex knitted form.

This paper will briefly review the earlier work of the research team, comprised of knit fabric specialists, computational designers and architects, in which grafted seams were used to produce an architecturally scaled dome using a craft/computational design methodology. It will highlight the successes and further questions revealed during the project and discuss how can the craft-based grafting of knitted seams can aid the development of complex knitted forms for architecture. This paper focuses on the textiles challenges of using knit for architectural formwork.

Background

40% of UK carbon emissions are related to the construction industry (RAEng, 2021) and the transition to net zero energy use in the building sector requires a fundamental rethink of material strategies. Knit can contribute to the positive environmental evolution of building materials by utilising the zero-waste construction capability of knit and using regenerative materials in its production.

The interest in the use of knit as an effective architectural material has grown significantly in the last decade, with use falling into two broad categories. Firstly, knitted fabrics have been used widely as tensioned canopies, utilising the extensibility of knitted fabrics, for example, the Isoropia canopy, part of the Danish Pavilion at the 16th Venice Architecture Biennale. (La Magna et al, 2018). Alternatively, the use of knit has been explored as a formwork for composite construction processes with either internal support provided by a steel scaffold (Popescu, 2022) or external support, provided by biomaterials (Scott et al, 2023) or concrete (Underwood 2009). Tensioned knit canopies and knit formwork are lightweight compared to conventional construction material, but more importantly, can produce complex geometries that would be difficult to produce using orthodox production methods.

Knit is an additive manufacturing process where the basic construction unit is the knitted loop. Fabric production can be specified at four levels, fibre, yarn, loop and fabric. In weft knitting, loops are formed sequentially in horizontal courses and depending on the equipment used for production, courses can be produced in reversing lateral directions to produce a plane or in a helical form to form a tube.

The capability to change the number of loops produced in each course and therefore the dimensions of the knitted fabric, affords knit the ability to produce fabric to a specified 2D or 3D form without further post-production processes. Thus, the zero-waste production capabilities of knit can reduce environmental and economic costs through a reduction in the material requirements when compared to subtractive manufacturing processes in the production of complex forms.

At a material level, the yarn requirements for a knitted fabric are length, strength and flexibility. These can be provided by a variety of fibres and filaments, either natural or synthetic. By using regenerative materials such as wool, which are also biodegradable post-use, as opposed to oil-based synthetic fibres, the environmental impact of material use can be reduced. This research recognises that all fibre and yarn production methods have some environmental impact during material processing.

Configured Knitting

Design and production of complex 3D knitted forms is enabled through the use of an integrated workflow including computational modelling and CNC 3D knit technologies. This poses challenges in the manufacture of complex 3D form at an architectural scale. CNC 3D knit technologies are designed for manufacturing at an apparel scale and fabric production is limited to widths of approximately 185cm resulting in either the need for the use of individual panels within an application or joining panels to produce a single large-scale structure. However, conventional knit seaming technologies alter the mechanical properties of knitted fabrics at the join and so disrupt accurate computational modelling.

In earlier work by the research team, Configured Knitting (2021) (Fig. 1) investigated the use of grafting, a traditional craft-based knit seaming process, as a joining mechanism. The project questioned *how can grafting be adapted as a joining technology for multiple textile components in architecture* and adapted the workflow to include computational modelling and craft practice.

Figure 1: Configured Knitting (2021).

The mechanical properties of knitted fabrics are well-documented (Spencer, 1989). Based on a loop structure, knitted fabrics demonstrate multi-directional extension, through loop deformation, at the application of low loads (Munden, 1959). Without material deformation, the fabric will relax to its original dimensions on the release of the load. The instability of knitted fabric was first discussed in the mid-twentieth century by researchers such as Chamberlain (1951), Doyle and Hurd (1953), Leaf and Glaskin (1955) and Munden (1959) who identified the individual knitted loop as the basic unit for fabric specification. Loop length is determined by needle action yarn tension and fabric takedown (Spencer, 1989) which is controlled accurately in CNC production and defines the deformation parameters of the fabric. Later work (Choi and Lo, 2006, Liu et al, 2017) recognised the effect of material use on loop deformation in terms of fibre content (which affects surface friction and bending rigidity), spinning processes, (which affects yarn diameter and torsional rigidity), tensile modulus and yarn compression (Choi and Lo, 2006). The complexity of predicting fabric extension under tension is compounded by dimensional differences in varying stitch structures, which are used to produce different mechanical fabric properties within a fabric to extend its form generation potential.

The need to seam knit fabrics to achieve an outcome at an architectural scale cause increasing complexity in terms of loop deformation. Configured Knitting reviewed three processes used to join knitted fabrics with unsecured edges, at the top and bottom of the fabric. Most seaming for architectural knit outcomes has been adapted from the two main knitwear seaming technologies; linking and overlocking. Both of these technologies use additional threads in a looped configuration to increase the extensibility of the seam (Brackenbury, 1992). Linked seams with one (Fig. 2) (stitch class BS100) or two (stitch class BS400) threads produce accuracy but restrict extensibility (ibid). Overlocking (stitch class BS500) with three or more threads closely resembles the knitted loop however is less accurate and produces bulky seams (ibid). Seams can be manipulated to replicate the extensibility of a knitted fabric more closely by increasing the run-in ratio of the seam (length of thread absorbed by the seam: length of the seam) (Choi et al, 2015) however, this reduces the integrity of the seam strength and may result in grinning between elements (Jones and Stylios, 2013).

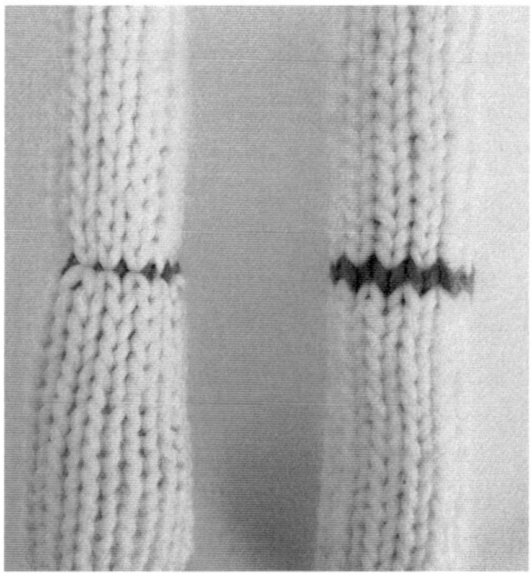

Figure 2: Image of tubes - Technical face of a linked seam (left) and a grafted seam (right).

The final class of seam reviewed, and later adopted was the traditional craft-based grafting, in which the joining thread replicates the structure of the knitted fabric exactly (fig. 2). Grafted seams demonstrated no reduction in extensibility and maintained the same tensile strength as the knitted fabric (Scott et al, 2021). To demonstrate the efficacy of grafted seams in architectural-scale practice using computational modelling, the design team produced an experimental architectural prototype using tubular branching topology. Based on processes developed in the production of 3D knitwear, the Configured Knitting design team produced a series of hand-knitted, tubular, pre-form modules which became the basis for computational exploration of branching dome topologies. Visual programming through Grasshopper revealed a series of dome topologies of increasing complexity, defined by the number of pre-form modules in each stratum of the dome and the number and position of branches within the module. The individual branches were produced integrally to the module and seaming occurred where two branches met to form a single beam. This improved the structural integrity of the final dome and simplified production.

Larger-scale fabrics can be produced by either knitting with a high number of courses and wales or as in this case using materials with more volume (untwisted Cheviot British wool approximately 1.5 cm in diameter) which enabled very clear visual analysis of fabric behaviour. The findings showed that the structure of the grafted seam replicated the behaviour of the knitted fabric exactly and did not disrupt the visual appearance of the structure.

Column module production

Whilst successful in assessing the integrity and deformation of grafted seams in complex 3D knitted assemblages, Configured Knitting presented further challenges. The digital production of architectural fabrics is conventionally produced at stitch densities of multiple stitches/cm. Successful seaming requires accuracy at the individual stitch level and currently, there is no mechanised process for producing grafted seams. Additionally, mechanised sewing and linking technologies are designed to create each stitch using the thread nearest the previously produced stitch. Grafting, like hand sewing, requires the full length of seaming thread to be pulled through each intersection of fabrics, expending additional energy and limiting the length of seam that can be produced without joining seaming threads. It is also vital that the loop length of the grafted seam is identical to the loop length used to create the original fabric; as a hand process, grafting presents the challenge of loop length control.

The second iteration of the modular knitted architecture project asked *is grafting an applicable seaming technology for fabrics produced through digital manufacture*. Following the successful design methodology employed by Configured Knitting, it extended the workflow to include computational modelling, craft practice and CNC 3D knit technologies. Whilst the overarching aim of the project was to test the use of grafting on fine gauge digitally produce knit modules, it was also an opportunity to extend the understanding of module geometries achievable using knit-tube configurations. Additionally, the project provided the opportunity to develop the computational form generation capability using knitted preforms and investigated the use of the column structure, observed widely in architectural practice (Fig. 3).

Figure 3, The Arndale Centre, Manchester.

Design Methodology

The design of the module evolved iteratively through a combination of craft and digital processes employing an interdisciplinary approach. Figure 4 shows the temporal organisation of design iterations however, each iteration wasn't completely discrete. The following section discusses the parameters explored during the project and their development within the workflow.

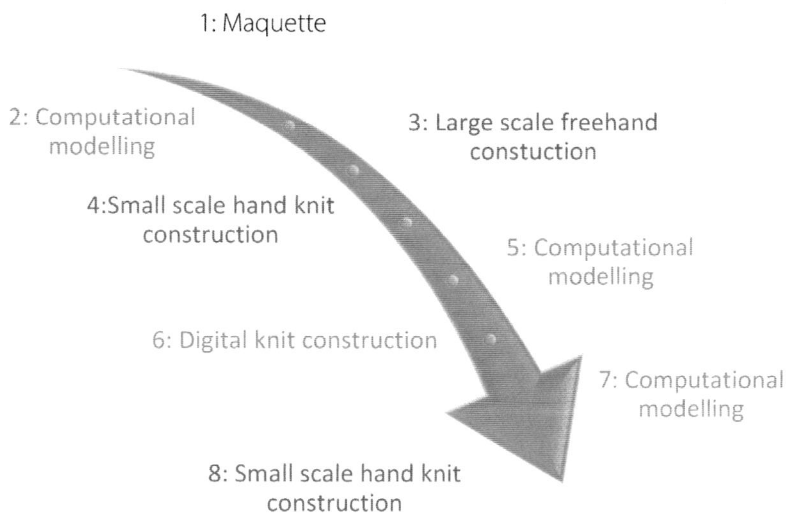

Figure 4 Project workflow demonstrating the relationship between design iteration and craft/digital process. Blue text denotes hand process, green text denotes digital process.

Module and global form

Previous work demonstrated that the global form of the architectural prototype was controlled through the branching topology, particularly the number of branches in each preform and the length of each branch. In the domed form, the branching topology was reductive, including fewer branches in each stratum of the structure. For the column form, the number of branches in each stratum was consistent and the global form was achieved through manipulation of the branch length. Initial craft-based experimentation in hand-knit produced a simple column based on three pre-forms in each stratum. The individual modules were designed as a simple X form, reducing the complexity of the module form from previous work. In each stratum, the modules were transposed laterally by half a module width to form a grid structure when grafted together (Fig. 5). The edges of the 2D grid were joined to form a 3D column with a helix structure.

In formwork structures manufactured in established materials such as steel or concrete, modular form panels are conventionally solid increasing the weight and cost of the column. In lattice columns, which reduce material use, the formwork consists of two elements with separate roles, compression and tension, working in opposition to provide stability (Stroud Foster and Greeno, 2017). Following the successful use of inflatable beams to provide internal support for Configured Knitting, the Column was also designed to incorporate the use of inflatable beams. Three X-shaped modules in each stratum produced six channels for inflatable beams, three in the S direction and three in the Z direction. It was anticipated that the compression force acting on the column would result in the deformation of the beams. To counter this, a further point of difference to Configured Knitting was the introduction of beams that specifically act as tensioning bracers with no load-bearing responsibility.

Figure 5: The configuration of CNC produced modules prior to construction.

Material and scale

Following the successful design methodology employed in previous iterations of the project, early form was generated using the large-scale freehand construction processes employed by Configured Knitting (Scott et al, 2021). Modules were constructed from untwisted Cheviot British wool roving (approx. 1.5 cm diameter). The compression beams had a diameter of fifteen loops which equated to approximately 10 cm diameter and tensioning bracer beams had a six-loop diameter. Each module was 70 courses in length. The double width join was 10 courses and the beams above and below the join were 30 courses each. Three modules were produced, which formed one stratum of a column at a 1:1 scale and were constructed from 20 kg of yarn. Whilst this weight of yarn was supported by inflatable beams in the dome-shaped Configured Knitting, inflatable beams were less successful in supporting a column form and the structure was unstable. However, this scale would have been more successful with external support, particularly when strengthened with bio-composites.

Reducing the weight of the structure was achieved through material use. In the second iteration of the column, a hand-knit craft process (Fig. 6) was employed using two ends of a softly twisted 1/1 Nm merino wool yarn knitted on a set of 4, 7.5 mm double-pointed needles. Using the same loop configuration as the large-scale free-hand construction model, a three-strata model was made from nine modules produced at a 1:4 scale. This was supported by inflatable beams with a 10 cm diameter inflated to 1 bar. The structure was held vertically with overhead support.

Figure 6: Hand Knit Column.

The successful form of the hand-knit version was translated into a digital programme for production on a 12-gauge Shima Seiki SSR machine and Apex 3 programming system (Fig 6). To use the two-bed machine for 3D production the machine was programmed to knit at half-gauge and the modules were knitted in two ends of 2/30 Nm merino wool. In order to produce individual modules efficiently and to accommodate the limitations of machine knit, the module design was adapted whilst maintaining the X configuration of beams and the use of additional bracer beams. The digitally knitted model was produced at a stitch density of 3,850 stitches/10 cm^2 (compared to the hand-knit version at 150 stitches/10 cm^2) therefore using the same loop configuration as the two previous iterations would have made the outcome too small to assess appropriately. As a result, the number of courses and wales in each beam was increased.

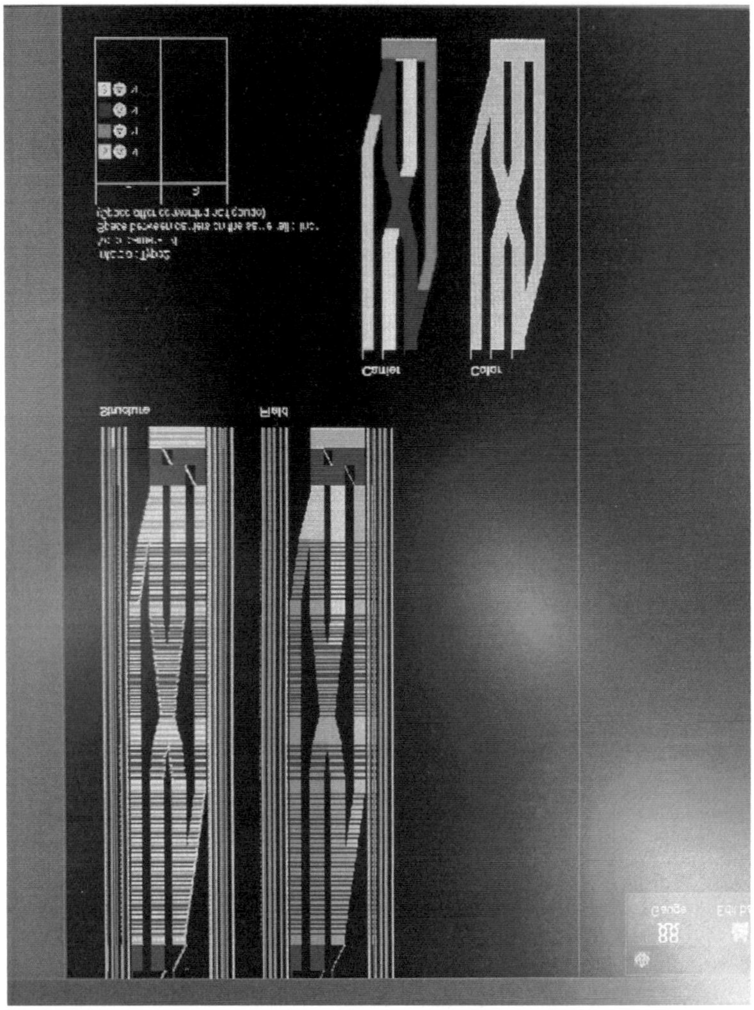

Figure 7: Apex 3 programming of CNC module production.

Centre Join Angle

To produce a secure, lateral join in two knitted tubes, for example, at a sleeve head in whole-garment knitwear, it is usual to transfer the loops forming one tube to overlap the loops forming a second tube and knit the overlapped loops together in the next knitting cycle (Brackenbury, 1992). Doing this allows precision in the angle of the join through control of the rate of transfer. However, this joining method would result in the centre of the X having the same width as the legs of the X and so restrict space at the point of crossover of two inflatable support beams. It would also be a potential point of weakness when translated into a bio-composite. In further models, the join at the centre of the X was formed by knitting the loops at the end of two adjacent beams as one double-width beam. This method of joining prevented precise angle placement at the join however at a model scale the inherent extensibility of the knitted fabric revealed the structure of the column. In the digitally knitted version, it was not possible to produce two adjacent tubes due to machine limitations and it was necessary to programme the machine to transfer the two tubes forming the beams below the join together and then transfer the further apart after the join. This resulted in a true X-shaped configuration, similar to the original maquette (Fig. 7). As this version was supported by expandable foam rather than inflatable beams there was less impact of the reduced width at the join however in further work this will need to be addressed.

Bracer Position

In the original maquette, the bracer beams originated at the end of the bottom right and top left compression beams however introducing or removing additional stitches to form the bracer beam in the first and last courses of knitting disrupted the top and bottom edge profile of the relevant beams which in turn disrupted the grafting process. Placing the bracer beams at least 2 cm from the end of the compression beams removed this distortion and a perfect grafted seam was achievable (Fig. 8). As this paper discusses the potential of grafting as an assembly process the structural integrity of bracer placement has not been discussed.

In the hand-produced models, the bracer beams were knitted to the required length.

Programming limitations resulted in the digitally knitted bracer beams being knitted to the same length as the compression beams and were later cut to the correct length. However, this added to the module's design potential during computational modelling, where the bracer could be positioned both within the same strata or across two or more strata.

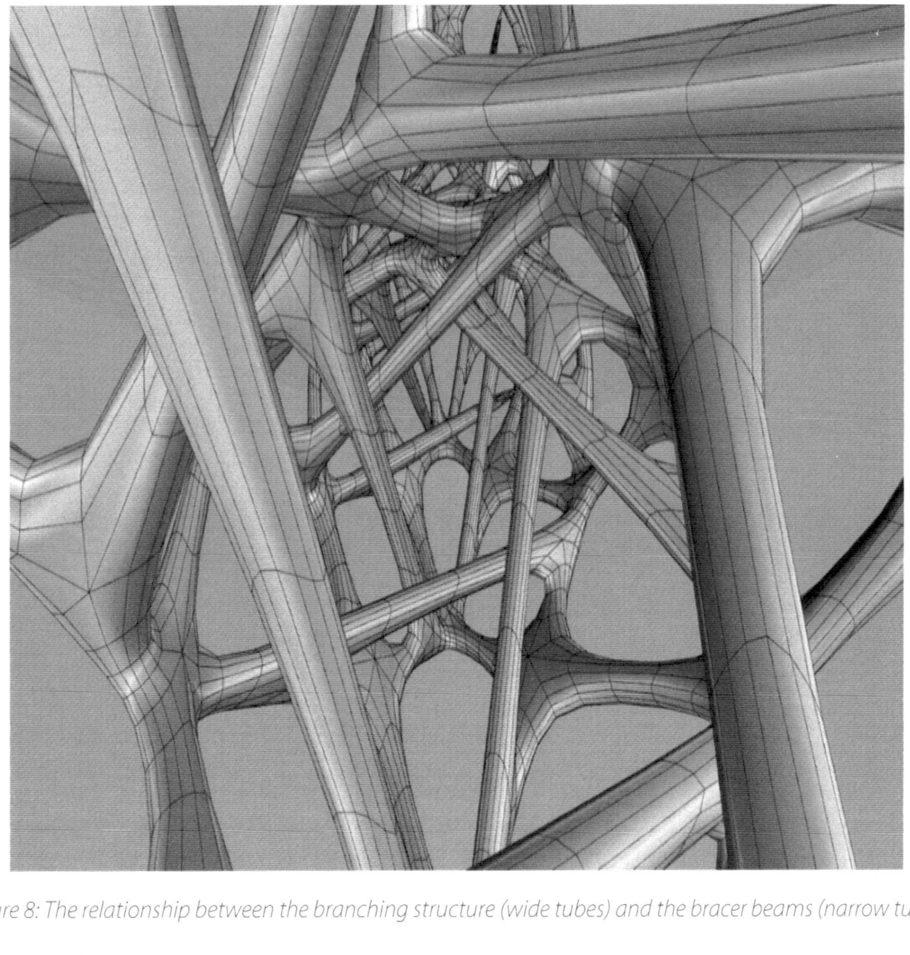

Figure 8: The relationship between the branching structure (wide tubes) and the bracer beams (narrow tubes).

Computational Design

Figure 4 positions the computational design aspects of the modelling system as discrete activities however, the team's interdisciplinary design approach moved between the material practice of knitting and digital modelling where each process informed and was informed by an understanding of each other. The modelling system designed to support fabrication by digital 3D knit technologies employed a multinodal system with variable components used to develop increasing complexity within modular knitted preforms. The system can be adapted to create design specifications for a diverse series of branching knit structures using sliding variables to control quantities, dimensions and variance of individual components within the system. Combinations of these variables compound to manipulate the geometry of the overall structure.

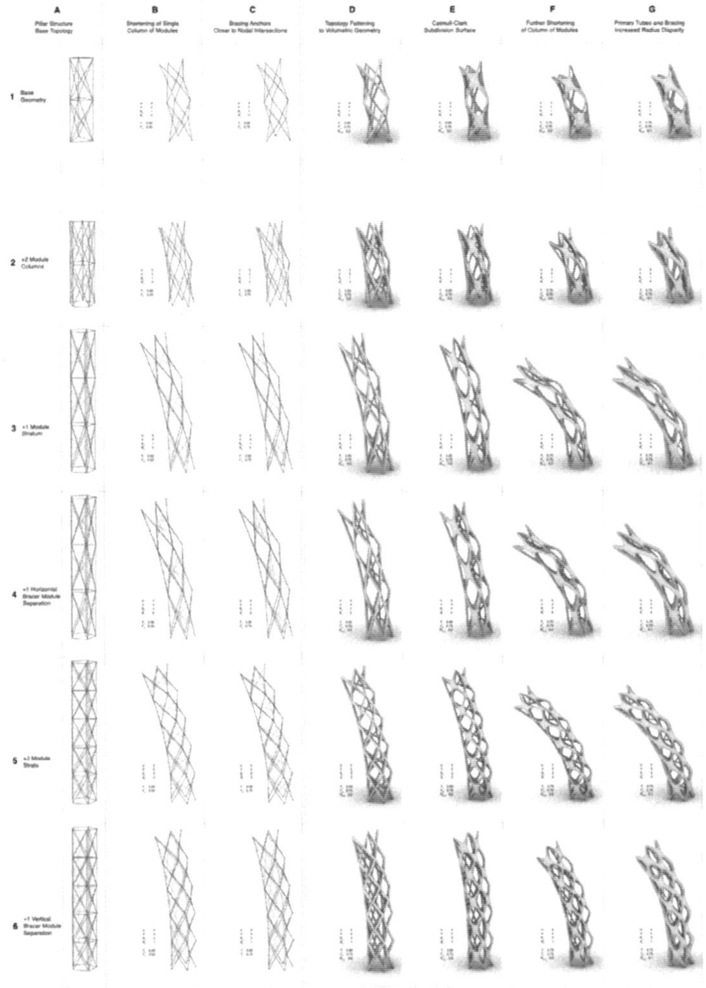

Figure 9: A directory of computationally generated topologies demonstrating bending behaviour.

Using Grasshopper, each module was originally described as a simple polygon which became one side of a 3D form. The number of sides of the form equated to the number of modules in each stratum. Each polygon was subdivided into four triangular panels to create a mesh and exert more control over module dimensions; in order to manipulate the overall column form. Bracer beams were introduced to start points on one compression beam within each module. The endpoint of the bracer is defined as a corresponding point in a neighbouring module of a given degree of separation either within or between strata. Input variables in modelling include beam length, number of modules per strata and bracer position. Once a wireframe topology is output from combinations of these values, the entire geometry is run through a physics simulation through the Kangaroo plugin.

At this stage, an extensibility variable is introduced to anticipate the behaviour of the knit elements under tension. Combined with the initial input variables and various constants, the resultant column bending form is created, with a precise measurement of each beam reported to assist in the physical production of the knitted module. Plugins were used to "fatten" the wireframe into a tubular solid to replicate the knitted form and report volumes and areas to help estimate required material quantities. All variables can be manipulated in real-time (Fig. 9) to generate inexhaustible forms. A fuller explanation of the programming will be available in a separate paper.

Results and discussion

Grafting was used as a seaming technology at the three scales and the results were compared in terms of seam quality defined by seam strength integrity, extensibility and ease of production.

At all three scales, the material integrity of the structure was achieved using the module construction material as the grafting thread. For the large-scale freehand construction column, only one stratum was produced, restricting seaming to the bracer beams. The large-scale production enabled a very clear visual analysis of the mechanical properties of the grafted seam. The seam demonstrated no visible difference in width-ways extension between the loops produced during the knit process and the loops produced during seaming. As a large-scale production process, seaming by hand was straightforward and produced a seam with good seam integrity.

In the model produced using a conventional hand-knit process, the seaming process was undertaken using a blunt-pointed darning needle. This allowed precision placement of each seam loop as it passed through the loops produced during knitting, particularly as the knitted loops presented a clear space in the centre of the loop. At a hand-knit scale, it was possible to control the loop length of the seam to replicate that of the knitted loop. Again, the scale of the fabric enabled a very clear visual analysis of the mechanical properties of the grafted seam. Again, the seam demonstrated no visible difference in width-ways extension between the loops produced during the knit process and the loops produced during seaming and again, the seaming process was straightforward and produced a seam with good seam integrity.

At the scale of digital production seaming was more complex. The hand grafting process was undertaken using a blunt-nosed sewing needle. Whilst a grafted seam was achieved, the process presented more challenges. Control of loop length is increasingly difficult as the loop length decreases. Whilst it was possible to achieve an even loop length in all loops of the grafted seam, this required greater skill, developed tacitly through repeated execution. Additionally, there was minimal clear space at the centre of the knitted loop which resulted in difficulty in positioning the needle as it passed through. This resulted in one of three outcomes; the needle passed through the loop as intended, the needle passed through part of the knitted loop or the needle passed through part of the waste yarn which marks the end of the knitted module as part of the production process. Unintentionally catching elements of either the fabric of the module or the fabric of the waste yarn in the seam resulted in difficulty unravelling the waste yarn to produce a clean seam. Despite the challenges of production, once executed, the grafted seam demonstrated no visible difference in width-ways extension between the loops produced during the knit process and the loops produced during seaming and was considered a successful seaming mechanism.

The computational modelling developed to generate complex 3D structures was extremely successful in terms of viable form generation and dimensional calculation. This was demonstrated by the production of a final structurally curved column where the curve was formed through variation in beam length using dimensions created in the Grasshopper programme (Fig 10).

Figure 10: The development of physical 3D form using computational analysis.

Conclusion and further work

This paper reports on the development of a knitted modular form used to develop a column structure at three scales. It focuses on the use of grafting as a seaming technology used to join individual modules. Findings demonstrate that grafting is a successful technology for seaming knit at all three scales in terms of seam integrity, however, in finer fabrics, in this case, 6 loops/cm, the hand grafting process requires enhanced skill and is a time-intensive process, therefore potentially of less use at a 1:1 architectural scale. Whilst demonstrably a successful seaming technology in terms of seam integrity, to be an effective seaming mechanism for 1:1 architectural scale construction, it will be necessary to develop a mechanised process for grafting.

In moving between the material practice of knitting and a digital modelling environment, the research explored digital design as a collaborative practice that informed and worked alongside tacit knowledge generated through making towards achieving sustainable material solutions for construction. Grafting enabled accurate computational modelling which is essential in 1:1 scale architectural modelling. The project explored global structure using a modular knitted form, developed computationally. The Grasshopper programme developed allowed innovative and versatile form generation that could be translated at multiple scales and production processes. The project clearly demonstrated that a transdisciplinary understanding of all processes allows fluid development of form in architectural research.

Acknowledgement

This project was made possible with funding from Northumbria University and Newcastle University.

References

Brackenbury T (1992), Knitted Clothing Technology, Oxford, Blackwell Scientific Publications.

Chamberlain J (1951), The Principles of Machine Knitting, Manchester, The Textile Institute.

Choi KF and Lo TY (2006), The Shape and Dimensions of Plain Knitted Fabric: A Fabric Mechanical Model, Textile Research Journal, 76(10), 777-786.

Choi W, Kim Y and Powell NB (2015), An investigation of seam strength and elongation of knitted neck edges on complete garments by binding-off processes, Journal of the Textile Institute, 106(3), 334-341.

Doyle PJ and Hurd JCH (1953), Fundamental Aspects of the Design of Knitted Fabrics, Journal of the Textile Institute Transactions, 44(8), 561-578.

Jones I and Stylios GK (eds) 2013, Joining Textiles Principles and Applications, Cambridge, Woodhead Publishing.

La Magna R, Fragkia V, Längst P, Lienhard J, Noël R, Baranovskaya Y S, Tamke M, Ramsgaard Thomsen M (2018) Isoropia: an Encompassing Approach for the Design, Analysis and Form-Finding of Bending-Active Textile Hybrids, Proceedings of the IASS Symposium 2018 Creativity in Structural Design July 16-20, 2018, Boston, MIT.

Leaf GAV and Glaskin A (1955), The Geometry of a Plain Knitted Loop, Journal of the Textile Institute Transactions,46(9), 587-605.

Liu D, Christe D, Shakibajahromi B, Knittel C, Castaneda N, Breen D, Dion G and Kontsos A (2017), On the Role of Material Architecture in the Mechanical Behavior of Knitted Textiles, International Journal of Solids and Structures, 109, 101-111.

Munden DL (1959), The Geometry and Dimensional Properties of Plain-Knit Fabrics, Journal of the Textile Institute Transactions, 50(7), 448-471.

Popescu M (2022), Knitnervi, Technoscape: The architecture of engineers, MAXXI National Museum of 21st Century Arts, Rome.

Royal Academy of Engineering (2021), Construction sector must move further and faster to curb carbon emmissions say engineers, Royal Academy of Engineers news. Available from https://raeng.org.uk/news/construction-sector-must-move-further-and-faster-to-curb-carbon-emissions-say-engineers [Accessed 2nd March 2023].

Scott J, Bridgens B, Ozkan D, Kaiser R, Agraviador A (2023), The Living Room, The Terry Farrell Centre, Newcastle upon Tyne.

Scott J, Gaston E & Agraviador A (2021), Configured Knitting: Grafting as an assembly process for knitted architecture. in V Stojaković & B Tepavčević (eds), Proceedings The 39th Conference on Education and Research in Computer Aided Architectural Design in Europe: eCAADe 2021 Towards a New, Configurable Architecture. Education and Research in Computer Aided Architectural Design in Europe (eCAADe), pp. 473-482, Novi Sad, eCAADe 2021.

Spencer DJ (1989), Knitting Technology (second edition) A Comprehensive Handbook and Practical guide to Modern Day Principles and Practices, Oxford, Pergamon Press.

Stroud Foster J and Greeno R (2017) Mitchell's structure & fabric Part 1 Seventh edition, London, Routledge.

Underwood J, (2009), The design of 3D shape knitted preforms, PhD Thesis, RMIT University, Melbourne, RMIT.

Figures

Figure 1 and 7 - photo Jane Scott.

Figure 2, 3, 5 and 6 photos Elizabeth Gaston.

Figure 8 and 9 images Armand Agraviador.

Figure 10 photo Elizabeth Gaston, image Armand Agraviador.

Microfibre Research

USING A LAUNDRY BOOSTER TO REDUCE THE ENVIRONMENTAL FOOTPRINT AND ENHANCE CLOTHING LONGEVITY

Helen Anne Hill[1], *Peter Slorach*[2], *Anna Marchiani*[3], *Dora Zamuner*[3] *and Susan Harris*[2]

[1]*Nottingham School of Arts & Design, Nottingham Trent University, Dryden St, Nottingham, NG1 4EY, U.K*
[2]*Anthesis Group, 355 Euston Road, London, NW1 3AL, U.K*
[3]*Reckitt Italy S.p.A, R&D Center, Piazza San Nicolò, 12/3 30034 Mira (VE), Italy*

Abstract

This research investigates the role laundry boosters play in extending the life of clothing and the impact this has on the environmental footprint.

Clothing has a significant lifecycle footprint, representing over five per cent global carbon emissions (WEF, 2021). The use phase represents up to 30% of lifecycle impact for garments (mainly in the laundering process (WRAP, 2017a)), with global studies showing only 12% of clothing is reused or recycled (EMF, 2017). Extending the active life of clothes during the use phase represents a key intervention for reducing impacts (BFC, 2021).

Research indicates laundering practices can influence the key reasons consumers dispose of clothes including discolouration of whites and stains.

Using methodology based on extensive wash testing and a consumer survey, results were reviewed, and comparisons were made on the impact of using detergent alone or detergent with a laundry booster. The results evidenced that using a laundry booster can enhance the longevity of garments through stain removal and by offering '360-degree protection'. Reduced environmental impacts are achieved by laundry boosters which provide superior stain and odour removal at a low temperature and gentle cycle, protecting the physical properties of the garment.

Keywords

Clothing, Longevity, Laundering, Circularity, Innovation, Consumer Education.

Introduction

The aim of this research was to investigate the role laundry boosters can play in extending the life of clothes, with a view to assessing the impact that adding a booster to the conventional laundry process can have on clothing longevity and the environmental footprint.

A laundry booster is a product that is added to the wash, with detergent, to help make the cleaning process more effective. By adding a laundry booster, it is possible to achieve the same or better results at a lower wash temperature and milder cycle. This results in a double benefit: superior stain and malodour removal from the action of the booster, alongside enhanced colour maintenance and fabric integrity and reduced pilling from the lower temperature and cycle (lower mechanical stress). Consequently, clothes stay in good condition for longer, which can extend their active life. This has many environmental benefits, including reduced energy use in the laundering process, extending garment lifetime, and reducing textile waste.

This is a new area of research and quantification, so a transparent methodology was developed, based on existing protocols and research. Extensive wash and testing trials were conducted based on the developed Method Statement, with the results reported in this paper. Results of the comparative lifecycle assessments conducted are reported in a separate paper anticipated to be published in 2023.

Fashion's Environmental Footprint and the Laundering Process

Clothing has a significant lifecycle footprint, representing over five per cent of global carbon emissions (WEF, 2020). The use phase represents up to 30% of lifecycle impact for garments, and most of that is in the laundering process (WRAP, 2017a). Additionally, textile waste is very high, with some global studies showing that only 12% of clothing is reused or recycled (EMF, 2017). Adopting a circular economy for clothing by extending the active life of clothes, and use/re-use represents a key intervention for reducing the environmental impacts of clothing (British Fashion Council, 2021).

Despite an observed decline in laundry temperatures between 1998 and 2001, and an anticipated further reduction (Dewaele et al., 2006), the average European washing temperature is still around 40 degrees Celsius (°C) (International Association for Soaps, Detergents and Maintenance Products (A.I.S.E.), 2020). This is despite campaigns promoted by detergent producers and industry associations advising consumer to wash at 30°C (e.g., A.I.S.E. 2015; Procter & Gamble, 2022). Behaviour change is difficult to achieve especially when the immediate benefit to the consumer is not clear (Unilever, 2021). Nearly two thirds (63%) of the European population still tend to wash above 30°C, with one of the main reasons preventing European consumers from washing at lower temperatures being a lack of confidence in cleaning efficiency (Electrolux Group, 2021). Furthermore, 47% of the people surveyed questioned the cleaning performance, 39% the stain removal, 29% the germ removal and 25% odour removal (Electrolux Group, 2021). This highlights the struggle consumers face – garments washed at lower temperatures retain stains and odours, whilst those washed at higher temperatures may be subject to pilling and clothes may become damaged (University of Leeds, 2020).

While detergent manufacturers suggest lowering temperatures, and communicate this message on-pack, there is still conflicting messaging around using a double dose of detergent, a higher temperature for whites or the use of different products in pre-treat. Lower temperatures can help to create a more environmentally friendly laundering process, but the products used still need to remove stains and odour. Laundry boosters can help achieve this result by being gentler on clothes, but tougher on stains creating a more efficient laundry process and helping to change consumer behaviour around laundering.

What is a laundry booster?

Laundry boosters are fabric care products designed to reinforce specific performance characteristics desired during laundering, for example, stain and malodour removal, colour and textile care, and whitening. They are added to the wash on top of regular laundry detergents and can be in the form of powder, liquid, or gel, or monodose (e.g., tabs).

This study was conducted with the laundry booster brand Vanish (Reckitt) and assessed the impact of two Vanish products available on the UK market (Pink Powder and White Powder), which both use a new catalyst made available on the market in early 2023 (Vanish, 2023).

Laundry boosters are intended to provide enhanced cleaning and fabric care of apparel in the washing process versus conventional wash patterns. However, the longevity benefits of using a laundry booster and the environmental impact reduction achieved have not previously been studied to this level of detail.

Temperature and Cycle

Studies (e.g., Cotton et al., 2020) that evaluate the impacts of washing temperature and washing time on garment colour loss, colour transfer and microfibre release show that washing at a lower temperature and with less mechanical agitation can significantly extend garment longevity and reduce environmental impact (Cotton et al., 2020). However, despite many consumer-facing campaigns such as the EU-wide A.I.S.E. '*I prefer 30*' campaign (A.I.S.E., 2015), studies show that as of 2021 over 58% of European households still wash at 40°C or above (Electrolux, 2021).

The Electrolux 'The Truth About Laundry' study shows that the primary reasons consumers do not wash at lower temperatures is due to concerns about whether garments will be clean, and specifically odour, stain, and germ free (Electrolux, 2021).

The '360-degree protection' concept

This research demonstrates that using a laundry booster can reduce environmental impacts and enhance the longevity of garments by offering '360-degree protection' – i.e., reducing stains and odour while retaining physical properties. The 360-degree protection concept (Figure 1) was developed by the authors of the paper and the wider Vanish R&D team.

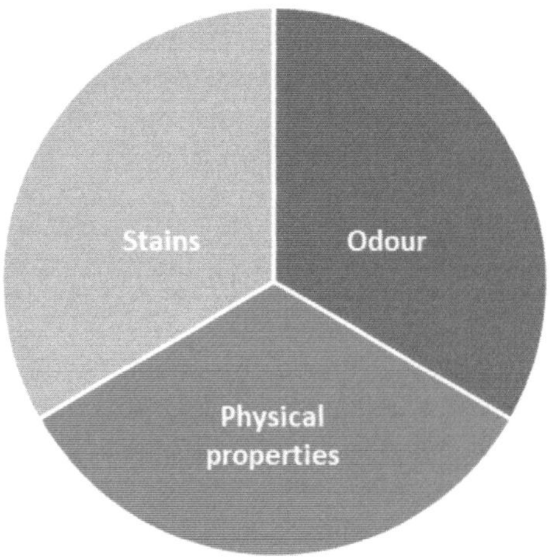

Figure 1 The Vanish concept of '360-degree protection' developed by the authors and Vanish R&D team. Helping clothes live longer lives.

This effect is achieved via a two-part action: not only does the formula provide superior stain and odour removal, but it can do so at a lower temperature and gentler cycle, which also protects the physical properties of the garment. This gives the same or better result at a lower wash temperature and lower mechanical agitation compared to washing with detergent alone at the most used EU wash temperature – that is, 40°C (Electrolux, 2021). Helping clothes live longer lives.

Methodology

No formal, publicly available, comprehensive standard for longevity in the apparel care and laundering sector exists (Cooper and Claxton, 2022). However, the Waste and Resources Action Programme (WRAP), a British charity working with consumers, businesses and communities to achieve a circular economy has developed widely-accepted guidance on physical durability of clothing and longevity in the apparel and textiles sector. The WRAP Sustainable Clothing Guide (WRAP, 2017b) identifies modes of failure that can be influenced during laundering. The WRAP Longevity Protocol (WRAP, 2014 and associated guidance) can be used to identify suitable testing to demonstrate improvement against these modes of failure (i.e., the longevity factors tested for).

The research team developed a methodology for assessing longevity, using existing garment longevity standards previously developed by Nottingham Trent University and WRAP as a basis for developing a method to assess the contribution of laundering products to garment longevity. A Method Statement (Figure 2) was created to ensure the process was as transparent and replicable as possible.

Test products against the threshold	Identify relevant longevity factors	Identify suitable testing modes	Identify suitable threshold	Test products against the threshold
Which modes of failure do Vanish's products seek to address?	What are the relevant aspects of longevity?	How can Vanish demonstrate enhanced performance on the longevity aspects?	From testing conducted so far, what would a reasonable "n x longer" threshold to test against?	Do products demonstrate threshold or above performance on each longevity parameter?
Using the WRAP Sustainable Clothing Guide (2017b) *Workshop with Vanish technical* *Consultation with academic partners U. of Leeds*	*Using Sustainable Clothing Guide (2017b) and industry testing norms* *Review by NTU academic*	*Using the WRAP Longevity Protocol and guidance, and industry testing norms* *Review by NTU academic*	*Using existing and early test results "2x longer" was selected*	*Via testing using the relevant industry tests conducted by Vanish selected test centres* *Review by NTU academic*

Figure 2 Method Statement.

Physical Longevity Factors

The focus of the research was physical longevity as affected by the washing process. Garment longevity is influenced by a range of physical and emotional durability factors (e.g., socio-economic cultural norms – see Chapman (2017)), during both washing and use, as well as by factors pre-determined by product design and production. Therefore, only the elements of physical durability influenced by laundering – washing, specifically – were considered the relevant factors, as these are the ones that can be directly influenced by the laundering process, and for which impact can be quantified. The WRAP Sustainable Clothing Guide (WRAP, 2017b) identifies common modes of failure regarding longevity (or how durable a garment is in practice).

Using this guidance as a starting point, the relevant longevity factors for this study were identified as:

- Stain removal.
- Colour fastness.
- Fibre integrity.
- Pilling.
- Odour removal.
- Whiteness.
- Tensile and tear strength.

Pilling or fuzzing, is the formation of balls of fabric fibres that have detached from the outer surface of the fabric and that tend to tangle forming lint / balls that are unsightly, and ruin the surface of the garment, reducing its shine.

These align with factors commonly identified with clothing longevity (WRAP, 2014); factors that can be impacted by the washing process (Cotton et al., 2021); and factors commonly identified as contributing to why consumers dispose of clothing (Cooper and Claxton, 2022).

Consumer survey

Data on why consumers dispose of clothing and how they launder clothes is limited (Cooper and Claxton, 2022). The most widely cited data is from a UK study conducted by Langley et al. for WRAP (2013), which is considered a seminal study methodologically in the garment durability field. To obtain more up-to-date UK data, the Reckitt Vanish team commissioned an Ipsos MORI consumer survey as part of this piece of research (Vanish, 2021).

The survey sought to replicate the Langley et al. (2013) clothing lifetime study. To do so, the relevant questions from the Ipsos Mori poll conducted for WRAP in 2013 were re-applied, along with additional supplementary questions where necessary. Questions were reviewed by Reckitt's Consumer Insights team.

Survey details are outlined in Figure 3:

Market: *UK*

Methodology: *Online omnibus survey (this is a survey which reaches a nationally representative population in the geographical market)*

Sample composition: *Nationally representative of the population (aged 16-75 in the UK)*

Sample size: *n=2000*

Questions: *Standard demographics (age, sex, social class, region, etc.); bespoke closed questions; and no open ended questions.*

Figure 3. Methodology overview from Ipsos Mori Poll, conducted by Reckitt's Consumer Insights team.

Performance testing

Physical tests were undertaken to assess whether washing with laundry boosters at 30°C or less could extend the lifespan of garments based on stain removal, colour fastness, fibre integrity, odour removal, whiteness and tensile and tear strength by two times compared to washing at 40°C with detergent alone. The laundry booster products tested were: Vanish Pink Oxi Powder and Vanish White Oxi Powder with a unique catalyst technology, available on the UK market since early 2023.

Appropriate performance tests for each longevity factor were identified, using the WRAP Longevity Protocol (2014) where the Protocol specified these, and standard industry tests for any factors where the protocol did not specify a testing method. It should be noted that the test methods used are established, well-recognised standard tests.

Table 1 shows modes of failure, related longevity factors, and the testing methods used. Tests are aligned with International Association for Soaps, Detergents and Maintenance Products (A.I.S.E.) test protocols and relevant International Organisation for Standardisation (ISO) or American Society for Testing and Materials International (ASTM) standards.

Table 1 Performance tests conducted against modes of failure

Mode of Failure	Longevity Factor	Testing Methods
Staining	Stain removal	Industry-standard to domestic laundering comparative test A.I.S.E. protocol for Stain Removal v.7 2020
Colour loss	Colour fastness	Industry-standard colour fastness to domestic laundering comparative test (as per Longevity Protocol; ISO 105 A – dE*)
Pilling and bobbling	Fibre integrity	Industry-standard pilling domestic laundering comparative test UNI EN ISO 12495
Malodour	Odour removal	Malodour removal domestic laundering comparative test developed by Reckitt, LabTest and Neotron
Discolouration of whites	Whiteness	Industry-standard whiteness domestic laundering comparative test. ISO Standard 4312: 1989 (E)
Loss of fabric integrity	Tensile and tear strength	Industry-standard bursting strength domestic laundering comparative ASTM D6797

Performance testing was performed by Reckitt Shared Capabilities laboratories at R&D Centre, Italy (for stain removal and whiteness tests) and an independent laboratory RITEX, ECOCHEM division, Vicenza Italy (for other colour fastness, bursting strength and pilling tests). Odour removal assessment were conducted by Labtest (odour preparation and washing), and NEOTRON Spa Modena, Italy (analytical measurement). An appropriate threshold improvement level was determined (that is, x times longer lasting) that could be tested for in the process. Initial tests established that a 2x threshold improvement level, or twice as long was a suitable threshold. The remaining testing was conducted as normal, and results were analysed to establish whether they met the 2x longevity threshold.

Results

Consumer survey

As shown in Table 2, the consumer survey clearly indicates that over 50% of the reasons those surveyed dispose of their clothing is directly correlated to the longevity and cleaning performance factors that laundry boosters can influence. Stain removal, colour-fastness, discolouration (of whites), pilling and fuzzing, fabric integrity, odour and shrinkage can all be impacted by a laundry booster.

Table 2 Consumer survey conducted by Vanish showing the reasons why consumers dispose of their clothes (multiple garment types)

Disposal Reason	%	
Stain removal	4%	
Colour fastness	7%	
Discolouration	5%	
Pilling & fuzzing	7%	Influenced by laundering practices - 52%
Fabric integrity	20%	
Odour	2%	
Shrinking	7%	
Reduce clutter	14%	
No longer wearing	16%	
No longer suitable	9%	
Didn't like anymore	8%	
None of the options	1%	
Bought new clothes	0%	
Broken/ damaged/ falling apart	0%	
Moth eaten/ chewed	0%	
Donated	0%	Not influenced by laundering practices
Sold	0%	
Old	0%	
Hygiene	0%	
Lost the pair	0%	
Other reasons (Please specify)	0%	
No answer	0%	
Don't know	0%	

Shrinkage was subsequently excluded from the longevity factors tested, because externalities beyond the washing process (such as garment type variation and human factors) made it difficult to isolate the longevity benefits related specifically to the washing process for this factor. Fabric integrity was excluded as a longevity factor for environmental assessments as it is significantly impacted by wear and tear during use.

It is noted that sentiment surveys cannot always be assumed to accurately reflect the reality of disposal patterns in practice. However, other primary research also supports these results. A recent study by Cooper and Claxton (2022), which analysed types of garment failure observed by physical assessment of discarded items, shows a similar pattern. Pilling, colour fading, fabric breakdown, discolouration and loss of dimensional stability are all recorded as key garment failures, and pilling and colour fading are identified as the most common (Cooper and Claxon, 2022).

Performance testing

The performance testing (Table 3) showed that, in all cases, powder laundry booster products in cold and delicate wash cycle contributed to a 2x longevity improvement compared to washing with liquid detergent alone at conventional wash pattern, 40°C cotton cycle.

Table 3 Performance test results summary

Mode of Failure	Longevity Factor	Tests	Result
Staining	Stain removal	A.I.S.E. protocol; Y value instrumental evaluation. Laundry booster pink powder plus detergent at 20°C temperature and lower mechanical agitation compared to liquid detergent alone at **40**°C on standard cotton cycle.	Laundry booster pink powder demonstrated superior stain removal at lower temperature and lower mechanical agitation than liquid detergent in conventional wash pattern.
Colour loss	Colour fastness	Instrumental assessment of change in colour (dE* scale) after ten and 20 washes. Laundry booster pink powder plus detergent at lower temperature and lower mechanical agitation compared to liquid detergent alone at **40**°C on standard cotton cycle.	Laundry booster pink powder protects colour to a commensurate level for at least twice as long as washing with liquid detergent alone.
Pilling and fuzzing	Fibre integrity	Pilling and fuzzing evaluation after ten washes. Laundry booster powder plus detergent at lower temperature and lower mechanical agitation compared to detergent alone at **40**°C in standard cotton cycle.	Laundry booster pink powder demonstrated reduced pilling shown over ten washes compared to five washes with liquid detergent alone at **40**°C.
Malodour	Odour removal	Sweat and foot synthetic malodours. Laundry booster pink powder plus liquid detergent versus liquid detergent alone, all products tested at 30°C cotton cycle.	Laundry booster pink powder demonstrated superior odour removal compared to liquid detergent alone.
Discolouration (of whites)	Whiteness	Whiteness evaluation. Greying is measured spectrophotometrically through Y-value. Whiteness impacted by optical brightener is assessed via the GANZ. Laundry booster white powder with catalyst and liquid detergent versus liquid detergent alone.	Laundry booster white powder demonstrated superior whitening results at 50 washes compared to 25 washes with detergent alone.
Loss of fabric integrity	Tensile and Tear Strength	Test methods for bursting strength of fabrics. Laundry pink booster powder plus liquid detergent at lower temperature and lower mechanical agitation compared to liquid detergent alone at **40**°C on standard cotton cycle.	Laundry booster pink powder demonstrated parity or better mechanical resistance results over 20 washes compared to ten washes with detergent alone.

Staining: Stain Removal

For stain removal, Y values of A.I.S.E. tests results indicate that a laundry powder booster used with liquid detergent can give significantly greater stain removal at 20°C in a delicate cycle than liquid detergent alone at 40°C in a cotton cycle (Figure 4).

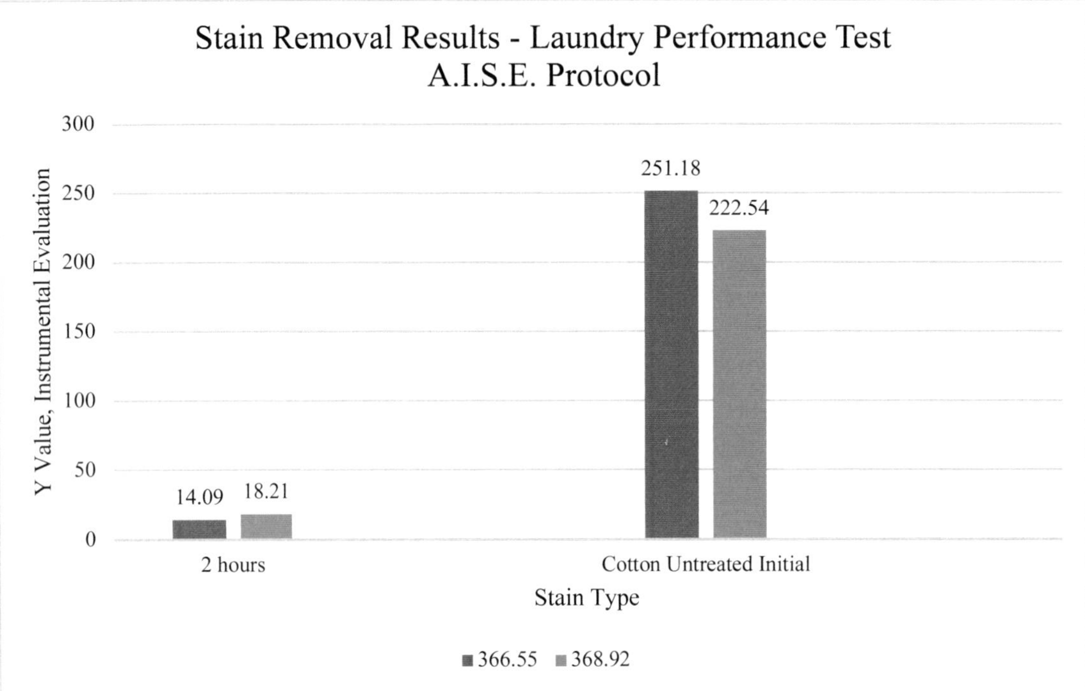

Figure 4 Stain removal performance results expressed as Y-value; means of eight measurements for each stain type. Outdoor stains: grass, grass and mud, unused motor oil. Beverage stains: tea, coffee, red wine, blueberry juice, chocolate. Food: tomato puree, balsamic dressing, French squeeze mustard, cooked beef sauce, soy sauce. Body soil: make up, blood. Instrumental reflectance measurements Y-value was taken using a Datacolor 650 Spectrophotometer, and four external replicates per stain were evaluated. Two measurements per stain were taken. The higher the Y value, the better the stain removal performance. A difference of two Y-value units is perceivable by the human eye. A liquid detergent representative of the UK market and laundry booster pink powder with catalyst, launched in the UK market in 2023 were tested. Products were dosed as recommended by manufacturer label instructions. The test was conducted in a European front load washing machine. Reckitt R&D Mira (Italy) Performance Test Report ShSe 117- 2020.

Colour Fastness: Colour Loss

Colour-fastness tests demonstrated that the addition of a laundry booster can give superior colour maintenance at a lower temperature and milder wash cycle (Figure 5). Independent laboratory test according to textile industry standard where conducted washing ten and 20 cumulative times a series of coloured fabrics representative of European market apparel. Results shows that thermal and mechanical stress applied on textiles during the laundering process have a large impact on colour maintenance causing higher fading versus the original unwashed fabric. While the addition of laundry boosters in a cold and delicate wash delivers better colour protection even after repeated washes, a superior level of stain and odour removal is achieved with a laundry booster under those mild washing conditions.

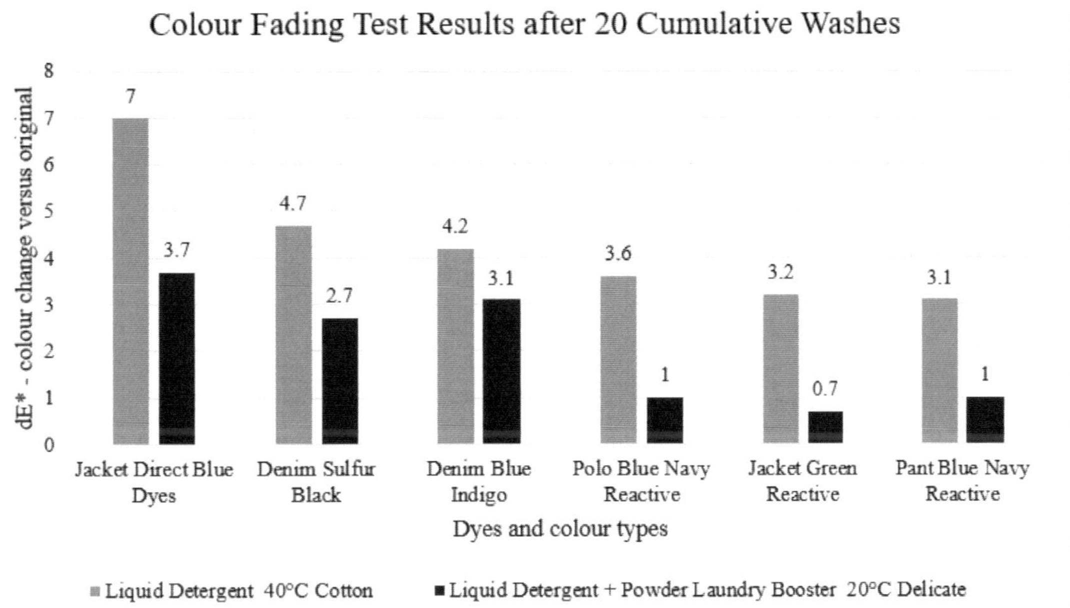

*Figure 5 Colour-fastness performance test conducted at an independent textile laboratory. Average of four measurements for each colour type. The measurement scale delta E dE * express the difference in colour of the washed textile versus the original unwashed fabric. The dE* value is determined instrumentally using a spectrophotometer using the formula CIElab. The higher the dE* correspond to a higher impact on colour fading and colour variation vs the original. When dE* is lower than 1.0 the difference between two colours is not perceivable by the human eye. A leading liquid detergent from the UK market and Vanish Pink powder with bleach catalyst were tested. Products dosed according to manufacturer label instructions. Test conducted by RITEX, ECOCHEM (Vicenza, Italy). Report number 20CR00137_138.*

All the values found after 20 washing cycle show an acceptable level of colour degradation. When comparing colour fading results for the ten washes with liquid detergent in 40°C cotton cycle (conventional European wash pattern) to cumulative washes with laundry powder booster, overall better colour maintenance is achieved in cold mild cycle with powder booster despite the double number of washes. Comparable results after ten washes can be found in Appendix 1.

Fibre Integrity: Pilling and bobbling

For pilling, testing in accordance with standardised textile industry methodology demonstrated that washing at a lower temperature and milder cycle when a laundry booster is added can significantly reduce pilling, whilst still delivering the same level of stain and odour removal.

In this test, conducted at an independent textile laboratory, a 100% cotton knitted fabric with pigment and a 50% cotton - 50% polyester knitted fabric both with pigment (red, blue, black, green), were subjected to five and ten washes in a domestic laundry machine. The effects of washing at 40°C with high mechanical action (cotton cycle) using a liquid detergent alone were compared to an alternative wash at 20°C with low mechanical action (delicate cycle) by washing with liquid detergent plus laundry pink powder booster, both products available on the UK market.

The visual evaluation highlighted in Figure 6 show that there was significant pilling formation in high-mechanically action washing at 40°C with liquid detergent only, while cold-washed fabric with a laundry powder booster underwent only slight pilling formation. Washing with a laundry booster in gentle cold cycles helps to better maintain the physical integrity of the fabric, extending its durability over time.

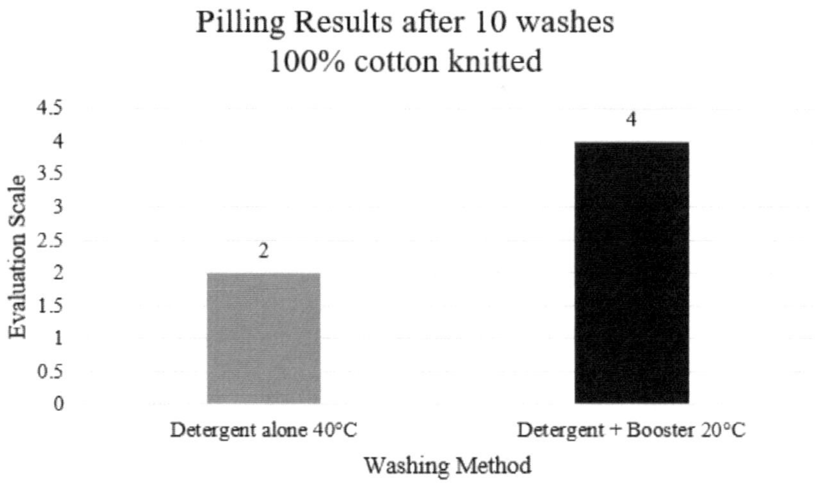

Figure 6 Pilling test results, visual evaluation. Average of two panellists. Evaluation scale: 5 – no pilling; 4 – slight surface pilling; 3 – moderate pilling; 2 – significant change, distinct pilling; 1 – severe change, dense pilling.

Tested fabrics were evaluated after five and ten cycles in a European front-loading domestic laundry machine (Table 4). Comparing the pilling and fizzing results after five washes with liquid detergent alone to ten washes with powder booster, the pilling formation is much lower when textiles are washed with laundry booster in cold mild delicate cycles. Even doubling the number of washes with a laundry booster delivers lower pilling, therefore better garment physical durability is achieved.

Table 4 Pilling and performance test results after five and ten cumulative washes. The higher the scale, the lower the pilling, corresponding to lower deterioration versus the original textile (unwashed), thus better garment physical durability. Test conducted by RITEX, ECOCHEM (Vicenza, Italy). Report number 20CR00137_138

Pilling visual evaluation	Liquid Detergent 40°C Cotton	Liquid Detergent + Powder Laundry Booster 20°C Delicate
100% Cotton Knitted (Sweater) – 5 washes	2	4/5
100% Cotton Knitted (Sweater) – 10 washes	2	4
50% Cotton 50% Polyester Knitted (Sweater) – 5 washes	2	4
50% Cotton 50% Polyester Knitted (Sweater) – 10 washes	2	3/5

Malodour: odour removal

Malodour removal test result show that washing with powder laundry booster on top of liquid detergent removes foot (Figure 7) and body (Figure 8) odour better than washing with detergent alone. The addition of a laundry booster protects the textile from malodour better than conventional wash pattern.

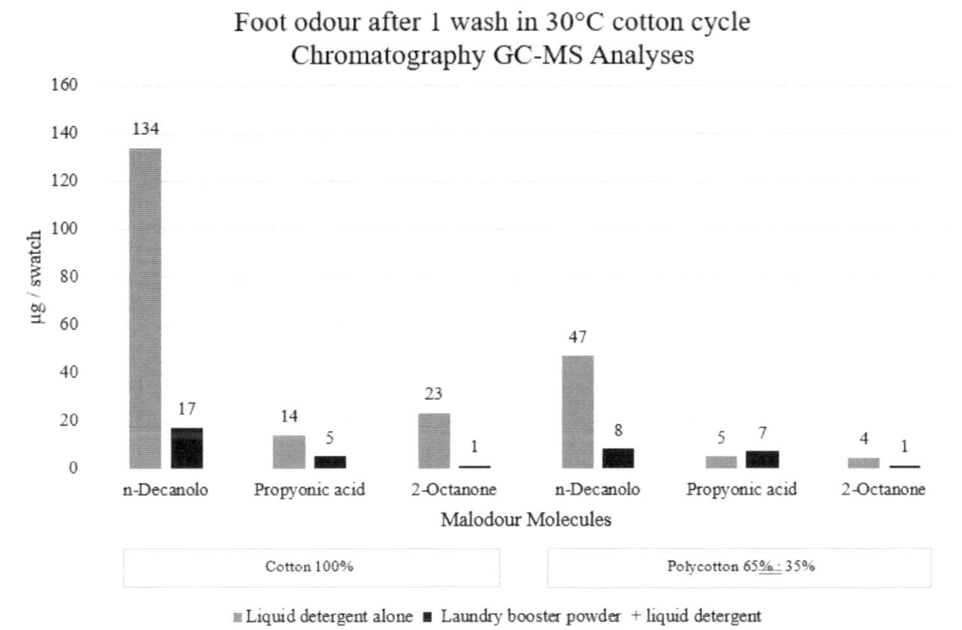

Figure 7
Amount of molecules present in foot synthetic malodour found on textiles after washing. Malodour molecule measured on cotton and polycotton fabrics using a chromatography GC-MS method. The higher the number, the higher the amount of odoriferous molecule left in swatches after wash.

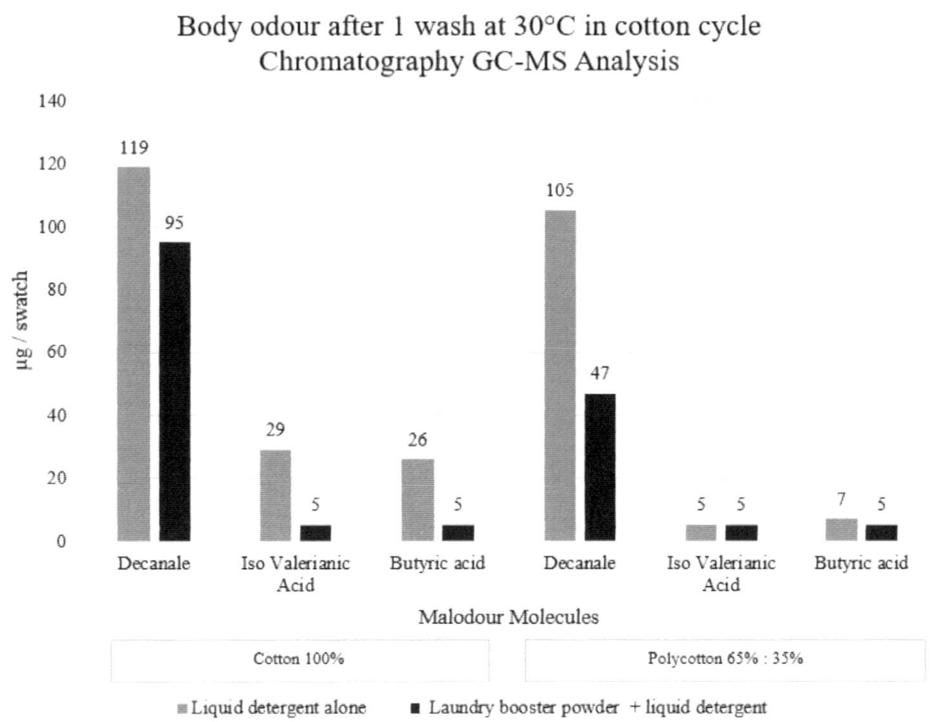

Figure 8 Amount of molecule present in body synthetic odour found on textiles after washing. Malodour molecule measured on cotton and polycotton fabrics using a chromatography GC-MS method. The higher the number, the higher the amount of odoriferous molecule left in swatches after wash.

Swatches were applied with synthetic odour mixtures to simulate foot and body (sweat) malodour; the odour mixtures were prepared by Lab-test (Venice, Italy). Three drops of foot odour, and six drops of body odour were applied to 15 cm by 15 cm cotton 100 % and polycotton 65% : 35% swatches. Swatches were then washed in European front load washing machine, using a 30°C cotton cycle. Liquid laundry detergent and powder laundry booster were dosed according to the manufacturer's recommended dosage on labels. Analytical measurement via chromatography GC-MS were performed at NEOTRON Modena (Italy). Test protocols details are reported in Reckitt R&D Mira report FA-2021-0010.

Discoloration of white: whiteness

Evaluation of the impact on whiteness maintenance using laundry booster white powder formula with catalyst and launched on the UK market in early 2023 was assessed. A comparison was made when laundry booster is used in-wash at 20°C on top of UK liquid detergent after 50 multiple washes versus the use of liquid detergent alone at 20°C after 25 multiple washes.

The whiteness degree was measured taking into consideration the whiteness improvement given by the deposition of optical brightener and maintenance of basic whiteness given by the overall formula composition. Deposition of optical brightener was measured through GANZ value, while basic whiteness was measured through Y-value.

Greying was determined by measuring the basic degree of whiteness (Y-value) after 25 and 50 washes (liquid detergent and laundry booster white powder) versus 25 washes of detergent alone. Y-values of detergent alone after 50 washes are also reported to prove superior results of laundry booster at the same number of washes. Measures of each swatch was carried out using a Datacolor 650 Spectrophotometer with UV-Filter at 420 nm to exclude the influence of optical brighteners. Mean and standard deviation of four measurements (four measurements each swatch, one internal replication, one external replications) were calculated: the higher the value, the higher the whiteness degree.

In addition, whiteness degree was measured with a Datacolor 650 Spectrophotometer including UV-light to register the influence of optical brighteners (GANZ value) after 25 and 50 washes. For complete information, GANZ values of detergent alone after 50 washes are also reported to prove superior results of laundry booster white powder at the same number of washes. Mean and standard deviation of four measurements (four measurements each swatch, one internal replication, one external replication) were calculated: the higher the value, the higher the whiteness degree.

The washing trials were carried out by Reckitt's R&D lab in Mira, Italy (Report Shse-perf-2022-036). Whiteness maintenance of basic whiteness results are reported in Table 5.

Table 5 Mean of Y-value (four measurements) after 25 and 50 washes for liquid detergent alone and after 50 washes for liquid detergent plus laundry booster white powder

Fabric Code	Liquid detergent 25 washes	Liquid detergent 50 washes	Laundry booster white powder + liquid detergent 50 washes
WFK 10A 100% cotton	86	81	90
WFK 20A 65% polyester, 35% cotton	88	84	91
Terry Cloth 100% cotton	85	80	89
Empa 221 100% cotton	86	80	92
Honeycomb, 100% cotton	82	78	88

Note: A difference >2 unit in the Y-value scale is visible to the human eye.

Whiteness degree Ganz results given by the deposition of optical brightener are reported in Table 6.

Table 6 Mean of Ganz value (four measurements) after 25 and 50 cycles for liquid detergent alone and after 50 cycles for laundry booster white powder plus liquid detergent

Fabric Code	Liquid detergent 25 washes	Liquid detergent 50 washes	Laundry booster white powder + liquid detergent 50 washes
WFK 10A 100% cotton	171	161	221
WFK 20A 65% polyester, 35% cotton	145	140	188
Terry Cloth 100% cotton	219	206	238
Empa 221, 100% cotton	178	162	231
Honeycomb 100% cotton	202	189	232

Note: A difference of GANZ > 20 unit is visible to the human eyes.

Based on the results reported in Tables 5 and 6, it can be concluded that the use of laundry booster white powder gives superior results on whiteness maintenance (measured both by Y-value and by GANZ) on the tested fabrics after 50 cumulative washes versus the 25 multiple washes with detergent alone.

Loss of fabric integrity: tensile and tear strength

Mechanical stress tests were conducted via bursting strength measurement on a 100% cotton knitted blue navy textile used for the production of casual apparel (e.g., polo T-shirt). This fabric was subject to ten and 20 cumulative washes in a domestic laundry machine in liquid detergent alone at a 40°C cotton cycle and compared to washing in laundry powder booster plus liquid detergent in a cold 20°C delicate cycle. The instrumental results shows that bursting strength of fabric after 20 cumulative washes with powder laundry booster in mild conditions is parity to ten washes with liquid detergent in conventional wash pattern commonly used by European conditions.

Table 7 Measurement of the bursting strength in N= Newton. Average of five measurements.

Bursting Strength of Fabrics, Newton N	Liquid detergent 40°C Cotton	Liquid detergent + powder laundry booster 20°C Delicate
100% cotton knitted, blue navy 10 washes	260	270
100% cotton knitted, blue navy 20 washes	240	260

The mechanical resistance properties such as bursting are to be considered acceptable for all tested products. Laundry booster plus detergent results at 20 cycles are parity to ten cycles with liquid detergent alone. Test conducted by RITEX, ECOCHEM (Vicenza, Italy). Report number 20CR00137_138.

Environmental impacts

The Life Cycle Assessment (LCA) analysis carried out as part of the wider project considered the carbon and water impacts of using a laundry booster across different UK washing cycles compared to a leading detergent only wash at 40°C. The goal was to investigate the lifecycle environmental impacts for a cotton T-shirt and to compare the impacts for a standard lifespan to an extended lifespan achieved using a laundry booster (Vanish pink powder). The per wear lifecycle carbon footprint impact of a standard T-shirt worn in the UK and washed at 40°C with liquid detergent only is estimated at 0.13 kg CO_2e. Considering a 30% extension in lifespan and washing at a lower intensity and cold temperature with Vanish, in line with consumer survey findings (Ipsos), there are carbon footprint and water savings of up to 23% per wear.

Full results quantifying the life cycle benefits and displacement of impacts will be presented in a separate paper later in 2023.

Conclusions

This research shows that by using a laundry booster stains and odour can be removed, whilst also enhancing fibre care, at a lower temperature and gentler cycle. The active life of clothing can be extended 2x in comparison to washing with detergent alone at the temperature and cycle most regularly used by European consumers (40°C). Adding a laundry booster can overcome common barriers to washing at lower temperature and cycle by ensuring a higher level of stain and odour removal, cleaning even in milder washing conditions.

In addition, results have shown how the use of the laundry booster in these cycles allows a better maintenance of the colour and a better physical durability of the garment.

Helping clothes live longer lives. Improving the physical durability of the fabric and promoting garment reuse, extending the life of the fabric, can result in a smaller amount of clothing thrown to waste into landfills or incinerated.

The use of laundry boosters could be the accelerator for gentle and cold washing, they can promote a more sustainable washing process and more responsible and sustainable use of clothing.

This can bring significant environmental benefits by reducing energy use in wash and extending the active lifetime of clothes. The work carried out also quantifies the life cycle benefits and displacement of impact that can be delivered and will be presented in a separate paper later in 2023.

Acknowledgments

WRAP, RITEX, ECOCHEM, Reckitt Consumer Insights Team [CMI], Susan Harris

Appendix 1

Table 7 Colour Fastness: Colour Loss results – comparable results after ten and 20 washes

dE* Instrumental Evaluation	Liquid detergent 40°C Cotton	Liquid detergent + powder laundry booster 20°C Delicate
Denim Sulphur Black Woven 98% cotton 2% elastan 10 washes	3.0	1.1
Denim Sulphur Black Woven 98% cotton 2% elastan 20 washes	4.7	2.7
Denim Blue Indigo Woven 98% cotton 2% elastan 10 washes	2.3	1.4
Denim Blue Indigo Woven 98% cotton 2% elastan 20 washes	4.2	3.1
Polo Blue Navy Knitted 100% cotton 10 washes (reactive dye)	2.5	0.4
Polo Blue Navy Knitted 100% cotton 20 washes (reactive dye)	3.6	1.0
Jacket Direct Blue Dyes Woven 100% cotton 10 washes	4.8	2.0
Jacket Direct Blue Dyes Woven 100% cotton 20 washes	7.0	3.7
Jacket Green Reactive Woven 100% cotton 10 washes	2.4	0.6
Jacket Green Reactive Woven 100% cotton 20 washes	3.2	0.7
Pant Blue Navy Woven 98% cotton 2% elastan 10 washes (reactive dye)	1.9	0.4
Pant Blue Navy Woven 98% cotton 2% elastan 20 washes (reactive dye)	3.1	1.0

References

A.I.S.E. (2015). I Prefer 30° campaign: Close out report October 2015. Available from: https://www.aise.eu/documents/document/20151102163916-ip30-report-2015_full_version_no_annexes.pdf

A.I.S.E. (2020). Pan-European Habits Survey 2020. A.I.S.E. Available from: https://www.aise.eu/documents/document/20200917130851-aise_consumerhabitssurvey_2020_highlights.pdf

A.I.S.E. (2023). I Prefer 30° campaign. A.I.S.E. Available from: https://www.aise.eu/our-activities/sustainable-cleaning-78/engaging-with-consumers/low-temperature-washing-campaign.aspx

British Fashion Council Institute of Positive Fashion (2021). The circular fashion ecosystems: a blueprint for the future. P. 39. Available from: https://instituteofpositivefashion.com/uploads/files/1/CFE/Circular_Fashion_Ecosytem_Report.pdf.

Chapman, J. (2017). Handbook of Sustainable Product Design, Routledge.

Cooper, T. and Claxton, S. (2022). Garment failure causes and solutions: Slowing the cycles for circular fashion. Journal of Cleaner Production. Volume 351, 1 June 2022, 131394

Cotton, L., Hayward, A. S, Lant, N.J, and Blackburn, R.S. (2020). Improved garment longevity and reduced microfibre release are important sustainability benefits of laundering in colder and quicker washing machine cycles. Dyes and Pigments. Volume 177, June 2020, 108120. Available from: https://doi.org/10.1016/j.dyepig.2019.108120

Dewaele, J., Pant, R., Schowanek, D. and Salducci, N. (2006). Comparative Life Cycle Assessment (LCA) of Ariel "Actif à froid" (2006), a laundry detergent that allows to wash at colder wash temperatures, with previous Ariel laundry detergents (1998, 2001). Procter & Gamble, Brussels Innovation Center, Central Product Safety-Environmental, Brussels. pp. 1-57.

Electrolux (2021). The Truth About Laundry. Electrolux. Available from: https://admin.betterlivingprogram.com/wp-content/uploads/2021/02/Electrolux_TheTruthAboutLaundry_WhitePaper-1.pdf

Ellen Macarthur Foundation (2017). A New Textiles Economy: Redesigning fashion's future. Ellen Macarthur Foundation. P. 20. Available from: https://emf.thirdlight.com/file/24/uiwtaHvud8YIG_uiSTauTlJH74/A%20New%20Textiles%20Economy%3A%20Redesigning%20fashion%E2%80%99s%20future.pdf

Langley, E., Durkacz, S., Tanase, S. (2013). Clothing Longevity and Active Use. Unpublished manuscript. WRAP, Banbury.

Procter & Gamble. (2022). Leading detergent brands to European consumers: Wash cold.

Procter & Gamble. Available from: https://www.politico.eu/sponsored-content/leading-detergent-brands-to-european-consumers-wash-cold/

Unilever. (2021). Unilever Sustainable Living Plan 2010 to 2020. Unilever. Available from: https://www.unilever.com/files/92ui5egz/production/16cb778e4d31b81509dc5937001559f1f5c863ab.pdf

University of Leeds. (2020). Quicker and cooler is best for clothes. University of Leeds. Available from: https://www.leeds.ac.uk/news-science/news/article/4524/quicker-and-cooler-is-best-for-clothes

Vanish (2021). Ipsos Mori Consumer research – Clothing Longevity.

Vanish (2023) Oxi Action Multi Purpose Powder. Available from: https://www.vanish.co.uk/products/laundry-boosters/oxi-action-multi-power-powder/

World Economic Forum (2021). Net Zero Challenge: the supply chain opportunity. p.12-13. Available from: https://www3.weforum.org/docs/WEF_Net_Zero_Challenge_The_Supply_Chain_Opportunity_2021.pdf.

WRAP (2013). Designing for Longevity – Guidance on increasing the active life of clothes

WRAP (2014). Clothing Longevity Protocol Extending clothing life protocol. WRAP. Available from: https://wrap.org.uk/resources/guide/extending-clothing-life-protocol

WRAP (2017a). Valuing Our Clothes – The Cost of UK Fashion. p.12. Available from: https://wrap.org.uk/resources/report/valuing-our-clothes-cost-uk-fashion

WRAP (2017b). Sustainable Clothing Guide. Available from: https://wrap.org.uk/resources/guide/sustainable-clothing-guide

Education & Policy

DESIGNING A CURRICULUM TO DEVELOP A SUSTAINABLE WORKFORCE FOR THE TEXTILE/CLOTHING INDUSTRY

Rohana U Kuruppu[1]

[1]Department of Fashion, Textiles and Clothing Technology, Sri Lanka Technological Campus, Research University, Ingiriya Road, Padukka, Sri Lanka.

Abstract

Clothing industry has become complex and fragmented. It has been influenced by product variety, significant uncertainties and short product life cycle (Chan et al. 2017). This is a matter of concern to those who trying to develop sustainability in the textile industry. Customers 'are beginning to demand sustainable textile products from textile/clothing manufacturers. There is less evidence that clothing manufacturing can adopt sustainability initiatives to produce sustainable products. No visible plans within the industry to develop a social, & environmental understanding workforce to manage a sustainable supply chain (SSC) (Vivek et al. 2020). Hence, a need to educate youth and those in the industry in sustainable development.

The accessibility to market is becoming increasingly difficult due to competitiveness. Therefore, efforts are taken to develop different methods to increase market penetration and among such efforts sustainable development seems getting noticeable prominence in the global market. Sustainable development is important because intelligent consumers are concerned about environment.

H & M has a sustainable supply chain in which it has embraced the entire spectrum of eco material to manufacturing and distribution to retailing (Shen, 2014). Clothing consumption have risen. Unacceptable working conditions have caused to design a fashion education that acknowledges ethical and sustainable aspects of fashion (Radcliffe-Thomas, 2018). The sustainable supply chain is a new phenomenon. This is the background to develop a curriculum to educate the undergraduates and those in the industry to manage sustainable supply chain effectively.

Fashion is changing fast but it is being disposed of at the same speed. Industry is operating under trying conditions to keep at low cost and zero waste to landfill. Textile business has an impact either socially or environmentally. Therefore, sustainable impact to the society is the responsibility of the businessman. The employees need to be educated and trained to implement a successful sustainable development process. In this paper the author has brought in some useful ideas that could be incorporated into the curriculum. There is information to develop a successful sustainable curriculum.

The customers are looking for sustainable products. Service organizations are offering certification programmes and training. Manufacturers are learning sustainable development pathways. Future is transforming to sustainable manufacturing. The industrial sector's success depends mostly on the talent and the strength of the workforce. Therefore, the purpose of this study is to develop a curriculum to develop the workforce in achieving a sustainable workforce in the textile and clothing industries.

Keywords

Sustainable Development, Supply Chain, Competitiveness, Sustainable Standards, Certification of Sustainability, Maintenance of Chain of Custody, Market Accessibility. Innovation and Newness.

Introduction

In today's context, unlike in the past, many countries are engaged in textile/clothing businesses. There are variety of products at very competitive prices. The whole textile/clothing business has become very competitive.

The potential for economic growth shall depend according to the market accessibility. Penetration into the world market is not an easy task. There are so many barriers e.g. competitiveness, lack of training and development, innovation and newness, market understanding and strategies, ability to negotiate etc. Amongst these factors, competitiveness and innovation and newness can be useful when trying to penetrate into markets. Whilst competitiveness is a strong contender to obtain market accessibility, the innovation and newness is also worth exploring its usefulness. The innovation and newness can be a new approach such as sustainable development. Perhaps, this could be used as a tactic to enter the competitive world market.

The world's exports have become the buzz word whilst sustainable development is becoming the marketing tool to increase exports. The contemporary consumers, demand sustainable textile/clothing products from the textile/clothing manufacturers. Sustainable textile/ clothing products are beginning to expose and being seen in department stores. Anyway, there is no evidence of plans among textiles/ clothing manufacturers to adopt sustainability initiatives to produce sustainable products in large scale in the foreseeable future (Vivek et al 2020).

Nevertheless, the countries that have challenges to penetrate the global market could explore sustainable products to develop their global market share. Sustainable products can have a new approach like organic/ reduced carbon foot print, new products of renewable energy, less water consumed clothing products, recycled and up cycled products, fair trade products etc. Since, not many countries have taken advantage of sustainable development to create accessibility to develop their markets, this would be a good opportunity to such developing countries.

Bangladesh garment industry is making an effort to reduce pollution and ecological footprint. They are hoping to bring a new paradigm of doing business called green business (GB). They want to maintain environmental safety and business competitiveness. Green strategy can act as a crucial tactic for gaining a competitive advantage over the potential competitors (Sarkar et al 2020).

Manufacturing sustainable products is a complex process. Therefore, factories need to be transformed into fibre to fabric to clothing products dedicated to sustainable development. A sustainable supply chain must be linked to the factory to complete the cycle. Above all, it is the employees that need to understand clearly sustainable development. Training and development must take a major role in the transformation process from conventional manufacturing to sustainable development process in the clothing industry. Produce a sustainable workforce. At present, the clothing factories are not practicing full scale sustainable development activities except for reduction of carbon foot print. As such, there is a need to draw a full scale learning and development programme for employees to produce sustainable products at competitive price.

Sustainability

Organic clothing is about sustainability. The consumers of contemporary times have realized the necessity of protecting themselves from chemicals, and as such, a number of clothing brands have taken the responsibility of keeping the earth green and are manufacturing eco-friendly clothing. Each brand has their own story, strategic reason for being organic and differentiate their brands from other labels. There is work being done by the brand owners to make organic clothing popular but still remains whether it generates the expected popularity thus earn the desired revenue for the clothing companies, need to be examined. In this context, is sustainable clothing products a profitable venture in the export industry? Sustainable clothing product would mean a product that meets the needs of the present, without compromising current and future generations to meet their own sustainable lifestyles (United Nations, 1987, p.8). Sustainable development refers to social, economic and environment. Sustainable development in the corporate sector would mean that a company must contribute to economic growth, social progress and promote environmental sustainability. Most companies would like to rank the importance in following order, environmental conservation, economic development and social development.

The environment aspect is the commitment to protect the environment by reducing risks of environment impacts of company activities. The challenges for companies would be; saving and preserving natural energy, assessing their footprint and reducing greenhouse gas emissions, and control waste of water thus availability of water for current and future generations. Therefore, renewable energy and less water consumption techniques in the textile/clothing industries must be applied to develop sustainability. The economic aspect of the sustainability is to promote the protection of the environment by controlling the risks posed by their production. Recycling of raw material and products and upcycling are important in the economic development of sustainability. The social aspect of sustainable development is about values that make equality and respect for individuals. This would include, gender equality, difference in salaries, promoting training, encouraging dialogue, fair trade products which guarantee a reasonable income for farmers, encouraging transparency and giving access to people with reduced mobility.

To create a sustainable development company, the three important pillars such as environment, economic and social must be systematically incorporated into the clothing business. The industrial sector's success depends mostly on the talent and the strength of the workforce (Narkhede and Gardas 2018). Therefore, the purpose of this study is to develop a curriculum incorporating the above three pillars to develop the workforce in achieving a sustainable workforce in the textile/clothing industries and produce sustainable products.

Problem

Although sustainable development is a fairly old terminology, yet there is room to impress upon clothiers to adopt sustainable practices thus lack of knowledge among employees has hindered the progress of sustainable products.

Aim

There is a need to train employees in manufacturing sustainable development products using the three pillars engaging the entire workforce in a clothing manufacturing unit and to perform at competitive level.

Methodology (Brief)

The success and development during a competitive environment of clothing industry of developing countries depend upon the strength of its workforce. The method of this study is to develop a curriculum for textile/clothing undergraduate students as well as already employed. This curriculum shall pay emphasis on sustainable development work so that workforce can intelligently adopt the practices of sustainable development which in turn, as a result, will be able to produce sustainable products. This approach would enable to create a pathway to have accessibility to the global clothing market. In such circumstances it will be possible to overcome the challenges and increase the clothing exports growth potential of developing countries.

At the very outset it would be necessary to develop a sustainable supply chain. All supplies must be certified for sustainability. At every stage in the supply chain there must be a certified scope certificate. It is a stage that has been audited and found to be in conformity with the sustainable standard. Then move on to the next stage until it reaches its final stage where it is the sustainable product.

The most important aspect is to maintain the sustainability standard at every stage and also to have chain of custody in place. This needs to be maintained in order to produce sustainable product. However, there are many sub activities to fulfil an audit to be successful e.g. under the product handling it is necessary to ensure there is no contamination. In this case not only it should be seen but also must keep all documents in this regard intact.

There are regulations and requirements to follow. All need to be complete to obtain sustainability certification.

Curriculum Development

The curriculum development will focus on sustainable development in a clothing manufacturing setting so that students will have an understanding of processes involved in making a sustainable product. The undergraduate programme will cover most subjects that normally required in a textile/clothing technology degree but there will be emphasis in subjects that are dealing with sustainability in a manufacturing scenario. Areas such as sustainable supply chain, sustainable standards, certification methods (e.g. farmer certification, organic certification etc.), standards regulations such as USDA NOP, APEDA NPOP, EEC 834 etc., audit and certification, scope certificate, transaction certificates, understanding of chain of custody, tests on genetically modified organisms (GMO), various standard owners such as Textile Exchange and Global Organic Textile Standard (GOTS), accredited certification bodies and their functions will be some of the sustainability action stations that need to explain in the degree course. Students should also understand about International Federation of Organic Agriculture Movements (IFOAM). This is an internationally applicable organic standard that can be used directly for certification. It is mostly used by the farmers to certify that their products are strictly organic. At the same time GOTS means that textile products are grown and manufactured under strict ecological and toxicological guidelines without any harmful chemicals.

Whilst providing a comprehensive knowledge covering economic, environmental and social including fair trade and equality aspects on sustainability for university students, it will be process oriented approach would be suggested for the already employed. Employees will learn about standards and implementation methods to secure chain of custody in the supply chain. They need to learn to perform effectively and efficiently. It will be necessary to request from certificates issuing bodies to conduct training sessions from time to time for industry personnel. Some agencies are already doing training but need to be interactive and interesting.

University students will need to visit more often clothing factories that practice sustainability to understand the process and activities. They would need to work alongside with the third party accredited certificate issuing bodies to understand the auditing methods.

Literature Review

A literature review was carried out to broaden the knowledge of sustainable development and curricula developed to train staff in practicing sustainable development in manufacturing.

The UN Sustainable Development Goals and Textiles 2030 shall soon become reality. Therefore, people on this planet should strive for sustainable product design which reduces the impact on the environment thus emerge as a superhero as it paves the way for cleaner planet for generations to come for long lifecycle. In this context, when the sustainable product reaches its end-of-life period it should be able to systematically dismantled and separated for recycling or upcycle thus material usage in the original product shall lower carbon footprint. In this way, one could extend its lifecycle thus reduce the environmental impact throughout its lifecycle which is known as measurement of lifecycle analysis (LCA).

Clothing industry has become complex and fragmented (Chan et al 2017). It has been influenced by product variety, significant uncertainties and short product life cycle and impulsive buying (Bruce, Daly & Towers 2004). Garments are designed for the occasion and saleable only for that purpose thus short time in the market (Christopher & Peck, 1997). The characteristics of fashion products pose challenges to the clothing industry and make supply chain complicated (Christopher et al 2004). As a result, fashion products are usually discarded fairly quickly hence pollute the environment. In such circumstances, it is better to manufacture clothing products that have long life span which makes environment friendly i.e. a variety of sustainable products over conventional garments. Our efforts should realize that clothing products are long life cycle thus no hindrance to the environment and society.

Soft skills such as creativity, communication, leadership etc. are essential tools that help to achieve sustainable goals are not common in fashion oriented curricula (Rana & Ha-Brookshire. 2018). This study group argued that critical thinking, collaborative skills, active listening, interpersonal skills, emotional skills, and administrative skills are essential to meet the sustainability goals. Further, they have suggested to develop a conceptual model by incorporating these skills and that will lead to a curriculum for sustainable development (Rana & Ha-Brookshire, 2018).

Haigh (2005) believes that academics can create better and self-sustainable world. Therefore, the academics have the opportunity to lead sustainability efforts. To incorporate sustainability in clothing syllabus, it would be necessary to focus on raw material selection, manufacturing processes and fair labour initiatives. Interestingly, Rana and Ha-Brookshire (2018) raised two questions that help in the development of curricula for sustainability. They are, (a) what are the critical processes in the clothing industry to achieve sustainability and (b) what are the key skill sets the workforce must possess to achieve sustainability. They discussed the three key business stages and the skill sets required to incorporate sustainability followed by a conceptual framework that includes skill sets to help students to understand sustainability related business operational settings.

Reflecting on Haigh's suggestion on academics contribution on sustainability it is however, not very visible that academics' involvement with sustainability in the context of clothing industry of developing countries. Sustainable development as a module is being taught in textile/clothing courses but not as implementing sustainability techniques in order to produce 100% sustainable products are not known. A recent survey carried out by the author among the South Asian countries did not record inferences in this regard. The studies carried out and observations made by the author reveals that in Sri Lankan clothing industry has made a concerted effort to control greenhouse gases (GHGs) thereby reducing carbon footprint. Anyway, there is no evidence to say that these efforts are organized and standardized across the clothing industry of Sri Lanka.

Shen (2014) explains when garment producing company take on sustainable products, it must first develop a sustainable supply chain. A well reputed fast fashion company, H & M has developed its sustainable supply chain which include environment friendly raw materials and processes, conserve energy and natural resources, activities that are economically sound and safe for employees and consumers, reducing carbon emission in all process activities, green retailing and in short sustainable manufacturing (Shen et al 2014). Shen (2014) claims that the H & M's sustainable programme has created many new jobs in less developed countries (LDCs), recyclable resources are used in production and consumers are educated to be more ethical (conscious action sustainability reports 2012 and 2013). This has made market and supply chain to be more economically, environmentally and socially sustainable. Therefore, Shen's (2014) objectives were to identify the structure of the sustainable supply chain and then based on the structure to investigate sustainable operations and finally to discuss lessons from supply chain in the context of country perspective. Li et al (2014) have discussed the impacts of corporate social responsibility (CSR) behaviour on the sustainability performance of H & M. In this case, Li at al has attempted to demonstrate the impacts of CSR on sustainable performance.

Unacceptable working conditions have caused to develop a fashion education that acknowledges ethical and sustainable aspects of fashion (Radcliffe-Thomas, 2018). Further, it has emphasized that focus should be on imparting skills, introducing knowledge and techniques, where students will be able to cut patterns, construct garments and produce fashion communications. It should be more than transmission of facts, figures and opinions. How can one involve students in learning and foster the curiosity and creativity? Sustainability education to be successful, it should operate as a collaborative venture (Radcliff-Thomas 2018).

Fletcher and Williams (2013) wanted to start a course that contributed towards sustainability in the fashion industry. They built the course on participative and practical understanding of sustainability. They decided to recruit students from practice- based and theoretical backgrounds to engage them in a creative learning process that gave the participants a host of different experiences and different learning environment. It has given a progressive broadening of educational content and different models of leaning within the sustainability framework.

A team of US educators have published that only a few fashion programmes have had sustainable practice into their curriculum in spite of vast knowledge about climate change and advantage of sustainability (Lovinski, 2019). The researchers have asked why sustainable practice curriculum lacking in industry? Their study reveals that there is a correlation between all types of academics such as researches, curriculum developers and those who feel that academia is responsible for future of fashion. Yet, there is a huge gap or refusal in incorporating sustainability in fashion based courses. It has been mentioned in this paper that some say, there is no competency of sustainability is required for graduates (Lovinski, 2019).

Whilst some quarters argue that competency of sustainability is not required and others challenge the textile industry to achieve zero discharge of hazardous chemicals across all path ways. The textile industry is taking initiatives to address this challenge. It is being mentioned that sustainable chemistry will be an important solution in the context of sustainable development (Nimkar, 2017). Therefore, sustainable development will be part and parcel in our future industries and all employees must be fully aware of it implications.

Kozer and Connell (2010) found that when sustainability was included in the curricula, people of upper echelon class were generally informed about sustainable products and more likely to consider environmental and social issues when buying clothing products than people of lower class in society. Therefore, their study confirms that more students and educated on sustainable development, the more knowledge is retained and that leads to sustainable applications and future careers.

Landgren and Pasriche, (2011) have reported even if students are well informed regarding sustainable development through their curricula, additional classes to focus on sustainability shall increase their knowledge on sustainability and have corresponding action. This suggest that there is an impact on behavioral patterns of people due to sustainability. As such, the author of this article has ample testimony to support the importance of sustainability in the textile and clothing industry which should be developed not only to make our industry successful but also to make the three pillars of sustainability successful and bring future generations 'prosperity.

Understanding of Sustainable Clothing Production

Let us understand how a sustainable clothing product is made. A sustainable product means one that has been certified under sustainable textile standard. There are number of sustainable textile standards e.g. GOTS., Global recycled Standard (GRS) etc. The certificates are issued by third party accredited certification bodies.

There is always a beginning to a product. It is the raw material that turns into a finished clothing product. As such, start with fibres which is organic certified. This organic certification should derive from any one of the following regulations. USDA (NOP), APEDA NPOP, EEC 834.

Fibres are then bought by the ginnery. If the sustainable standards were to be continued, the ginnery should conform to the sustainable textile standards. Then the fibres in the cotton bale is conformed to sustainable standards because ginnery has got the scope certification. Then these fibres are purchased by the spinner. There is a transportation and a new buyer. It must be ensured that the goods have arrived at the spinning mill intact. At this point, a transaction certification (TC) is issued. TC is a document issued by the third party certification body that verifies the goods being transported from one place to the next conform to a given standard. No damage done or quantity change for the goods. At this juncture, the spinner must ensure their organization meets the sustainable standards. If this is true, then the yarn they produce is sustainable yarn. The process keeps moving forward ensuring its sustainability. From yarn to fabric making and colouring completes one cycle in the supply chain. At this juncture it must ensure it has maintained the chain of custody. This process continues forward until clothing product is produced. A such product is a sustainable product. The process owners must ensure there is no way to deviate from the sustainability pathway.

Discussion

To manufacture a sustainable textile/clothing product is not an easy task. Not that it is difficult to follow the method but it gives so much stress trying to maintain the standards. First of all, the employees must be fully aware of the purpose of making a sustainable product in their factory. The objective of sustainable production. To follow the standards to the best satisfaction, the employees must be a sustainable workforce. All staff must undergo rigorous training to be sustainable workforce.

The employees involved in sustainable production standards claim that there is no flexibility to take commercial decisions. For example, a certain certified company had identified a certified supplier to do the embroidery for their production. The accredited supplier failed to meet the demand as such, the client decided to get the balance done by a non-certified supplier. The accredited certification body decided this does not comply with the standard hence non-compliant certificate (NC) was issued. The client argument was to get the production and deliver the goods on time. One could see this as a trivial matter but its effect can be dangerous losing your customer.

The staff must not only follow the sustainable standards but they may have problems with lower cadres not meeting production targets. Whilst adhering to production targets, must learn to handle lower cadres in order to fulfil your production. One would need to be careful when dealing with such cadres in such situations. There is no excuse for low production because of the competition. Production supervisors must be vigilant about their line production and maintain targets. Shop floor employees find difficult to hold sustainable standards and production targets intact.

In South Asia there is no big interest for sustainable products as in other parts of the world. As such, there are not much of sustainable orders to perform. Sometimes, one could consider to remain in the conventional production lines rather than changing over to sustainable products. This could be a reason for not much interest in making sustainable products.

In today's context, there is a recession in the world. There is a shortage of orders in the clothing industry across the world. As a result, everyone is trying hard to grab orders. In this circumstances it can be a good strategic move to produce sustainable products to regain orders. However, still need to be very competitive because of economic crisis the world is facing at the moment. Certainly, sustainability can be a marketing tool and if it can be handled well, there will be plenty of opportunities for growth. According to industry studies it reveals that there is not much of enthusiasm moving into sustainable production mainly due to stress factors. It has brought staff moving over to conventional production in developing countries but yet if this situation can be changed there can have lot of benefits for the industry.

In order to make sustainability a profitable centre, the industry should not be left alone to transform it into sustainability. Most probably that tactic will bring success only to those factories that have facilities to overcome challenges. The whole industry must be successful. Therefore, the government involvement is necessary. The sustainability must be developed as a national project and it must be driven by the government giving an impetus to all the export oriented clothing factories to produce sustainable products. Training will be a major part in this project. Without government initiatives it will not have the courage for entrepreneurs to prepare their industries to the next generation.

The author has had opportunity to discuss with accredited certification bodies as well as industry personnel about the issues in developing a sustainable textile factory. It has been made to understand that there is lack of interest in factory personnel. When organized training sessions for factory personnel to develop their knowledge and skills on sustainability, they often avoid these training sessions. Similar sentiments can be expressed at university level. There is clear lack of interest about the subject. Efforts must be taken in the future to change this position and bring young talent into this sustainability textile industry to make it a growth centre.

Conclusion

Sustainability is a vast subject and it needs patient hearing and involvement. People should be encouraged to practice sustainability. There should be interest to it.

In response to the requirements of incorporating sustainability into the curriculum, this paper has attempted to offer a variety of important aspects of sustainability. It has provided basic requirements of sustainability for a university course that will be studied along with the textile and clothing subjects. To make this course more interactive and practical it has suggested that students shall visit factories that practice sustainability to broaden their practical knowledge. For those who are already engaged in the industry are also encouraged to follow organized courses on sustainability from time to time.

This paper has given information about sustainability in the textile industry and method to start a sustainability process to produce sustainable clothing products. It has discussed a few areas where employees have felt difficult to continue in a sustainability supply chain.

It has also shown a brighter side of sustainability for developing countries where they can use it as a marketing tool. If the respective governments can support this industry to develop sustainability, then most countries will have prosperity because they can capture the market.

Sustainability is a new phenomenon that needs careful handling. Since it has not caught up in the market place in big way, consumers may not know the product and its background. Lack of knowledge about the product can be stopping people showing interest. A concerted effort must be launch to make sustainable production for exports in big way.

References

Bruce, M., Daly, L., & Towers, N. (2004) *Lean or agile: A solution for supply chain management in the textile and clothing industry*. International Journal of operations and production management. 24, 151-170.

Chan, A.T.L. at al. (2017). The effects of strategic and manufacturing flexibilities and supply chain agility on firm performance in the fashion industry. European Journal of Operational Research, 259, 486-499.

Conscious Action Sustainability Report 2012.
http://about.hm.com/content/dam/hm/about/documents/en/CSR/reports [13-05-2023]

Christopher, M., & Peck, H, (1997). Managing logistics in fashion markets. International Journal of Logistics Management, 8, 63-74

Christopher, M., Lowson, R, & Peck, H. (2004). *Creating agile supply chain in the fashion industry*. International Journal of Retail and Distribution management, 32, 367-376.

Fletcher, K., and Williams, D., (2013). *Fashion education in sustainability in practice*. Research Journal of Textile and Apparel. 17. 81-88.

Haigh, M., (2005). Greening the university curriculum: Appraising an international movement. Journal Geography in Higher Education, 29, 31-48.

Landgren, T.M., & Pasricha, A (2011). *Transforming the fashion apparel curriculum to incorporate sustainability*. International Journal of Fashion Design, Technology and Education, 14, 187-196.

Li, Y.; Zhao, X.; Shi, D.; Li, X. (2014*) Governance of sustainable supply chains in the fast fashion industry*. European Management Journal. 32, 823–836.

Lovinski, N.,Copeland, L., Kim, J., (2019). *Perceptions of sustainability curriculum in US fashion academia*. International Journal of Fashion Design, Technology and Education. 12, 364- 373.

Kozer, J. M., & Connell Hiller, K.Y. (2010). *Socially responsible knowledge and behaviours: Comparing upper Vs Lower classmen*. College student journal, 44, 279-293.

Narkhede, B.E. & Gardas, B.B. (2018). Hindrances to sustainable workforce in the upper stream oil and gas industries-interpretive structural modelling approach. International Journal Business Excellence. 16, 61-81

Nimkar, U., (2017). *Sustainable chemistry: A Solution to the textile industry in a developing world.* Current opinion in green and sustainable chemistry. 9, 13-17.

Radclyffe-Thomas, N. (2018). Designing a sustainable future through fashion education. Clothing Cultures, Clothing Cultures. 5, 183-188.

Rana, Md. R. I., Ha- Brookshire, J. (2018). New conceptual frame work for fashion business ideation, negotiation, and implementation undergraduate curricula for sustainable development. International Journal of Fashion Design, Technology and Education. Taylor & Francis. 12, 140-148.

Sarkar, A., Qian,L., Peau, A.K., Overview (2020) Green business practices within the Bangladesh RMG industry: Competitiveness and sustainable development perspective. Environmental Science and Pollution Research. 27, 22888–22901.

Shen, B. (2014). *Sustainable Fashion Supply Chain. Lessons from H & M.* Sustainability in Fashion Business Operations. 6, 6236-6249.

United Nations (1987) Our Common future: Report of the world commission on environment and development. http// www.undocuments.net/ocf-02.htm# [13-05-2023]

Vivek, R. (2020). Reactive and proactive pathways to sustainable apparel supply chains: manufacturers perspective on stakeholder salience and organizational learning toward responsible management. International Journal of Production Economics. 227, 107672.

Textile Institute
8th Floor, St. James' Buildings
79 Oxford Street
Manchester, M1 6FQ
United Kingdom

ISBN 978-1-7138-9696-8

9 781713 896968

CURRAN ASSOCIATES INC.
proceedings.com